THE EMERGENCE OF THE FOURTH DIMENSION

The Emergence of the Fourth Dimension

Higher Spatial Thinking in the Fin de Siècle

MARK BLACKLOCK

OXFORD
UNIVERSITY PRESS

OXFORD
UNIVERSITY PRESS

Great Clarendon Street, Oxford, OX2 6DP,
United Kingdom

Oxford University Press is a department of the University of Oxford.
It furthers the University's objective of excellence in research, scholarship,
and education by publishing worldwide. Oxford is a registered trade mark of
Oxford University Press in the UK and in certain other countries

© Mark Blacklock 2018

The moral rights of the author have been asserted

First Edition published in 2018

Impression: 2

Published in the United States of America by Oxford University Press
198 Madison Avenue, New York, NY 10016, United States of America

British Library Cataloguing in Publication Data
Data available

Library of Congress Control Number: 2017953265

ISBN 978–0–19–875548–7

Printed and bound by
CPI Group (UK) Ltd, Croydon, CR0 4YY

Acknowledgements

A stroke of good fortune launched this project: Professor Steven Connor agreed to supervise it in its PhD incarnation (against, I suspect, his better judgement). At completion it has been unsurprising to realize the extent of the subtle changes of direction Steve's encyclopaedic guidance, detailed correction, and focused encouragement have wrought on the final work. I can think of no better indicator of his generous tutelage than the fact that he continues to allow me to pass off his spare ideas as my own. They are recognizable by being the best ideas you will read here; in his work they would be made to soar.

Roger Luckhurst informed me early in the piece that as my second supervisor he was 'in deep cover': to which I can only respond with the hope that Roger doesn't pursue a second career in espionage, so poor are his skills of self-concealment. His help and guidance continue to be generously and freely given and his work remains a source of inspiration.

The English Department at Birkbeck College, at which I was a student for eight years and in which I have been employed since, has been immensely supportive. It is a fine institution: collegiate, dynamic, generative, and welcoming. In practical terms, I am enormously grateful for the award of a studentship that paid my fees for four years. I should also like to thank every member of staff who has given invaluable advice, support, friendship, and even occasional paid work over the past decade, in particular Isobel Armstrong, Anthony Bale, Joseph Brooker, Carolyn Burdett, and Hilary Fraser. To Birkbeck teaching alumni Rebecca Beasley and Laura Salisbury I owe a debt of gratitude for providing impetus: while under their tutelage I first encountered the texts that became the seedbeds of this project.

Gerry Kennedy, an independent scholar and descendant of the Boole clan, has shared thoughts, chats, and information on Hintonian material: he's a top man. Anthony Ossa-Richardson gave generously of his expertise in early modern philosophy and provided invaluable comment on the More section; Christina Scholz has assisted me with German-language queries. Specific tips for sources later came from David Gillott and John Holmes: Edwin Abbott and May Kendall send their thanks! Many former Birkbeck postgraduate friends and colleagues are now to be found working throughout the Eng. Lit. academy and I cannot recommend highly enough their work and company, but I am particularly grateful to Dennis Duncan and James Emmott for numerous coffees and discussion in the British Library. Special thanks go to Henderson Downing, whose quite brilliant MA work opened the whole thing up for me. I look forward to his monograph with eager anticipation.

Members of staff of several libraries and archives have fielded inquiries, but particular thanks to the archivist at the Swan Sonnenschein Collection at Reading University who dug out for me a letter from Charles Howard Hinton to his publisher that still provokes in me a most unscholarly emotional response. The work has been honed at many conferences and in contact with numerous researchers, but

I would like to thank in particular the editors at several journals which have permitted me to develop ideas in their pages. My PhD examiners, Professors Alice Jenkins and Mark Turner, gave advice that assisted my research to pupate into the book you hold in your hands, and the anonymous readers of the manuscript gave the most valuable honest and constructive criticism: I sincerely hope that in my responses to their comments I have improved the book. I am very grateful to all those who have worked on the book, my editors at OUP and beyond, for their patience, support, and professionalism: Jacqueline Baker, Rachel Platt, Eleanor Collins, Aimee Wright, Kavya Ramu, Monica Kendall, and Ingalo Thomson. Any errors that remain are mine entirely.

Finally, my wife Katie has not only supported me financially for the duration of my studies as I took on less and less freelance work to concentrate on this project, but has also given birth to our three wonderful daughters during the same time period. In the fourth dimension, over which I claim dominion, Katie is a saint and empress. Back here in three dimensions she is yet more awesome. This book is dedicated to her.

Contents

List of Illustrations	ix
List of Abbreviations	xi
Introduction	1
1. Conditions of Emergence: Kant, Helmholtz, and Analogy	14
2. Knots: Topology, Conjuring, and the Spiritualist Fourth Dimension	41
3. A Square: *Flatland*, Play, and Tradition	72
4. Cubes: Hintonian Higher Space and its Thinking Subject	103
5. Through: The Theosophical Society, Authority, and Mediation	135
6. Fictions: The Spaces of Literature after *n*-Dimensions	166
Conclusion	206
Bibliography	209
Name Index	225
Subject Index	230

List of Illustrations

1. Edwin A. Abbott, *Flatland* (1884)   73
   Inside front illustration
2. Edwin A. Abbott, *Flatland* (1884)   74
   Inside back illustration

List of Abbreviations

ANE C.H. Hinton, *A New Era of Thought*
F Edwin A. Abbott, *Flatland*
I Joseph Conrad and Ford Madox Hueffer, *The Inheritors*
S C.H. Hinton, *Stella and an Unfinished Communication*
SR C.H. Hinton, *Scientific Romances*
THB Mary Wilkins Freeman, 'The Hall Bedroom'
TP Johann Carl Friedrich Zöllner, *Transcendental Physics*

Introduction

It is a curious physical fact that any point in space can be defined by three numbers. Length, breadth, and height, as seems self-evident, are the first three dimensions: x, y, and z. As our algebraic symbols for the axes indicate, these three derive from co-ordinate geometry, developed by Descartes in 1637 to describe the position of a point in a plane using two intersecting axes.[1] Analytic geometry, using this system of co-ordinates on perpendicular axes, bound algebra and geometry together.

The reason for the tri-dimensionality of space troubled Immanuel Kant, who could not accept Leibniz's claim that the multiplicity of dimensions derived from the fact that three lines meeting in a point could be drawn orthographically to each other. 'I have thought of proving the threefold dimension of extension from what is to be observed in the powers of numbers,' wrote Kant:

> Their first three powers are quite simple, and cannot be reduced to any others; whereas the fourth power, as the square of the square, is nothing but a repetition of the second power. But however satisfactorily this property of numbers may appear to account for the threefold dimension of space, it fails to stand the test of application. For in everything representable through the imagination in spatial terms, the fourth power is an impossibility.[2]

And yet this impossible fourth power, the elusive w-axis, has proved curiously resilient. August Möbius, writing in 1827, echoed Kant:

> It seems remarkable that solid figures can have equality and similarity without having coincidence, while always, on the contrary, with figures in a plane of systems of points on a line equality and similarity are bound with coincidence. The reason may be looked for in this, that beyond the solid space of three dimensions there is no other, none of four dimensions.[3]

---

[1] Fermat simultaneously and independently developed, but did not publish, a similar system extended to three dimensions.

[2] Immanuel Kant, 'Thoughts on the True Estimation of Living Forces (Selected Passages)', in *Kant's Inaugural Dissertation and Early Writings on Space*, ed. and trans. John Handyside (Westport, CT: Hyperion Press, 1979), pp. 3–18 (p. 10). Kant's early physical works are considered beyond the scope of the Cambridge edition of his complete works and translations are often fragmentary. Handyside's translation, although only a selection, is the most complete available. All subsequent references to Kant texts are given from the Cambridge edition.

[3] August Möbius, 'On Higher Space', in *Sourcebook in Mathematics*, ed. D.E. Smith, trans. Henry P. Manning (New York: McGraw Hill Book Company, 1929), pp. 525–6 (p. 526) (first published in *Der barycentrische Calcul* (Leipzig: [no publisher] 1827)).

Not only did Möbius insist that there was not space of four dimensions, but his consideration of 'equality and similarity' paralleled Kant's later pondering of incongruent counterparts, solid objects that are the same shape and occupy the same volume but cannot be made to coincide with each other by dint of being mirror images. Möbius, however, went on to demonstrate that he could make three-dimensional objects that were inversions of each other coincide, if only he could have access to this nonsensical space of four:

> It will be necessary, we must conclude from analogy, that we should be able to let one system make a half revolution in a space of four dimensions. But since such a space cannot be thought, so is also coincidence in this case impossible.[4]

It is ironic that Möbius thought such a space unthinkable, given that thinking it was precisely what he was doing, founding what became *n*-dimensional geometry after Arthur Cayley published 'Chapters in the Analytic Geometry of (n) Dimensions' in 1843. Cayley's *n* is a variable that can stand for any integer. It is instructive to think briefly about that meta-sign *n*. We first encounter it parenthesized, retreating within its own semiotic regime, bracketed off from surrounding language. It is frequently italicized, as it is in this book, placed askance. And yet we sound it as a letter and when we sound it in conjunction with its compound term, dimensional, we sound the word 'end', a word the meaning of which the symbol *n* assumes away. So too do we sound the Greek prefix 'en-', for inside. In these two sounds we are given a pre-sentiment of the imaginative conditions of *n*-dimensional space: elsewhere, expansive—indeed, infinite—yet simultaneously directing our thought inwards.

A study of higher space does not begin with space but with geometry and mathematics. Nineteenth-century studies have outlined the significance of Euclidean geometry for British culture. Alice Jenkins has described the shifting conditions of space in the first half of the nineteenth century, as Romantic abstract space became domesticated by geometric education—'democratized and concretized and—to the opponents of this development—fundamentally denatured'—before a 'profoundly dematerializing tendency' in the physical sciences of the mid-century exploded spatial certainties.[5] Mathematical historian Joan Richards identifies in the emergent non-Euclidean geometry, described by Bolyai, Lobachevsky, and Beltrami, later canonized by Gauss and Riemann, and the concurrently developed idea of geometry of *n*-dimensions, 'a radical mathematical development which threatened to drastically distort the Victorian intellectual tapestry in which geometrical study was so integrally woven'.[6]

The reformation of geometry had far-reaching epistemological ramifications. For Mary Poovey 'modern abstraction [...] derives from the imposition of a conceptual grid that enables every phenomenon to be compared, differentiated, and

---

[4] Möbius, 'On Higher Space', p. 526.
[5] Alice Jenkins, *Space and the 'March of Mind'* (Oxford: Oxford University Press, 2007), pp. 175, 177.
[6] Joan L. Richards, *Mathematical Visions: The Pursuit of Geometry in Victorian England* (Boston: Academic Press, 1988), p. 117.

measured by the same yardstick'.[7] That grid was Cartesian and Euclidean: the imposition of geometric formality onto space was both a signature movement in the creation of the modern subject, in Poovey's account, and in the creation of distinct disciplines:

> As the elusive heart of the epistemology associated with modernity, this form of abstraction—and its historical production as abstract space—facilitated the disaggregation of conceptual and spatial domains: the nineteenth-century institutionalization of separate social sciences (economics, anthropology, sociology) owes as much to the naturalization of abstract space as does the physical reordering of London.[8]

The Emergence inquires into the further abstraction of the primary object. The development of post-Euclidean thinking described here was both a reformation of knowledge and responsive to knowledge of that reformation. What processes accompanied such reformations? Where and how was geometric reformation registered?

The project that emerged from Poovey's research into this abstraction focused on the production of disciplinary difference, examining double-entry book-keeping, moral philosophy, political economy, and the discussions around the method of induction to describe the creation and formalization of the modern fact. She acknowledges here her kinship with the work of Bruno Latour, whose 'programmatic' statement in *We Have Never Been Modern* of a sociology that does away with the ruptures between nature and society and, founded upon this first separation, the splitting of the ancient and the modern, also informs the current work.[9]

Latour's book is now twenty-three years old but he has continued to nuance its themes in his recent work. The lead essay in *On the Modern Cult of the Factish Gods* focuses on the encounter of Portuguese traders with West Africans in the Gold Coast to describe the attempt by the cross-wearing Portuguese to dissuade the West Africans of their belief in their idols, tellingly exploring the shared etymologies of the words 'fact' and 'fetish' and proposing, in characteristic style, the hybrid term 'factish' for the objects we 'moderns' encounter.[10] If, as Latour argues, *We Have Never Been Modern*, then the process of purifying the realm of science by casting out apparently man-made things, of insisting upon the separateness of society and nature, was always an illusion. The continued imposition of purification results in the production of what Latour terms hybrids.

In 'Thinking Things', an essay responding to and summarizing contributions to a 'thingly turn' in critical theory, Steven Connor identifies Latourian Actor Network Theory (ANT) as 'the most important of the influences' in a 'new thingly

---

[7] Mary Poovey, 'Making a Social Body', in *Making a Social Body: British Cultural Formation, 1830–1864* (Chicago: University of Chicago Press, 1995), pp. 1–24 (p. 9).

[8] Mary Poovey, 'The Production of Abstract Space', in *Making a Social Body: British Cultural Formation, 1830–1864* (Chicago: University of Chicago Press, 1995), pp. 25–54 (p. 25).

[9] Mary Poovey, *A History of the Modern Fact: Problems of Knowledge in the Sciences of Wealth and Society* (Chicago: University of Chicago Press, 1998). Bruno Latour, *We Have Never Been Modern*, trans. Catherine Porter (Cambridge, MA: Harvard University Press, 1993).

[10] Bruno Latour, *On the Modern Cult of the Factish Gods* (Durham, NC: Duke University Press, 2010).

disposition' that sees things as 'intimately involved with and expressive of [the human]'. ANT 'follows up the hunch that objects and subjects may in fact be reciprocally constitutive'.[11] Connor stresses the acknowledged Latourian borrowing of the idea of the 'quasi-object' from Michel Serres. For Serres

> the quasi-object is not an object, but it is one nevertheless, since it is not a subject, since it is in the world; it is also a quasi-subject, since it marks or designates a subject who without it would not be a subject.[12]

Serres gives the example of a rugby ball in a game of rugby or the *furet* in a French children's game, pass the slipper.

> The quasi-object, when being passed, makes the collective, if it stops, it makes the individual. If he is discovered, he is 'it' [*mort*]. Who is the subject, who is an 'I,' or who am I? The moving furet weaves the 'we,' the collective; if it stops, it marks the 'I.'

Latour extends quasi-objects, freighting them with a heavier load than Serres's focus on 'intersubjectivity'. Latourian quasi-objects are hybrids of society and nature, intersubjective, certainly, but also churned up, messy:

> As soon as we are on the trail of some quasi-object, it appears to us sometimes as a thing, sometimes as a narrative, sometimes as a social bond, without ever being reduced to a mere being. Of quasi-objects, quasi-subjects, we shall simply say that they trace networks. They are real, quite real, and we humans have not made them. But they are collective because they attach us to one another, because they circulate in our hands and define our social bond by their very circulation. They are discursive, however; they are narrated, historical, passionate, and peopled with actants of autonomous forms.[13]

How, then, to deal with networks containing these quasi-objects? We need to occupy the position of the quasi-objects in order to trace these forms, 'to look at networks of facts and laws rather as one looks at gas lines or sewage pipes'.[14] In order to avoid becoming an 'uprooted, acculturated, Americanized, scientized, technologized Westerner [...] a Spock-like mutant' we must bring the network of human and non-human, the entire assemblage, including these hybrids, or quasi-objects, into symmetry.[15]

This injunction presides over *The Emergence*, and I hope to occupy multiple quasi-objectival positions. It takes its lead from Latour, and from Serres, in refusing the purification of the ancient from the modern, or the human from the non-human. It accords with Mary Poovey's remark that

> questions of epistemology encourage the historian to consider how domains of knowledge production overlap (whether or not the modes of knowledge production

[11] Steven Connor, 'Thinking Things', *Textual Practice*, 24 (2010), 1–20 (p. 5). An expanded version of the same essay published on Connor's website is referenced in Chapter 5, n. 92.

[12] Michel Serres, *The Parasite*, trans. Lawrence R. Schehr (Baltimore and London: Johns Hopkins University Press, 1982), p. 225.

[13] Latour, *We Have Never*, p. 89.   [14] Latour, *We Have Never*, p. 117.

[15] Latour, *We Have Never*, p. 115.

have yet to be stabilized as discourses) how knowledge about nature resembles (as well as differs from) knowledge about society, and how knowledge practices are gradually differentiated, codified, and institutionalized as different kinds of knowledge.[16]

This spirit is manifested in a close attention to the forms around which higher-dimensional knowledge and thought coheres, an attention I want to ally with Bruno Latour's description of Michel Serres's method as 'a structuralism of contents': in Chapter 2 this form is the knot; in Chapter 4, the cube. These are exemplary quasi-objects: forms that are both abstractions from nature and from thought. I aim not to dismiss on the grounds of a correctness imposed from considerable distance the claims of those, such as Johann Carl Friedrich Zöllner or a legion of Theosophists, that the forms they encountered were not purely man-made, but aim instead to permit them to describe the forms they saw. This spirit therefore also pays attention to the circulation of 'science's blood flow' identified by Latour in *Pandora's Hope*, a flow that loops in the form of a knot, binding together social participants in the creation of churned-up facts.[17] As Gillian Beer has argued, 'Scientists and writers dwell in the land of the living where multiple epistemological systems interlock, overlap, contradict, and sustain our day-by-day choices.'[18]

Not all thought after the interventions of Hobbes and Boyle, identified by Latour and Poovey, following Schaffer and Shapin, as the great purifiers of the modern contract, is characterized by an urge towards codification and purification. As Christopher Herbert has tracked in considerable detail, there was a rich tradition of relational thinking in the nineteenth century, in fields as distinct as physics and comparatist anthropology. Herbert's insistence that nineteenth-century relational thinking anticipated 'the intellectual condition defined by polyvalency, indeterminacy, constructivism, *différance*, and the ideological critique of knowledge' can rely upon a conflation and reduction of the specificities of each of these more recent currents and by a reading of 'relativity' that can seem to encompass even comparison.[19] Nevertheless, his point is well attested that thinkers as diverse as Herbert Spencer, Karl Pearson, and George Frazer were, at certain points in their published work, committed to thought that would today be characterized as relativism. What Poovey and Latour offer, I think, is a way of understanding how the epistemological condition in which these aspects of thinking might have been 'occluded' has come about; of reinstating some of the granularity that is lost in methodological or disciplinary purifications. What I hope to do with *The Emergence* is to bring into symmetry the occluded elements of the ideas of higher-dimensional space that demonstrate that space itself does not sit easily within any disciplinary framework.

The development of the new geometries—*n*-dimensional and non-Euclidean—in the second half of the nineteenth century was concurrent and they shared an

---

[16] Poovey, *The Modern Fact*, p. 19.

[17] Bruno Latour, *Pandora's Hope* (Cambridge, MA: Harvard University Press, 1999), p. 100.

[18] Gillian Beer, 'Translation or Transformation? The Relations of Literature and Science', *Notes and Records of the Royal Society of London*, 44 (1990), 81–99 (p. 97).

[19] Christopher Herbert, *Victorian Relativity: Radical Thought and Scientific Discovery* (Chicago: University of Chicago Press, 2001), p. xii.

assumption that Euclid should not be a limit for geometry, but they were distinct. Indeed, *n*-dimensional geometry might be either entirely Euclidean or entirely non-Euclidean, and non-Euclidean geometry may be assumed within any number of dimensions. For *n*-dimensional thought any number of contributions might be selected as a mark in the sand, a point upon the grid at which to plot a starting point: August Möbius's paper is particularly important for beginning to think what an *n*-dimensional space might do, what features it might have. This book plots J.J. Sylvester's address to the British Association in 1869, republished in *Nature*, as the intersection of axes. From this point on, English-language discussion of the fourth dimension took place beyond mathematical papers.

The range of the period under consideration is usefully circumscribed in the early twentieth century: Linda Dalrymple Henderson's magisterial 1983 study, *The Fourth Dimension and Non-Euclidean Geometry in Modern Art*, described the profound influence on artistic Modernists of a body of thought overlooked for much of the twentieth century. Henderson read in the papers of Boccioni, Picasso, Duchamp, and Kandinsky detailed discussions of new geometric ideas developed in the nineteenth century. Her updated 'Re-Introduction', published in 2013, describes a 'new phase of interest' in the fourth dimension—and a concurrent waning of interest in non-Euclidean geometry—in popular culture and art post-1984, the centennial year of the publication of *Flatland*, the year of the 'birth of "superstring" theory', and the publication of William Gibson's *Neuromancer*. Subsequent developments in computer graphics have maintained interest in the 'adaptable and suggestive fourth dimension of space'.[20]

Many of the figures or texts I deal with are encountered in the introductory chapters of Henderson's original work and it has been a conscious decision to work up to and into her main thesis, to unpack further some of the prehistories she indicates. I have shifted my attention towards literature: Henderson's work demonstrates that the geometries developed in the nineteenth century and their popular and occultist elaborations informed Modernist production in the visual arts and outlines potential lines of inquiry in the literary arts. I have followed these leads.

Faced in the twentieth century with an artistic Modernism profoundly informed by new geometries, and preceded by a Victorian Britain whose spatial imagination was fundamentally configured by the old geometries, a series of questions emerge from the nexus: what did the *n*-dimensional turn do to the spatial imaginary? How was culture altered after thought of higher space? To put it another way: what happened between *n*-dimensions and Modernism?

A one-word answer to the question may well be possible: occultation. Between Gauss, Cayley, and Sylvester, and Picasso, Kandinsky, and Duchamp are spiritualists and Theosophists in their thousands, locating intelligences, ghosts, and thought-forms in higher space. How to deal with these? How do they think higher space and what does the way they think it do to the form?

---

[20] Linda Dalrymple Henderson, 'Re-Introduction', in *The Fourth Dimension and Non-Euclidean Geometry in Modern Art* (Cambridge, MA and London: MIT Press, 2013), pp. 1–96.

Roger Luckhurst's *The Invention of Telepathy* identifies a *modus operandi* for investigating a hybrid construction theorized at disciplinary 'vanishing points'. Luckhurst aims to 'approach telepathy as if it were a possible formulation' rather than to dismiss the late Victorian interest in the supernatural as eccentricity.[21] I aim to approach higher space in the same spirit. The methodology of the entire project therefore responds to the challenges set by the object of inquiry at varying junctures. It seeks to understand how higher space was thought and what changes such thought worked on the cultural spatial imaginary. It therefore pursues significant events, works, and movements within the sphere of higher-dimensional thought wherever they lead before refocusing to consider the representation of the idea of higher space in the literature of the fin de siècle.

Steven Connor's afterword to a collection of essays on 'The Victorian Supernatural' notes a further 'twist' in the tale of the supernaturalist engagement with higher space, that it 'belonged to and perhaps even helped to condition the further development of the mathematics of shapes and spatial relations known as projective geometry, or topology, that was already underway by the 1870s'.[22] For Connor the Victorian supernatural 'is not an inert and finished shape in space, but a continuing potential for reshaping of the space it is in, and so partly includes us'.[23] In the case of resurgent higher dimensions of space this is undeniably so, and paying attention to the folds that bring different historical moments into contact with each other is important to this project.

The present work is also the study of an idea of mathematical origin, and must consider its relation to other such works, even if it risks suffering by comparison. The writing of Arkady Plotnitsky, a veteran of the Science Wars, provides in its account of those border disputes both reminder of the risks of a profligate approach to disciplinary boundaries, certainly as far as ideas from mathematics and physics are concerned, and a bulwark in defence of rigorous work with such ideas in the humanities. Jeremy Gray's histories of non-Euclideanism, meanwhile, provide a foundational background against which to consider the distinct but closely related geometry of *n*-dimensions. His account of the Modernist transformation of mathematics treats *n*-dimensional geometry as a case study of popularization, an account underlined here. His picture of Modernist mathematics as abstract and axiomatic interpolates mathematics and philosophy at the expense of broader Modernist culture, and our concerns diverge in that. Both Plotnitsky and Gray are able to draw on considerably greater mathematical and philosophical erudition than I can claim and I can only beg the patience of my readers in this matter.[24]

In spirit, if not in method or period, I would also like to be able to claim kinship with the research of Brian Rotman, another gifted mathematician and philosopher,

---

[21] Roger Luckhurst, *The Invention of Telepathy* (Oxford: Oxford University Press, 2002), p. 2.

[22] Steven Connor, 'Afterword', in *The Victorian Supernatural*, ed. Nicola Bown, Carolyn Burdett, and Pamela Thurschwell (Cambridge: Cambridge University Press, 2004), pp. 258–77 (p. 268).

[23] Connor, 'Afterword', p. 274.

[24] See Arkady Plotnitsky, *The Knowable and the Unknowable: Modern Science, Nonclassical Thought, and the 'Two Cultures'* (Michigan: University of Michigan Press, 2002); Jeremy Gray, *Plato's Ghost: The Modernist Transformation of Mathematics* (Princeton: Princeton University Press, 2008).

whose works on zero and infinity have focused on the semiotics of mathematical concepts. I have allowed the present project to be guided instead by the cultural work done by the idea of higher space, rather than the functioning of the sign co-opted for its signification. Rotman concludes that to continue to the end his project to pursue zero

> would require a semiotics that went beyond zero to the whole field of mathematical discourse. A semiotics which, in order to begin at all, would have to demolish the widely held metaphysical belief that mathematical signs point to, refer to, or involve some world, some supposedly objective eternal domain, other than that of their own human, that is time bound, changeable, subjective and finite making.[25]

A very similar conclusion might be reached about the fourth dimension: that without any measurable basis in the empirical it must be a product of thought alone. Yet I am less interested in the idea of an eternal signified behind higher space—or the need to demolish such a metaphysical notion—than in the ways such a notion or metaphysics has been thought.

In order to ground the methodological promiscuity indicated by this diverse and interdisciplinary body of informative work, I proceed first and foremost from primary research.[26] Chapter 1 describes the disparate conditions for the emergence of higher-dimensioned space as a cultural object. It takes as an inauguration point James Joseph Sylvester's 1869 Address to the British Association and in so doing marks this book as an English-language study. On publication, Sylvester's address launched a lively discussion of Kantian space. As Kant's philosophy was the context within which British intellectuals thought matters spatial, I describe Kant's original work on space, and particularly his thoughts on the dimensionality of space, in some detail. Isobel Armstrong considers Kantian space 'foundational' for the nineteenth-century novel, so this groundwork is essential for later considerations of how higher spatial thought works in fiction.[27]

Discussions over Kantian space in British scholarly journals in the 1870s revealed stark disagreements over the relationship of geometry to space and the nature of space itself. The structures of the symbolic languages under discussion were often obscured. The dimensional analogy was presented as a guaranteed representation of the problem at stake but its reliance on metaphor was not always made clear. Returning to scholarly discussions in British periodicals identifies the persistent use of analogy as a rhetorical device for explaining the ideas of dimensionality, and work on this formulation continues throughout the book. The dimensional analogy identifies the significance of modelling as a process of thought and the work done by the forms that mediate abstractions. It identifies,

---

[25] Brian Rotman, *Signifying Nothing: The Semiotics of Zero* (Basingstoke: The Macmillan Press, 1987), p. 107.

[26] I should note that digital archives and databases, from collections of nineteenth-century periodicals to online archives of Theosophical magazines, have been invaluable to this project. That is not to say that it is characterized by an approach that might be described as originating in the digital humanities, but that it is certainly post-digital.

[27] Isobel Armstrong, 'Spaces of the Nineteenth-Century Novel', in *The Cambridge History of Victorian Literature*, ed. Kate Flint (Cambridge: Cambridge University Press, 2012), pp. 575–97 (p. 575).

too, the fact that geometry itself is a model of the more abstract form that is space; it alerts us to a structural shift between domains early in the life cycle of the fourth dimension, as it leaves geometry—a domain of pure thought—to enter space, a phenomenon of the physical world.

The exchanges between algebraic geometry and descriptive geometry that accompanied the development of non-Euclideanism can be complicated in a period in which geometric theories proliferated. This chapter therefore shines some light on these interchanges, defining the relations between *n*-dimensional and non-Euclidean geometry, their shared heritage and the points at which they diverge. So, too, does it consider Henry More's 'spissitude', an example of higher-dimensional thought that was developed in immediate response to Descartes.

Chapter 2 concentrates on a crucially catalytic episode in the history of cultural higher space and on a particularly pregnant form at the heart of this episode. The series of experimental seances conducted by the Leipzig-based astrophysicist Johann Carl Friedrich Zöllner with the medium Henry Slade and the knot that Zöllner proclaimed as experimental evidence of the fourth dimension are encountered as soon as one starts researching the fourth dimension in the nineteenth century.[28] This chapter outlines Zöllner's theoretical position and its sources; his allegiances and feuds; the experiments themselves and their legacy. Zöllner drew higher-dimensioned space into occultist discourse, a field in which it can still be discerned. This shift requires the mobilization of different resources: attention to the historical phenomenon of popular spiritualism and its discourse networks; consideration of the relations between professionalizing science and spiritualism; and the negotiation of the knot, an object that is thing and idea, form and material, mediator and terminus.

As an historical case study of a pseudo-scientific experiment drawing on resources from literature, popular performance, occult texts and journals, and rigorously scientific work, Chapter 2 is best considered under the aegis of its guiding figure, the knot. It is nevertheless here that fraternity may be discerned with Roger Luckhurst's *The Invention of Telepathy* and, in shared nineteenth-century literary interest, Daniel Brown's work on the poetry of Victorian mathematicians.[29] I share with Brown—and, indeed, with Clerk Maxwell, author of a poem more capably read by Brown but also considered here—an interest in analogy. Brown uses Maxwell's essay for the Apostles, 'Are there Real Analogies in Nature?', as a lever for reading his poetry, highlighting Maxwell's insistence on the reciprocity of puns and analogies and noting that the essay's opening remarks 'define Maxwell's poised orientation to the problem of knowledge, which he recognizes here as both metaphysical and cultural'. Brown continues to remark that: 'Modern physics, a milieu of field theory, ethers, energy and mathematical models, lends itself to the use of analogies with their attendant risks of semantic duplicity, conflation,

---

[28] See Corinna Treitel, *A Science for the Soul: Occultism and the Genesis of the German Modern* (Baltimore: Johns Hopkins University Press, 2004) for a particularly rich account of Zöllner's experiments.
[29] Daniel Brown, *The Poetry of Victorian Scientists: Style, Science and Nonsense* (Cambridge: Cambridge University Press, 2013).

and indeed febrile imaginings.'[30] The current study supports this argument, particularly in relation to the Zöllner event but also more broadly.

Chapter 3 considers the first English-language higher spatial fiction, Edwin Abbott Abbott's *Flatland* (1884). *Flatland* marks a crucial juncture in the *n*-dimensional turn, a deceptively complex and playful literary response that inserted the palpably difficult idea of higher-dimensional space into the cultural sphere. *Flatland* casts a long shadow, appropriately enough for a two-dimensional text that explores the raised dimensional consciousness of its planar narrator. I disinter its satirical forebears and work through the criticism of the text in an effort to construct a usable critical paradigm, focusing on pedagogy, rhetoric, imagination, and space. I am particularly interested in the play with analogy of *Flatland*, and how it may consider its own functioning as a model.[31] My reading of *Flatland* seeks to stress the originality and staggering difference of this text, both to its peers and to its author's other work.

I consider at some length the continuation of the text by A Square—*Flatland*'s purported author—into post-publication correspondence. This meta-textual ploy works to expand the text dimensionally, constituting new dimensions of reading in distinct locations, making *Flatland* an active participant in modelled higher dimensionality. In working out what this text is, we discover that A Square may yet be our most capable guide into higher space.

Chapter 4 focuses on the work of Charles Howard Hinton, arguably the least well-known yet most influential theorist of higher space of the late nineteenth century. If there is one thinker around whose work this study is organized it is Hinton. Early encounters with his extraordinary, quasi-visionary work catalysed this project. Linda Dalrymple Henderson has done as much as anyone to recuperate Hinton's work in a scholarly context, but there has been sporadic interest from other critics and writers. Bruce Clarke, whose study of allegory focusing on thermodynamics devoted a chapter to Hinton, is the most notable among these and is a highly suggestive reader of Hinton's work. More recently Elizabeth Throesch has written a thesis on Hinton's *Scientific Romances*. Outside the academy, the SF author and critic Rudy Rucker has edited and engaged with his project in a thoroughly hands-on manner, but otherwise Hinton is not much read or considered.[32] Hinton's work is the heavy mass around which much else in this book orbits.

Indeed, the term higher space is itself Hinton's, a contraction of the phrase higher-dimensioned space that for Hinton carried with it the necessary implication of ascension to a higher consciousness for those who were able to educate themselves to visualize the fourth dimension. I prefer higher space to the more

[30] Brown, *Poetry of Victorian Scientists*, p. 40.

[31] See Andrea Henderson, 'Math for Math's Sake: Non-Euclidean Geometry, Aestheticism, and "Flatland"', *PMLA*, 124 (2009), 455–71; and Mark McGurl, 'Social Geometries: Taking Place in Henry James', *Representations*, 68 (1999), 59–83.

[32] Bruce Clarke, *Energy Forms: Allegory and Science in the Era of Classical Thermodynamics* (Ann Arbor: University of Michigan Press, 2001); Elizabeth Lea Throesch, 'The *Scientific Romances* of Charles Howard Hinton: The Fourth Dimension as Hyperspace, Hyperrealism and Protomodernism', doctoral thesis, University of Leeds, 2007; see, for example, Rudolf v. B. Rucker, *Speculations on the Fourth Dimension: Selected Writings of Charles H. Hinton* (New York: Dover Publications, 1980).

common hyperspace for the reason that it sidesteps the prepositional confusion that surrounds the fourth dimension. The American mathematician G.B. Halsted briefly toyed with the term 'pro-space', before settling on 'meta-space'. The radical journalist W.T. Stead coined the notably ugly 'throughth' and the dabbling columnist Rev. J.B. Bartlett suggested 'inwardness'.[33]

There is good reason for this prepositional confusion: many prepositions are spatial and all spatial prepositions are derived from the experience of lived space. They rapidly prove insufficient for describing relationships or movements in a space that has never been lived or perceived. What is the meaning of 'above' or 'behind' for a four-dimensional being? In short, all these phrases elide a term every bit as important as space in this idea: dimension.

Hinton's work hinges my own investigations. For Elizabeth Throesch, 'Hinton's hyperspace philosophy is [...] concerned with mediation, the ways in which the consciousness thinks and creates with and through the aesthetics of space.'[34] 'Hinton was an important mediating figure,' writes Steven Connor, 'because, like some of the physical scientists who investigated Spiritualism, his grasp of scientific principles was extensive and subtle.'[35] Indeed, his work fed into the literature of occult groupings, avant-garde art, Modernist poetry and fiction, and back into geometry and orthodox science.

I give a brief account of the work of Hinton's father, James, who developed an idiosyncratic philosophy of service, and a more detailed account of Hinton's own work, highlighting his acknowledged and implied sources, Kepler, Kant, and his father, before focusing on his invention of a system of cubes for training the subject in the visualization of higher space. This set of cubes are again read as quasi-objects, things that make fluid the distinction between thinking thing and thing thought on, between mind and material object.

Chapter 5 considers the development in popular occultism of higher spatial ideas, with an emphasis on the Theosophical Society's correlation of the fourth dimension with the astral plane. While Kant was canonical within the academy, philosophies of space which held geometry as an expression of the sacred informed the occult networks within which the fourth dimension provoked at least as much discussion as it did among establishment thinkers. Unpacking a brief history of Platonic geometry and its afterlives in the Neoplatonic tradition gives some context for the popular mystifications that will be discussed in more detail in later chapters. From the *Timaeus* and the geometrically formed view of nature described therein to the geometric cosmos of Johannes Kepler, the foundational and ritual aspects of geometry provide essential background for the development of higher space in occult contexts. The popular impact of the appropriation of higher spatial ideas by leading figures in the Theosophical Society and far briefer but no less important engagements on the part of figures whose influence in the period

---

[33] W.T. Stead, 'Throughth: Or, On the Eve of the Fourth Dimension: A Record of Experiments in Telepathic Automatic Handwriting', *Review of Reviews*, 7 (1893), 426–32 (p. 426); Rev. J.B. Bartlett, 'A Glimpse of the "Fourth Dimension"', *The Boy's Own Paper*, 12 (1890), 462.
[34] Throesch, 'The *Scientific Romances*', p. i.   [35] Connor, 'Afterword', p. 264.

has long been established—Edward Carpenter and W.T. Stead—describes the elaboration and cross-fertilization of various speculated supernatural phenomena with higher-dimensioned space and the elaboration of these relationships in overlapping social groupings and a popular but contested body of literature. This chapter stresses a shift in rhetorical emphasis: in these accounts we have moved beyond speculation and into the presentation as empirical of ideas that maintain no basis in the empirical.

Chapter 6 considers how cultural conceptions of space had been shifted by higher spatial thought in its various forms and how this was reflected in the popular and literary fiction of the fin de siècle. Investigating the production of space in fin de siècle literature, it focuses on embodiment, the senses, and particularly narrative voice and mood. A newly configured spatiality that owes its conception to higher space becomes a driving force behind certain techniques of narrative fiction in the period and plays directly into Modernism.

This chapter makes a return to some of the concerns of Chapter 3, in recreating the literary production of higher space in this period and considering what space did to fiction. It operates in the scholarly space opened up over a quarter of a century ago by Gillian Beer's *Darwin's Plots*, by seeking to read the ways in which a scientific theory '[h]as been assimilated and resisted by novelists who, within the subtle enregisterment of narrative, have assayed its powers. With varying degrees of self awareness they have tested the extent to which it can provide a determining fiction by which to read the world.'[36]

This is not, of course, to argue that the ideas of n-dimensional geometry should be regarded on a par with evolutionary theory. Far from it: the thinkers encountered in this book who try to make the fourth dimension an aspect of the human are the most metaphysical contortionists of all, although, as I hope to show in Chapter 1, the two theories were on familiar terms. Higher space remains a highly potent imaginative construction, to this day, perhaps because over a century later it retains the powers Beer ascribes to new theories: that they 'rebuff common sense. They call on evidence beyond the reach of our senses and overturn the observable world. They disturb assumed relationships and shift what has been substantial into metaphor.'[37]

This imaginative construction is what I am after. In terms of the account given in Stephen Kern's *The Culture of Time and Space* (1983), this book seeks to amplify higher spatial thinking, all but ignored there. In relation to the burgeoning body of criticism that orbits the definitively modern condition of agoraphobia, this book wants to make a new suggestion: that the material conditions typically thought of as the root cause of *peur d'éspace*—urbanization and the alienating experience of living in the city—might have a metaphysical corollary, if not competitor. Just as Linda Dalrymple Henderson has shown that there was a rich tradition of intellectual engagement with the idea of the fourth dimension in art before public engagement

---

[36] Gillian Beer, *Darwin's Plots: Evolutionary Narrative in Darwin, George Eliot and Nineteenth-Century Fiction* (London: Routledge & Kegan Paul, 1983), p. 4.
[37] Beer, *Darwin's Plots*, p. 3.

with the ideas of relativity theory, so by expanding that arena to include literature, we can see how that prior thinking was more prevalent than has been previously considered. Whether we view space as innate or learned, the current study argues that we should figure its imaginative power as significant as its lived experience.

Thought and the shapes it can take are therefore the thematic anchors of *The Emergence of the Fourth Dimension*. Analogy emerged as an intellectual theme of this study through thinking the relationship between geometry and space as indicated by the persistent use of this rhetorical construction in efforts to explain the idea to a popular audience. Thinking in terms of models and translations drew attention to another persistent feature of higher-dimensional thought: its frequent recourse to mediating its ideas through material things.

What then is the final work? Why work in this way? As an early career researcher with no formal mathematical training whatsoever I lay no claim to authority on the mathematical aspects of *n*-dimensional geometry. Yet *n*-dimensional thought so rapidly exceeded the mathematical—how could it not, this 'fairyland of geometry', introducing as it did such fantastic possibilities for mind and body— that reading it requires familiarities beyond the mathematical. The method pursued may frustrate some readers by being insufficiently literary critical. To such readers I can only respond that it is not my aim to criticize. I offer, instead, Gillian Beer's remark that 'more is to be gained from analysing the transformations that occur when ideas change creative context and encounter fresh readers'.[38]

I hope by tracking higher space, and playing with the forms it took, to be able to offer something in the way of an illumination of the shapes of this mode of thought, how they come about and how we then relate these forms to each other. In an introduction to an issue of *Critical Quarterly*, Steven Connor described the working out of an idea of the practice of cultural phenomenology that might describe the aims of this project. Such a practice would

> enlarge, diversify and particularize the study of culture. Instead of readings of abstract social and psychological structures, functions and dynamics, cultural phenomenology would home in on substances, habits, organs, rituals, obsessions, pathologies, processes and patterns of feeling. Such interests would be at once philosophical and poetic, explanatory and exploratory, analytic and evocative. Above all, whatever interpreting and explication cultural phenomenology managed to pull off might well be accomplished in the manner of its getting amid a given subject or problem, rather than the completeness with which it got on top of it. It would inherit from the phenomenological tradition an aspiration to articulate the worldliness and embodiedness of experience—the in-the-worldness of all existence.[39]

This seems to me a method worth aiming for, even if failing to achieve it might be a prerequisite.

---

[38] Beer, 'Translation or Transformation?', 81.
[39] Steven Connor, 'Making an Issue of Cultural Phenomenology', *Critical Quarterly*, 42 (2000), 2–6.

1

Conditions of Emergence
Kant, Helmholtz, and Analogy

In his inaugural Presidential Address to the Mathematical and Physical Section of the British Association at Exeter in August 1869, James Joseph Sylvester presented an overview of the continental developments in geometric thought and how they related to space.[1] He introduced the idea of n-dimensional space in an analogy:

> for as we can conceive beings (like infinitely attenuated bookworms in an infinitely thin sheet of paper) which possess only the notion of space of two dimensions, so we may imagine beings capable of realising space of four or a greater number of dimensions.

Sylvester's analogy was apparently borrowed from Gauss, via his biographer, and embedded within an appreciation of the recent non-Euclidean work of Bernard Riemann: 'Like his master Gauss, Riemann refuses to accept Kant's doctrine of space and time being forms of intuition, and regards them as possessed of physical and objective reality.'[2]

When his address was published in *Nature* as 'A Plea for the Mathematician', Sylvester added copious footnotes. In these he expanded on his positions on both higher space and Kant, referring in the case of the former to the work of W.K. Clifford, his colleague at University College, who had

> indulged in some remarkable speculations as to the possibility of our being able to infer, from certain unexplained phenomena of light and magnetism, the fact of our level space of three dimensions being in the act of undergoing in space of four dimensions (space as inconceivable to us as our space to the suppositious bookworm) a distortion analogous to the rumpling of the page.[3]

His critique of Kant was expanded to draw in a number of influential British thinkers:

> It is very common, not to say universal, with English writers, even such authorised ones as Whewell, Lewes or Herbert Spencer, to refer to Kant's doctrine as affirming

---

[1] For a definitive account of the life of Sylvester see Karen Hunger Parshall, *James Joseph Sylvester: Jewish Mathematician in a Victorian World* (Baltimore: Johns Hopkins University Press, 2006).

[2] J.J. Sylvester, 'A Plea for the Mathematician', *Nature*, 1 (1869), 237–9, 261–3. Sylvester was responding to T.H. Huxley, who had recently published an article in which he had distanced mathematics from the natural sciences, writing that 'mathematics is that study which knows nothing of observation, nothing of experiment, nothing of induction, nothing of causation'. T.H. Huxley, 'The Scientific Aspects of Positivism', *The Fortnightly Review* (1869), 653–70 (p. 667).

[3] Sylvester, 'A Plea', 238.

space 'to be a form of thought' or 'of the understanding.' This is putting into Kant's mouth (as pointed out to me by Dr. C. M. Ingleby), words which he would have been the first to disclaim, and is as inaccurate a form of expression as to speak of 'the plane of a sphere', meaning its surface or superficial layer, as not long ago I heard a famous naturalist do at a meeting of the Royal Society.[4]

G.H. Lewes, working at the time on his overview of psychology and perception, *Problems of Life and Mind* (1874), wrote to defend himself against the accusation of inaccuracy. Suggesting that those who claimed to have read Kant outnumbered those who actually had, and implying that Sylvester might be among their number, Lewes argued that 'there is no discrepancy at all in also saying that he taught space to be a "form of thought," since every student of Kant knows that intuition without thought is mere sensuous *impression*'.[5] A fortnight later Huxley, Ingleby, and Sylvester all wrote to challenge Lewes's inaccurate turn of phrase and to cite passages from the original to clarify Kant's categories. Lewes dug in over the course of a heated correspondence, as various writers challenged him. The final word, before the editor closed down the debate, was allowed to the Belfast philosopher W.H. Stanley Monck, who directed attention towards the 'Transcendental Aesthetic' and the 'Transcendental Analysis' sections of Kant's *Critique* to provide a final, contextual counterargument:

> The criterion by which Kant distinguishes between Intuition and Thought (under which term he includes both the understanding proper and the reason proper) is that, in the former the mind is passive (receptive) while, in the latter, it is spontaneously active; and it is precisely on this ground—the passive reception of them by the mind—that he refers Space and Time to Sensibility rather than Thought.[6]

Sylvester's tipster C.M. Ingleby, meanwhile, threw back at the mathematician his remark that although he himself could not conceive of it, 'if Gauss, Cayley, Riemann, Schalfli [*sic*], Salmon, Clifford, Krönecker, have an inner assurance of the reality of transcendental space, I strive to bring my faculties of mental vision into accordance with theirs'.[7] Ingleby responded: 'It would be more satisfactory to unbelievers like myself if the gifted author of the address were to assure the world that he had an insight into, or clear conception of, this transcendent space.'[8] Pushing through the vexed and pedantic discussions of Kantian categories, Ingleby indicated an element just as crucial to Sylvester's argument: conception of this idea could not be delegated or taken on authority.

Sylvester's address and the debates it founded indicate the terrain for the emergence of higher space as a cultural object in Britain. From this starting point we can work outwards: as Sylvester wrote, he, Salmon, and Cayley had 'all felt and given evidence of the practical utility of handling space of four dimensions as if it were a conceivable space' in their mathematical work; his address was not the origin in the English language of the idea of higher space, but it was the inauguration of a discourse around it that exceeded the confines of those papers.

---

[4] Sylvester, 'A Plea', 238.　　[5] G.H. Lewes, 'Kant's View of Space', *Nature*, 1 (1870), 289.
[6] W.H. Stanley Monck, 'Kant's View of Space', *Nature*, 1 (1870), 386.
[7] Sylvester, 'A Plea', 238.　　[8] Dr. C.M. Ingleby, 'Transcendent Space', *Nature*, 1 (1870), 289.

That this discourse began with a high-profile squabble over Kantian categories of thought makes clear the persistent influence of Kantian philosophy over space in the British academy of the late nineteenth century. Kant was foundational but not universally agreed upon. Kant's insistence on space as an a priori form of thought was brought to the fore by the thought of a space that certainly was not received a priori and the tensions between the innate and the empirical in perception were one axis around which higher spatial debate revolved. That Lewes's error roused biologist, mathematician, and philosopher alike illustrates the interstitial landscape into which we must strike: the contestation of space occurred at the intersection of numerous disciplines and drew scholars of all hues into turbulent terrain.

It might be observed that higher space flourished in the cracks of Kantian metaphysics. Kantian space will be an orientation point throughout this book, precisely because it was for theorists of higher space throughout the period under investigation. Kant is essential background but his work does not provide the only set of conditions for the emergence of higher spatial thought. Another set is brought into play by a parallel tradition of philosophical thought—Neoplatonism—that survived within esoteric discourse and drew on pre-Kantian philosophies. Indeed, these alternative traditions of both geometric and spatial thought were particularly popular within the occult revival of the late nineteenth century, and their relationship to higher spatial thought is no coincidence: these more mystical traditions provided a fertile seedbed for the seemingly sterile constructions of abstract geometry by maintaining pre-Kantian ideas of space that were congruent with aspects of the suggested fourth dimension.

Spatial philosophies provide underlying conditions for a spatial imaginary but the specific set of ideas that prompted their revision were developed within a mathematical context. Sylvester's address indicates the contemporary geometric sources that will need fleshing out. He refers to several fellow mathematicians—Riemann, Gauss, Bolyai, Lobachevsky—and indicates innovations. He also highlights 'remarkable speculations' that applied this new geometry in the physical sciences. There was extraordinary mobility in geometry in this period and this mobility destabilized fixed ideas of space and the relationship between mathematical and physical thought. What were the competing currents? To what extent was n-dimensional geometry related to non-Euclidean geometry? How were these intermingled but distinct currents received?

If we are to ask how these ideas were received, we should ask also how they were delivered. The language of geometric and spatial discourse is crucial, particularly the rhetorical device of analogy. Analogy had been a commonplace of scientific argument since Aristotle and became particularly significant in a period of engaged scientific popularization as practitioners sought to explain to lay audiences the concepts with which they were working. This construction was central to higher-dimensional reasoning throughout the late nineteenth century and beyond. I want to consider how the spatial imagination responds to this kind of rhetorical construction and more broadly to think about the types of language, and consequently the types of images, used to describe spatial and higher spatial concepts.

KANTIAN SPACE

The outbreak of Kantian discussion following the publication of Sylvester's 'Plea' in *Nature* indicates the extent to which the idea of higher space drilled directly into Kant's writing. The subsequent English-language discourse surrounding the idea of higher space frequently seems to be an extended negotiation of Kantian spatial philosophy.

A survey of Kant's developing thinking on space and geometry is required orientation and such a survey reveals the source of some of the differences in the late nineteenth century. Kant's thinking changed a great deal between 'The True Estimation of Living Forces', his first published work written when he was only twenty-two, and the mature transcendental philosophy of the *Critique of Pure Reason*, but his concern with the subject of space remained central throughout.

'The True Estimation' explicitly addressed dimensionality. The prevailing philosophical view of space in 1747 was derived from the work of Christian Wolff, in which it was held that objects possessed no force of their own. Kant argued against this, claiming that space was relational and derived from the force objects possessed in relation to each other: 'It is easily proved that there would be no space and no extension, if substances had no force whereby they can act outside themselves.'[9]

Kant wrote that he had considered proving tri-dimensionality through 'what is to be observed in the power of numbers' and insisted that the fourth dimension was simply a reiteration of an earlier dimension because it was two squared: 'For in anything representable through the imagination in spatial terms, the fourth power is an impossibility.' Space was assumed to be three-dimensional and force was the root of tri-dimensionality:

> The threefold dimension seems to arise from the fact that substances in the existing world so act upon one another that the strength of the action holds inversely as the square of the distances.[10]

Ultimately, though, there was a power greater than that of numbers. Failing to produce a fixed answer, the young Kant deferred to an omnipotent creator:

> This law is arbitrary and [...] God could have chosen another, for instance the inverse threefold relation; and lastly, that from a different law an extension with other properties and dimensions would have arisen. A science of all these possible kinds of space would undoubtedly be the highest enterprise which a finite understanding could undertake in the field of geometry.[11]

Kant's early engagement with the idea of dimensionality therefore provides potential for future respondents to locate him as a prophet of higher dimensionality, a Super-Geometric seer issuing an invitation to speculative geometers. Despite its theistic wriggle, we also read the seeds of what would become a cornerstone of his critique of metaphysics, his insistence on the *a priorism* of space, in a construction

---

[9] Immanuel Kant, 'Thoughts on the True Estimation of Living Forces (Selected Passages)', in *Kant's Inaugural Dissertation and Early Writings on Space, trans. and ed. John Handyside* (Westport, CT: Hyperion Press, 1979), p. 10.

[10] Kant, 'Thoughts on the True Estimation', p. 11.

[11] Kant, 'Thoughts on the True Estimation', pp. 11–12.

that anticipates the latterly more developed claim of correlation, the lighthouse around which contemporary critique of Kant's project navigates:

> The impossibility, which we observe in ourselves, of representing a space of more than three dimensions seems to me due to the fact that our soul receives impressions from without according to the law of the inverse square of distances, and because its nature is so constituted that not only is it thus affected but that in this same manner it likewise acts outside itself.[12]

In the short essay 'Concerning the Ultimate Ground of the Differentiation of the Directions in Space', Kant began to describe 'an absolute and original space' that was subjective but not reliant upon the senses. He highlighted a shortcoming in Leibniz's *analysis situs*: that it did not distinguish between objects of opposite orientation, considering equilateral triangle A, pointing to the left, exactly the same as equilateral triangle B, pointing to the right.

Kant still assumed the tri-dimensionality of space: 'Because of its three dimensions, physical space can be thought of as having three planes, which all intersect each other at right angles.'[13] These planes not only defined the body, but also cognition of objects outside the body. He identified a curiosity with the vertical plane that split the body into left and right: based on subjective physical experience there were objects that were either left-handed or right-handed or rotated clockwise or anticlockwise. Kant identified many such examples: the swirl of hair on the crown of the head; hops growing up a pole; snail shells. He enlisted these naturally occurring objects to prove a point:

> The ground of the complete determination of corporeal form does not depend simply on the relation and position of its parts to each other; it also depends on the reference of that physical form to universal absolute space, as it is conceived by geometers.[14]

Objects, such as right and left hands, which could not be made to occupy the same limited space—'can be exactly equal and similar, and yet still be so different in themselves that the limits of the one cannot also be the limits of the other'—he termed 'incongruous counterparts'. These proved that relationality was not between objects but towards a geometric space and put an insurmountable barrier in the path of *analysis situs*, 'according to which space simply consists in the external relations of the parts of matter which exist alongside each other':

> Our considerations make it plain that the determinations of space are not consequences of the positions of the parts of matter relative to each other. On the contrary, the latter are consequences of the former [...] Finally, our considerations make the following point clear: absolute space is not an object of outer sensation; it is rather a fundamental concept which first of all makes possible all such outer sensation.[15]

---

[12] Kant, 'Thoughts on the True Estimation', p. 12.

[13] Immanuel Kant, 'Concerning the Ultimate Ground of the Differentiation of the Directions in Space', in *Cambridge Edition of the Works of Immanuel Kant: Theoretical Philosophy 1755–1770*, ed. and trans. David Walford in collaboration with Ralf Meerbote (Cambridge: Cambridge University Press, 1992), pp. 361–72 (p. 366).

[14] Kant, 'Concerning the Ultimate Ground', p. 369.

[15] Kant, 'Concerning the Ultimate Ground', p. 371.

In these thoughts we read the emerging development of the philosophy that reached its fruition in the *Critique of Pure Reason*. His 'Inaugural Dissertation' built on the idea of space as a priori, the ground of external sense. In Section 15, he gave a series of propositions 'On space' that serve as the defining features of Kantian space:

> A. The concept of space is not abstracted from outer sensations [...] B. The concept of space is a singular representation embracing all things within itself [...] C. The concept of space is a pure intuition [...] D. Space is not something objective and real, nor is it substance, nor an accident, nor a relation; it is, rather, subjective and ideal; it issues from the nature of the mind in accordance with a stable law as a scheme, so to speak, for co-ordinating everything which is sensed externally.[16]

Under the banner of this fourth proposition, Kant considered again the view of relational space. He argued that those who supported it were 'in headlong conflict with the phenomena themselves, and with the most faithful interpreter of all the phenomena, geometry'.[17] Such thinkers reduced geometry to an empirical science based on experience and induction and denied its 'necessity'. If they were correct 'we might hope, as happens in empirical matters, one day to discover a space endowed with different fundamental properties'.[18] This was not a matter of leaving the door ajar for such a possibility: as he had stated, Kant believed such a view to be incorrect.

The 'Inaugural Dissertation' displays significant revisions and progression from the view he had put forward in 'Living Forces'. He insisted with his fifth proposition that, despite being 'ideal', space was nevertheless 'a concept which is in the highest degree true'. He went on again to discuss the relationship between geometry and space, making clear that he saw the two as indivisible and in this binding creating a ground for much of the discourse that would surround the idea of a geometry that did not correlate to lived space and how an unsensed space might result from this.

Since space is a prerequisite, a mediator of experience, then everything that comes before the senses 'conforms with the fundamental axioms of space and its corollaries (as geometry teaches) [...] Accordingly, nature is completely subject to the prescriptions of geometry, in respect of all the properties of space which are demonstrated in geometry.' The development of non-Euclidean and *n*-dimensional geometries would, after Kant, seemingly sanction the existence of such spaces.

> Assuredly, had not the concept of space been given originally by the nature of the mind (and so given that anyone trying to imagine any relations other than those prescribed by this concept would be striving in vain, for such a person would have

---

[16] Immanuel Kant, 'On the Form and Principles of the Sensible and the Intelligible World [Inaugural Dissertation]', in *Cambridge Edition of the Works of Immanuel Kant: Theoretical Philosophy 1755–1770*, ed. and trans. David Walford in collaboration with Ralf Meerbote (Cambridge: Cambridge University Press, 1992), pp. 373–416 (pp. 395–7).
[17] Kant, 'On the Form and Principles', p. 397.
[18] Kant, 'On the Form and Principles', p. 398.

been forced to employ this self-same concept to his own fiction), then the whole of geometry in natural philosophy would be far from safe.

This sentence indicates the fictive nature, on Kantian grounds, of *n*-dimensional spaces. Anyone trying to imagine such spaces would be using the concept he was trying to imagine as the basis for his 'fiction'. Kant continues: 'For one might then doubt whether this very concept of space, which had been derived from experience, would agree sufficiently with nature, since the determinations from which it had been abstracted might perhaps be denied.' The coils of his reasoning drawing ever tighter, Kant declared that space derived from experience would necessarily be on shaky ground.

This section of the 'Inaugural Dissertation' demonstrates the development of Kant's thinking regarding dimensionality from his earliest published work to his mature critical philosophy. In 'Living Forces', still cited by mathematicians as anticipating *n*-dimensional geometry, Kant allowed the possibility of such space through a theological dodge: he bowed before an assumed creating power. By the time of his 'Inaugural Dissertation', as his project to recast metaphysics was gathering momentum, Kant reasoned that the human imagination could only falsely produce such spaces, that Euclidean geometry was, like space, a priori to thought. Nevertheless, his tight coupling of geometry and space dictated that when internally coherent geometries were developed that were *n*-dimensional or non-Euclidean, that did not seem to subject nature to their prescriptions, the *logos* would have to be recast: one would have to entertain that space was similarly non-Euclidean or *n*-dimensional.

The 'metaphysical exposition of the concept of space' cast this reasoning within the terms developed for the 'transcendental aesthetic':

> Geometry is a science that determines the properties of space synthetically and yet *a priori*. What then must the representation of space be for such a cognition of it to be true? It must originally be intuition; for from a mere concept no propositions can be drawn that go beyond the concept, which, however, happens in geometry [...] But this intuition must be encountered in us *a priori*, i.e., prior to all perception of an object, thus it must be pure, not empirical intuition. For geometrical propositions are all apodictic, i.e., combined with consciousness of their necessity, e.g., space has only three dimensions.[19]

The problem we might identify is precisely the *a priorism* of space. In the 'Prolegomena', Kant addressed himself to those who had mistakenly assumed his idealism to imply that there was no objective reality. Again, he insisted upon the tri-dimensionality of space as intuitively necessary:

> That full-standing space (a space that is itself not the boundary of another space) has three dimensions, and that space in general cannot have more, is built upon the proposition that not more than three lines can cut each other at right angles in one point; this proposition can, however, by no means be proven from concepts, but rests

---

[19] Immanuel Kant, *Critique of Pure Reason*, ed. and trans. Paul Guyer and Allen W. Wood (Cambridge: Cambridge University Press, 1998), p. 176.

immediately upon intuition, and indeed on pure *a priori* intuition, because it is apodictically certain.[20]

The mature Kant of the Critical Philosophy and beyond was convinced both of the certainty of three-dimensional space and of the tight coupling of geometry with this space. The development of *n*-dimensional geometry therefore placed before Kantian spatiality a series of problems to which nineteenth-century thinkers responded with vigour. Lewes's suggestion that there were more who claimed to have read Kant than actually had should perhaps be modified to allow also for those who, like Lewes, had *mis*read him.

SPISSITUDE: A CURIOSITY?

In 1881 the Austrian historian of philosophy Robert Zimmermann published an account of a seventeenth-century curiosity: the Cambridge Neoplatonist Henry More's notion of 'spissitude' or, as both More and Zimmermann termed it, a 'fourth dimension'.[21] Zimmermann was a student of Herbart, whose description of the correspondences between psychological and intelligible space relied upon a description of the three-dimensional manifold and informed Hermann von Helmholtz's speculative essays (discussed in this chapter, in 'Three-Space in Four-Space'). His commentary on More's 'spissitude' registered a response to the currency of ideas of the fourth dimension in German-language spiritualist circles, described in detail in Chapter 2: the following year he would publish *Anthroposophie* (1882), the book that would give Rudolf Steiner a philosophical blueprint for his spiritual system following his split from the Theosophical Society. More's 'spissitude' was unorthodox but directly responsive to the theories of space that dominated the philosophy of the period and developed in correspondence with Descartes.

Writing in 1926, mathematical historian Florian Cajori summarized historical approaches to the idea of higher dimensionality from Aristotle to the seventeenth century, approaches that were, time and again, 'repulsed':

> We have now cited the judgments of eight thinkers, distributed in time over 2000 years and geographically over Greece, Egypt, Italy, Germany, and France. These men rejected the possibility of a space of more than three dimensions as at variance with our external sense-perception. Theirs were arguments based on experience, much like that of the discouraged fat man who was certain there was no fourth dimension, for if there was one, he surely would have it.[22]

---

[20] Immanuel Kant, 'Prolegomena to Any Future Metaphysics', in *Cambridge Edition of the Works of Immanuel Kant: Theoretical Philosophy after 1781*, ed. Henry Allison and Peter Heath, trans. Gary Hatfield, Michael Friedman, Henry Allison, and Peter Heath (Cambridge: Cambridge University Press, 2002), pp. 29–170 (p. 80).

[21] Robert Zimmermann, *Henry More und die vierte Dimension des Raumes* (Vienna: Carl Gerold's Sohn, 1881).

[22] Florian Cajori, 'Origins of Fourth Dimension Concepts', *The American Mathematical Monthly*, 33 (1926), 397–406 (p. 399).

Aristotle had been one of those who had rejected the idea of higher dimensions. Considering the difficulties of 'place' in Book IV of the *Physics*, Aristotle focused his reasoning on dimensionality and the impossibility of co-presence: 'Now it has three dimensions, length, breadth, depth, the dimensions by which all body is bounded. But the place cannot *be* the body; for if it were there would be two bodies in the same place.'[23] Over the first five sections of Book IV Aristotle demonstrated to his satisfaction the existence and nature of place, assessing the concept in terms of bounding, and contrasting place to Plato's space, which he read as the same as matter. He came to the conclusion:

> Hence the place of a thing is the innermost boundary of what contains it [...] For this reason place is thought to be a kind of surface, and as it were a vessel, i.e. a container of the thing. Further, place is coincident with the thing, for boundaries are coincident with the bounded.[24]

Over sections six to nine he considered competing theories of the void, 'a sort of place or vessel, which is thought to be full when it holds the bulk which it is capable of containing, void when deprived of that'.[25] Highlighting cases for interpenetration that were used as arguments for the void, Aristotle systematically argued against distinctions between void and place, writing that every definition of the void required the same bodies it suggested were vacated from the void, and using examples of volume displacement as illustrations. Addressing arguments that claimed 'the existence of rarity and density', a key problem of ancient Greek physics introduced by Anaximenes, to demonstrate the existence of void on grounds of compression and contraction, he concluded: 'From what has been said it is evident, then, that void does not exist either separate (either absolutely separate or as a separate element in the rare) or potentially, unless one is willing to call the cause of movement void, whatever it may be.'[26]

These Aristotelian concepts of space and opposition to the idea of the void dominated spatial thought in Europe throughout the early Middle Ages and oriented the scholastic tradition of the medieval universities. Edward Grant suggests that the Aristotelian denial of the existence of a void space 'must form the point of departure for any consideration of the history of spatial concepts from the Middle Ages to the Scientific Revolution'. Grant notes considerable consistency in the terms in which space was thought in the seventeenth century:

> Although scholastics almost unanimously rejected an infinite three-dimensional space whereas numerous nonscholastics accepted it, there was otherwise a surprising degree of agreement on certain spatial properties such as infinity, incorporeality, penetrability, indivisibility, lack of resistance, ability to coexist with bodies, homogeneity, immutability, and especially the exclusion of space from the traditional categories of substance and accident.[27]

---

[23] Aristotle, 'Physics', in *The Complete Works of Aristotle: The Revised Oxford Translation*, ed. Jonathan Barnes, 2 vols (Princeton: Princeton University Press, 1984), I, pp. 315–446 (p. 356).

[24] Aristotle, 'Physics', p. 361.   [25] Aristotle, 'Physics', p. 362.

[26] Aristotle, 'Physics', p. 369.

[27] Edward Grant, *Much Ado About Nothing: Theories of Space and Vacuum from the Middle Ages to the Scientific Revolution* (Cambridge: Cambridge University Press, 1981), p. 221.

That is not to say that spatial thought was static during this period. Given the scholastic context in which these arguments were taking place, Aristotle was required to cohabit with Christian theology, and the fusion effected between the two bodies of thought was often painful. Where could a Christian God exist within the various universes of the ancient Greeks: not only the Aristotelian cosmology, but the pagan, Pythagorean tradition that informed the geometric universe described by Timaeus? Was God co-extensive with space? Was God the same thing as space, or nature, as suggested by Spinoza? To assign to God three-dimensionality would be to limit Him to the conceptual framework of body from which the dimensionality of space had been extrapolated. For this reason dimensionality became a central issue in the spatial imaginary of the period. For the Jesuit scholars at Coimbra university in Portugal, authors of a series of significant commentaries on Aristotle, space was non-dimensional because nothing could be eternal and infinite but God.[28]

For Amos Funkenstein, the seventeenth century witnessed the emergence of a secular theology. Barriers between traditions that had been central to the Aristotelian and Scholastic traditions were eroded from the fourteenth century onwards as mathematical approaches were essayed in physics and then also in social theory, notably in Hobbes. Reading Galileo, Descartes, Leibniz, Newton, and Hobbes, Funkenstein proceeds by examining key themes, including the divine predicate of omnipresence, a particularly thorny issue for spatial theorists in the period. Where Descartes distinguished between *res extensae*, extended bodies, and non-extended spirits, Henry More insisted that spirits must also be extended. For More, the fact that the geometrical physics of Descartes could not explain different states of density was a problem that could be solved with a theological solution. God was his 'Spirit-in-Chief', but 'spirits' were no mere metaphysical entities for More: 'Forces, properties, spirits are often interchangeable terms,' writes Funkenstein.[29] More's spirits possess 'plastical power' as opposed to the mechanical power that characterizes the Cartesian idea of quantity of motion. More's spirits could penetrate each other *and* bodies, could expand and contract and in moments of interpenetration were intensified. In this way, spirits could account for states of density and, like God, they were 'all-penetrating': omnipresent.

More writes:

> And, that I may not dissemble in any way, although all material substances considered in themselves are measured [*contentae* = contained or enclosed] only in three dimensions, a fourth however is to be admitted in the universe, which can, I think, be sufficiently called essential spissitude. Which, although it refers most properly to those spirits which can contract their extension into a less Ubi, can however by an easy analogy be referred further to the mutual penetration of spirits, both of matter and of themselves, so that, wherever either many essences or more of essence is contained in

---

[28] See Grant, *Much Ado*.
[29] Amos Funkenstein, *Theology and the Scientific Imagination from the Middle Ages to the Seventeenth Century* (Princeton: Princeton University Press, 1989), p. 78. See also Alexander Koyré, *From the Closed World to the Infinite Universe* (Baltimore: Johns Hopkins Press, 1979).

some Ubi than that which is adequate to its amplitude, there is acknowledged this fourth dimension which I call essential spissitude.[30]

'Spissitude' was a quality of spirit, not an extension into higher Euclidean dimensions—indeed, *contra* Descartes, not an extension at all—and it was characterized by 'self-penetration'. This was a deliberate move beyond the Cartesian grid. As Alexander Jacob writes: 'This feature of self-reduplication allows spiritual extension to be absolute, that is, at once infinite and eternal.'[31] The distinction between extended, Cartesian, mathematical space was underlined by More in a shared analogy that described his notion of 'plastical power'; considering the malleability of wax, the same material considered by Descartes in his second meditation, More argued:

> For, unless one wishes to consider that a piece of wax extended, say, to an ell's length, and afterwards gathered and rolled up into the form of a globe, would lose some of its original extension on account of this globulation, it would be necessary for one to acknowledge that a spirit has not lost anything of either its extension or essence in its contraction of itself into a less space, but, as in the case of the above-mentioned piece of wax, its diminution of longitude is compensated by the present increment of latitude and profundity, so, in the spirit contracting itself, the recent diminutions of its longitude, latitude, and depth are compensated by the essential spissitude which it acquires by this contraction of itself.[32]

These features of More's spissitude make it an intriguing case in the prehistory of the spatial imagination. We find a theory of space that allows for spiritual interpenetration, for co-location—More wrote in *The Immortality of the Soul*: 'For I mean nothing else by Spissitude, but the redoubling or contracting of Substance into less space then [*sic*] it does sometimes occupy. And Analogous to this is the lying of two substances of several kinds in the same place at once.'[33] The development of such a theory in correspondence with Descartes's geometric purification of space and, indeed, Hobbes's Euclidean account of the social locates an alternative spatial tradition to the theories that came to dominate spatial epistemology at the moment of their conception. It illustrates also the close proximity of metaphysics and physics in the period that we would do well to recall when reading speculative spatial theories of the nineteenth century.

While Henry More was never an explicit source for late nineteenth-century theorists of higher space—although he was later an occasional addition to the Theosophical canon, possibly through Steiner—we find a key feature of the fourth dimension mapped out in his work, a heritage for the conceptual nexus of the

[30] Alexander Jacob, *Henry More's Manual of Metaphysics: A Translation of the Enchiridium Metaphysicum (1679) with an Introduction and Notes*, 2 vols (Zurich: Georg Olms Verlag Hildesheim, 1995), I, p. 121. Variant translations suggested by Anthony Ossa-Richardson.
[31] Jacob, *Henry More's Manual of Metaphysics*, p. xxv.
[32] Jacob, *Henry More's Manual of Metaphysics*, p. 121.
[33] Henry More, *The Immortality of the Soul*, ed. Alexander Jacob (Dordrecht and Lancaster: Nijhoff, 1987), p. 27.

fourth dimension as it was at the fin de siècle.[34] His theory was of interest to historians of science and scholars of early modern philosophy and was maintained through publications in these fields.

More's desire to return God to the space from which Descartes had effectively exorcized Him, to reanimate mechanical, gridded space, is what most closely aligns him to the occultists who adopted the mathematical fourth dimension as a legitimate space in which to locate the spiritual forms—Christian or otherwise—that were increasingly purified from the scientific world view of the late nineteenth century. Ernst Cassirer also notes that More's attempt to respiritualize the mechanistic space of Descartes informed Newton's absolute space.[35] The gridded space of European thought in this singular but historically significant British example was extended and interpenetrated by 'spirit'. This interpenetration was well repressed, however, during a period when Euclid gained supremacy in distinct fields of knowledge. As an historical artefact, 'spissitude' exemplifies the hybridity of space, its resistance to disciplinary purification.

THE DEVELOPMENT OF *N*-DIMENSIONS

Recent nineteenth-century studies have restated the position of unimpeachable prestige held by geometry in the first half of the century. Joan Richards has described how central geometry was to the educational institutions of the period and how the inclusion of a geometry paper in the Cambridge Tripos examinations made it a core body of knowledge for at least half of Britain's intellectual elite. She writes: 'Triumphantly accurate in both the subjective and the objective realms, geometry was the *summum bonum* of human knowledge.'[36] Alice Jenkins concurs, noting Whewell's wholesale borrowing from Kant and the lofty position in which Coleridge and Wordsworth held the science of space, complicating the picture with an account of the democratizing moves against foundation geometry of Baden Powell and Dionysus Lardner. Jenkins describes a shift in the perception of space and geometry in the mid-century, from 'Romantic abstract space: not space imagined as a thing, but the condition for imagining things', to an abstract space 'under threat not only of eradication from the education experience of increasing numbers of people but also of diminution into physical space, the kind of space which could be investigated with tools'.[37]

That the non-Euclidean geometries developed on the Continent were also instrumental in the turbulence surrounding geometry in this period is universally agreed, and forms the core of Richards's research: 'As first presented by the writings

[34] 'Henry More (1614–1687), one of the Cambridge Neo-Platonists, is sometimes added to this list [the Theosophical corpus of the seventeenth century].' Antoine Faivre, *Theosophy, Imagination, Tradition: Studies in Western Esotericism* (New York: State University of New York Press, 2000), p. 10.

[35] See Ernst Cassirer, *The Platonic Renaissance in England*, trans. James P. Pettegrove (Edinburgh: Nelson, 1953), p. 146.

[36] Joan L. Richards, *Mathematical Visions: The Pursuit of Geometry in Victorian England* (Boston: Academic Press, 1988), p. 2.

[37] Alice Jenkins, *Space and the 'March of Mind*' (Oxford: Oxford University Press)', pp. 152, 174.

of Helmholtz, Riemann and Clifford, non-Euclidean geometry was a radical mathematical development which threatened to drastically distort the Victorian intellectual tapestry in which geometrical study was so integrally woven.'[38] As the American mathematician G.B. Halsted noted in his 'Bibliography of Hyper-Space and Non-Euclidean Geometry': 'Hyper-Space [...] though springing at first from a purely analytical basis, has become intimately connected with the former [Non-Euclidean Geometry].'[39] An overview of the development of *n*-dimensional geometry is required to show the intimacy of this connection and allows us to begin to put some space between the two conjoined currents in geometric thought.

Sylvester's 1863 paper in *The Philosophical Magazine*, 'On the centre of Gravity of a truncated triangular pyramid, and on the principles of barycentric perspective', although it did not name the author of a 'well-known geometrical construction for finding the centre of gravity in a plane quadrilateral', built on the barycentric calculus of August Möbius. Möbius, quondam student of Gauss, and Professor of Astronomy at the University of Leipzig, had first speculated on *n*-dimensional geometry in a section of his 1827 paper *Der barycentrische Calcul*, entitled 'On Higher Space':

> It seems remarkable that solid figures can have equality and similarity without having coincidence, while always, on the contrary, with figures in a plane of systems of points on a line equality and similarity are bound with coincidence. The reason may be looked for in this, that beyond the solid space of three dimensions there is no other, none of four dimensions [...] For the coincidence of two equal and similar systems, A, B, C, D,...and A', B', C', D',...in space of three dimensions, in which the points D,E,...and D', E',...lie on opposite sides of the planes ABC and A'B'C', it will be necessary, we must conclude from analogy, that we should be able to let one system make a half revolution in a space of four dimensions. But since such a space cannot be thought, so is also coincidence in this case impossible.[40]

Möbius's conclusion that 'such a space cannot be thought' echoed Kant but his work on coincidence, or congruence, suggested that problems could be conceived that would question Kant's observations on handedness. By indicating that the impossible space of four dimensions would enable the inversion of solid figures, Möbius began to sketch the possibilities for the fourth dimension.

Sylvester's citation of Clifford and Cayley gave notice of British mathematical thought on the subject. Cayley's 1843 paper 'Chapters in the Analytical Geometry of (n) Dimensions' was originally published in the *Cambridge Mathematical Journal*.[41] This article was traditionally algebraic, but in 1846 Cayley returned to the subject, publishing in *Crelle's Journal* 'On Some Theorems of Geometry of Position', a more detailed piece of analytical geometry in which he explored the advantages of extending the spatial manifold:

---

[38] Richards, *Mathematical Visions*, p. 117.

[39] George Bruce Halsted, 'Bibliography of Hyper-Space and Non-Euclidean Geometry', *American Journal of Mathematics*, 1:3 (1878), 261–76 (p. 262).

[40] August Möbius, 'On Higher Space', in *Sourcebook in Mathematics*, ed. D.E. Smith, trans. Henry P. Manning (New York: McGraw Hill Book Company, 1929), pp. 525–6 (p. 526) (first publ. in *Der barycentrische Calcul* (Leipzig: [n. pub.], 1827)).

[41] Arthur Cayley, 'Chapters in the Analytical Geometry of (n) Dimensions', *Cambridge Mathematical Journal*, 4 (1843), 119–27.

We can, in fact, without having recourse to any metaphysical notion in regard to the possibility of a space of four dimensions, reason as follows (all of this can also be translated into language purely analytical): In supposing four dimensions of space it is necessary to consider *lines* determined by two points, *half-planes* determined by three points, and *planes* determined by four points (two planes intersect in a half-plane, etc.). Ordinary space can be considered as a plane, and it will cut a plane in an ordinary plane, a half-plane in an ordinary line, and a line in an ordinary point.[42]

Higher spatial terminology was loose—Cayley's use of the terms 'half-plane', to refer to an ordinary plane, and 'plane', to refer to what would become a hyperplane in projective geometry, was not carried forward—but this paper developed a four-dimensional geometry by considering three-dimensional space in four-dimensional terms: it was effectively a system of notation. Cayley justified the reasoning behind the application of this *n*-dimensional geometry in a 'Memoir on Abstract Geometry', received by the Royal Society on 14 October, two short months after Sylvester's address, and read on 16 December. He wrote: 'The science presents itself in two ways: as a legitimate extension of the ordinary two- and three-dimensional geometries, and as a need in these geometries and in analysis generally.'[43] The justification for thinking abstract higher space as a mathematician was utilitarian. If embedding certain geometric problems in a putative four-dimensional space made them easier to solve, why not so embed them? The strictly mathematical deployment of these techniques was largely unproblematic and in such papers their practice remained distinct from the non-Euclidean geometries in which the parallel postulate was assumed away.

The new geometries were rapidly interpolated with another developing practice: projective geometry. Derived from the techniques of perspective drawing, projective geometry was the study of a projective space, considering geometric properties that were invariant under the conditions of projection. It was non-metrical, uninterested in distance, and assumed that parallel lines met at infinity. Using this framework, mathematicians were able to work with higher dimensions in projection, reducing their dimensionality, and the results were startling.

In 1875 Felix Klein published a paper on closed space curves, 'Bemerkungen über den Zusammenhang der Flächen'.[44] Klein summarized his discovery in his own later account of nineteenth-century mathematics:

This result was that the presence of a knot can be considered an essential (i.e., invariant under deformations) property of a closed curve only if one is restricted to move in three-dimensional space; in four-dimensional space a closed curve can be unknotted by deformation. Hence knottedness is no longer a property of *analysis situs* once our considerations have gone beyond the usual space.[45]

---

[42] Arthur Cayley, 'Sur quelques théorêmes de la géometrie de position', *Crelle's Journal*, 31 (1846), 213–27 (pp. 217–18).

[43] Arthur Cayley, 'A Memoir on Abstract Geometry', *Philosophical Transactions of the Royal Society of London*, 160 (1870), 51.

[44] Felix Klein, 'Bemerkungen über den Zusammenhang der Flächen', *Mathematische Annalen*, 9 (1876), 476–82.

[45] Felix Klein, *Development of Mathematics in the 19th Century*, trans. M. Ackerman (Brookline: Math Sci Press, 1979), p. 157.

Similar projective work was underway in the USA, where J.J. Sylvester had decamped to found the mathematics department at Johns Hopkins University. His colleague Simon Newcomb opened the inaugural issue of the *American Journal of Mathematics* with a 'Note on a Class of Transformations which Surfaces May Undergo in Space of More Than Three Dimensions'. Newcomb extended Klein's conclusions to all closed material bodies:

> If the material bodies which surround us were placed in a space of more than three dimensions, their kinematic susceptibilities would be increased in a manner which, at first sight, would seem very extraordinary [...] If a fourth dimension were added to space, a closed material surface (or shell) could be turned inside out by simple flexure; without either stretching or tearing.[46]

Within the frameworks of analytical geometric scenarios that mathematicians knew to be unrepresentative of physical space, results regarding higher-dimensioned space were produced.

THREE-SPACE IN FOUR-SPACE

The interest in n-dimensional geometry of William Kingdon Clifford coincided with that of Sylvester and began in similarly mathematical engagements. Clifford's first published work on the subject was in his own solution to a problem he had set for the *Educational Times* of January 1866. Having provided an initial solution, Clifford went on to write: 'Now consider the analogous case in geometry of n dimensions. Corresponding to a closed area and a closed volume we have something which I shall call a confine.'[47]

Again we see the early mobility of terminology in n-dimensional work: from Cayley's half-planes to Clifford's combines. Clifford's n-dimensional innovation, however, was to put the idea to work in the physical sciences. Sylvester's citation of Clifford's 'remarkable speculations' suggests that the two had corresponded on the subject of Clifford's work on Riemannian space, or that Sylvester had been given insight into the contents of a paper read in Cambridge in 1870, 'On the Space-Theory of Matter', for which only an abstract remains. In this Clifford proposed to demonstrate:

1) That small portions of space *are* in fact of a nature analogous to little hills on a surface which is on the average flat; namely, that the ordinary laws of geometry are not valid in them.
2) That this property of being curved or distorted is continually being passed on from one portion of space to another after the manner of a wave.

---

[46] Simon Newcomb, 'Note on a Class of Transformations which Surfaces May Undergo in Space of More Than Three Dimensions', *American Journal of Mathematics*, 1 (1878), 1–4 (p. 1).

[47] W.K. Clifford, 'Problems and Solutions from The Educational Times', in *Mathematical Papers*, ed. Robert Tucker (London: Macmillan and Co., 1882), pp. 565–627 (p. 603).

3) That this variation of the curvature space is what really happens in that phenomenon which we call the motion of matter, whether ponderable or ethereal.

4) That in the physical world nothing else takes place but this variation, subject (possibly) to the law of continuity.[48]

The clear indication that in Clifford's work non-Euclidean space was no longer assumed as a purely mathematical abstraction was developed throughout the 1870s and described in a number of his lectures, reaching a non-specialist audience in the pages of various journals. *The Postulates of the Science of Space* was first read at the Royal Institution of London in March 1873 and published the following year in the *Contemporary Review*. In this Clifford identified four new rules, or postulates, of space. He argued that the tri-dimensional nature of space was not itself a postulate: 'The science of space, as we have it, deals with relations of distance existing in a certain space of three dimensions, but it does not at all require us to assume that no relations of distance are possible in aggregates of more than three dimensions.'[49]

His third postulate, of superposition, equivalent to Kant's congruence and Möbius's coincidence, by which figures remain constant in size or shape when moved through a Euclidean space or on a surface, was read topologically. A figure on a surface would remain constant despite crumpling or bending of the surface:

> This property of the surface, then, could be ascertained by people who lived entirely in it, and were absolutely ignorant of a third dimension [...] The supposed people living in the surface and having no idea of a third dimension might, without suspecting that third dimension at all, make a very accurate determination of the nature of their *locus in quo*.[50]

In the same year he provided the first English translation of Riemann's paper 'On the Hypotheses which Lie at the Bases of Geometry'.[51]

Complementary and contemporary to Clifford's work on non-Euclidean geometry was that of Hermann von Helmholtz, whose articles on 'The Axioms of Geometry', first in *Academy* in 1870 and expanded in the inaugural issue of *Mind* in 1876, provoked a number of correspondents from within the British scientific community. Helmholtz's influence also reached beyond the academy, and responses to his work would later emerge in more popular media. The attention he gave to higher-dimensioned space was scant, but the structure of his arguments, his insistent working by analogy from two dimensions to three, demands attention.

[48] W.K. Clifford, 'On the Space-Theory of Matter', *Transactions of the Cambridge Philosophical Society*, 2 (1876), 157–8 (repr. in *Mathematical Papers*, pp. 21–2).

[49] W.K. Clifford, 'The Philosophy of the Pure Sciences III: The Postulates of the Science of Space', in *Lectures and Essays*, ed. Leslie Stephen and Sir Frederick Pollock, 2 vols (London: Macmillan and Co., 1879), I, pp. 295–323 (p. 304).

[50] Clifford, 'The Postulates', pp. 314, 315.

[51] Bernhard Riemann, 'On the Hypotheses which Lie at the Bases of Geometry', trans. W.K. Clifford, *Nature*, 8 (1873), 14–17.

Published during a period when he was researching for the third edition of his *Handbuch* on visual perception, Helmholtz's articles on non-Euclidean geometry present a sustained, detailed, and targeted attack on the Kantian idea of space as an innate function of the mind. R. Steven Turner situates Helmholtz's position as dictated by a rivalry with the Leipzig-based physiological researcher Ewald Hering over the innate or learned nature of visual perception. Both had conducted research into depth perception and the location of the horopter, but each held firmly opposed opinions regarding the basis for visual perception.[52] For Helmholtz, it was learned, while Hering believed that depth perception was entirely innate. Helmholtz was much concerned to make his case and perceived the need to address the foundational arguments for this immediate and primitive perception as read in the work of Kant. Gary Hatfield argues that Helmholtz the physiologist attributed Kant's mistake to 'faulty psychology'.[53]

Helmholtz turned to the emergent geometries to make a case against the innatism of Kant and Hering. He worked primarily with the mathematics of Riemann and Beltrami to undermine the assumption of the axioms of geometry as transcendentally true. The reasoning of his first article for *Academy* requires unpicking. As Joan Richards writes: 'Helmholtz's non-Euclidean papers contain physiological, philosophical and mathematical arguments entwined so closely as to be virtually indistinguishable from one another.'[54]

Describing the 'more simple case of the geometry of two dimensions', Helmholtz argued: 'There is no logical impossibility, in conceiving the existence of intelligent beings, living on and moving along the surface of any solid body, who are able to perceive nothing but what exists on this surface and insensible to all beyond it.'[55] Sketching how these creatures might understand geometry on different types of surface, Helmholtz moved on to describe Gauss's work on curved surfaces, and Beltrami's on pseudo-spherical surfaces, demonstrating how the axiom of parallel lines did not hold in these non-Euclidean geometries. In extrapolating the same arguments to spaces of more than two dimensions, Helmholtz shifted his methodology from analogical reasoning to analytical geometry:

> These results regarding surfaces of spaces extended in two dimensions only can be illustrated, as we have tried to do, because we live in a space of three dimensions and can represent in our ideas, or model in reality, other surfaces than the plane (on which alone the geometry of Euclid holds good). When, however, we try to extend these researches to space of three dimensions, the difficulty increases, because we know in reality only space as it exists, and cannot represent even in our ideas any other kind of

---

[52] 'The curved surface of points in space that, for a given degree of ocular convergence, are projected on to corresponding retinal points, all points on the horopter being perceived as the same distance away as the point being fixated.' Andrew M. Colman, *A Dictionary of Psychology* (Oxford: Oxford University Press, 2008).

[53] Gary C. Hatfield, *The Natural and the Normative* (Cambridge, MA and London: MIT Press, 1990), p. 5. Curiously, Hatfield is currently the Adam Seybert Professor in Moral and Intellectual Philosophy, the position at the University of Pennsylvania funded by the estate of Henry Seybert on the condition of the completion of the Seybert Commission Report. See Chapter 2.

[54] Richards, *Mathematical Visions*, p. 77.

[55] Hermann von Helmholtz, 'The Axioms of Geometry', *Academy*, 1 (1870), 128–31 (p. 128).

space. This part of the investigation, therefore, can be carried on only in the abstract way of mathematical analysis.[56]

Helmholtz went on to use the concept of congruence, implying 'the possibility of motion of bodies of invariable form', and to describe a set of experiments through which bodies of invariable form could be shown to move freely in non-Euclidean spaces, thereby demonstrating the reality of Riemann's hypothetical non-Euclidean space.

His article prompted immediate discussion. William Stanley Jevons accused Helmholtz of 'an *ignoratio elenchi*'.[57] Sylvester's conceivability had come home to roost. Jevons argued that the inconceivability of non-Euclidean spaces did not make Euclidean geometry any less apodictically certain. Tupper added his support to Jevons, claiming that Helmholtz was begging the question he proposed to ask by assuming that his intelligences had any conception of a triangle at all. Helmholtz responded in *Academy* clarifying that he was not arguing for the existence of *n*-dimensional space but simply putting these alternate spaces to work in a thought experiment:

> Where I say that geometrical axioms are true or not true for beings living in a space of a certain description, I mean that they are true or not true in relation to those points, or lines, or surfaces, which can be constructed in these spaces, and which can become objects of real perception to those beings [...] No mathematician ever came to the conclusion that a fourth dimension of space exists, even though he find it convenient to write his equations as if it existed.[58]

Helmholtz returned to the subject in 1876 with an expanded, revised, and clarified version of his paper on the axioms. His first article was presented as the basis of his expanded arguments with a second section describing how various non-Euclidean spaces 'would appear to an observer whose eye-measure and experiences of space had been gained like ours in Euclid's space'.[59] These descriptions of worlds as seen through convex mirrors and lenses added a powerful visual element to Helmholtz's argument. Such an observer

> would think he saw the most remote objects round about him at a finite distance, let us suppose a hundred feet off. But as he approached these distant objects, they would dilate before him, though more in the third dimension than superficially, while behind him they would contract.[60]

The conceivability of higher space was an entirely different case, however:

> As all our means of sense-perception extend only to space of three dimensions, and a fourth is not merely a modification of what we have but something perfectly new, we

[56] Helmholtz, 'The Axioms', 129.

[57] William Stanley Jevons, 'Helmholtz on the Axioms of Geometry', *Nature*, 4 (1871), 482. *Ignoratio elenchi*: 'a logical fallacy which consists in apparently refuting an opponent, while actually disproving some statement different from that advanced by him; also extended to any argument which is really irrelevant to its professed purpose' (*OED*).

[58] Hermann von Helmholtz, 'The Axioms of Geometry', *Academy*, 3 (1872), 52–3.

[59] Hermann von Helmholtz, 'The Origin and Meaning of the Axioms of Geometry', *Mind*, 1 (1876), 301–21.

[60] Helmholtz, 'The Origin and Meaning', 316–17.

find ourselves by reason of our bodily organisation quite unable to represent a fourth dimension.[61]

On this occasion, Helmholtz was taken to task by the Dutch philosopher J.P.N. Land, who argued that because 'science has no suspicion of a distinction between "objectivity" and "reality"', Helmholtz was incorrect to move so freely between analytical geometry and philosophy.[62] Land also challenged Helmholtz's idea of what was imaginable: 'Even admitting for a moment that our mind is capable of imagining different sorts of space, it might still be maintained that the only possible form of actual intuition, for a mind like ours, as affected by real things outside of it, is Euclidean space.'[63]

Hatfield's assessment of the success of Helmholtz's argument is mixed. Helmholtz was 'on solid ground in his criticisms of Kant for accepting as universal and necessary what can receive only empirical support'. More fundamentally, however, 'the notion that Kant's point about the status of Euclid's axioms could be distilled into a thesis of psychological nativism revealed Helmholtz's misunderstanding of the notions of transcendental knowledge and a priori form'.[64]

The success or otherwise of Helmholtz's arguments is of secondary significance in this account. What is clear is that the concepts of higher space were first disseminated in the English language through discussions of the nature of lower space, and specifically the visual perception of space. Helmholtz had to depart from psychology and enter the field of epistemology to counter the work of his rival Hering, and in so doing he found the pure mathematics of Riemann and the theorization of non-Euclidean geometry his most useful weapons in insisting upon an empirical basis for space. As Kant had warned, perhaps, he had entered the world of fiction, describing speculative spaces in terms that made him closer to a romancer than a scientist. The rhetorical device of analogy, for example, insistently used by Helmholtz, became a near-constant in accounts of higher-dimensional thought. As a linguistic construction it was doing a great deal of significant work, work we should put on open display.

ANALOGY AND THE DIMENSIONAL MENAGERIE

Analogy has been a preoccupation of scholars working in science and technology studies and nineteenth-century literature and science for some fifty years. Foundational to both reason and argument, it mediates between the domains. Gillian Beer's account of analogy in *Darwin's Plots* described this essence:

> The activity of making analogies is essential to human perception as much as to argument. Meaning presupposes analogies. It would not be possible to describe a thing

[61] Helmholtz, 'The Origin and Meaning', 318–19.
[62] J.P.N. Land, 'Kant's Space and Modern Mathematics', *Mind*, 2 (1877), 38–46 (pp. 38–9).
[63] Land, 'Kant's Space', 42.   [64] Hatfield, *Natural and the Normative*, p. 224.

which was totally *sui generis*. We understand the new by reference to the already known. We cannot do without comparison.[65]

Beer also warned of the close relationship of analogy to story and its tendency to overreach itself: 'The speculative, argumentatively-extended character of analogy ranges it closer to narrative than to image.'[66] Over the course of this book we will read of the inadvertent fusion of two such scientific stories—*n*-dimensionality and evolutionary progress—and their consequent re-emergence in hybridized form in occultist and literary narratives.

n-dimensional geometry was instantiated in an analogy. Having described geometrically the conditions for congruence in two dimensions as a half-revolution of a two-dimensional figure in space of three, August Möbius concluded that space of four dimensions would be required to enact the same congruence for a solid figure in space of three dimensions: 'It will be necessary, we must conclude from analogy, that we should be able to let one system make a half revolution in a space of four dimensions.'[67] Addressing Kant's much-pondered congruence, Möbius described a space that made the inversion of solid objects possible. Further, reasoning strictly by analogy—as from A to B, so from C to D, or A/B = C/D—he was working within the most ancient traditions of descriptive geometry, indeed was reworking the moment of instantiation of geometry itself: ἀναλογία (*analogia*), proportion, being the discovery of Thales in the shadow of the pyramid, the recognition of ratio, the geometric model.

The use of the ratio in geometry was enshrined in Book V of Euclid: '3. A ratio is a sort of relation in respect of size between two magnitudes of the same kind [...] 6. Let magnitudes which have the same ratio be called proportional.'[68] The particular scale model of Thales—of two alike triangles—was described in Book VI, concerned with similar figures: 'Similar rectilineal figures are such as have their angles severally equal and the sides about the equal angles proportional.'[69]

The possible uses of analogy were broadened by Aristotle, applied to biology, metaphysics, justice, and poetic metaphor. In this process analogy became significantly less rigid. In the *Poetics*, Aristotle defined metaphor based on analogy:

> Metaphor [...] from analogy is possible whenever there are four terms so related that the second is to the first, as the fourth to the third; for one may then put the fourth in place of the second, and the second in place of the fourth. Now and then, too, they qualify the metaphor by adding on to it that to which the word it supplants is relative. Thus a cup is in relation to Dionysus what a shield is to Ares. The cup accordingly will be described as the 'shield of Dionysus' and the shield as the 'cup of Ares'. Or to take another instance: As old age is to life, so is evening to day. One will accordingly

---

[65] Gillian Beer, *Darwin's Plots: Evolutionary Narrative in Darwin, George Eliot and Nineteenth-Century Fiction* (London: Routledge & Kegan Paul, 1983), pp. 82–3.

[66] Beer, *Darwin's Plots*, p. 80.   [67] Möbius, 'On Higher Space', 526.

[68] Euclid, *The Elements*, trans. Sir Thomas Heath, 3 vols (New York: Dover Publications, 1956), II, p. 114.

[69] Euclid, *The Elements*, II, p. 188.

describe evening as the 'old age of the day'—or by the Empodoclean equivalent; and old age as the 'evening' or 'sunset of life'.[70]

Aristotle's expansion took the concept of analogy some distance from its roots in mathematical proportionality. Substitution became a possibility, the transference of metaphor. In the *Rhetoric*, Aristotle reiterated the reciprocity of the terms of the analogy:

> The proportional metaphor must always apply reciprocally to either of its co-ordinate terms. For instance, if a drinking bowl is the shield of Dionysus, a shield may be fittingly called the drinking-bowl of Ares.[71]

The argument by analogy was therefore similarly supple and flexible. A:B::C:D could be the basis for the formation of constructions that elided or transferred elements of the construction: A's D, or C's B.[72]

Considered in light of Aristotelian definitions, Sylvester's version of the dimensional analogy is more complex than it at first seems, and of a different order to that of Möbius: 'for as we can conceive beings (like infinitely attenuated bookworms in an infinitely thin sheet of paper) which possess only the notion of space of two dimensions, so we may imagine beings capable of realising space of four or a greater number of dimensions'.[73] Sylvester noted a debt to Gauss for this argument, acknowledging Gauss's autobiographer, Sartorius von Walterhausen. The passage in Gauss's biography to which Sylvester referred was one of few pieces of evidence that Gauss, Möbius's erstwhile teacher, had employed any form of dimensional analogy in his work:

> According to his frequently expressed convictions Gauss regarded the three dimensions of space as a specific characteristic of human beings. People who could not understand this he humorously called Boeotians. 'We can think of creatures who are conscious of themselves in only two dimensions,' he said. 'Higher above us in like manner would stand those who look down on us. Certain problems pertaining to this', he continued jestingly, he had 'put aside to deal with later through geometry, in a higher state of existence.'[74]

Like Gauss, Sylvester placed a stable human subjectivity in the centre of the operation, eliding its own dimensional conditions and making conceivability the proportional structure of comparison: as A (to whom B) can conceive C to whom D, so can A (to whom B) conceive E to whom F. Deviating from Möbius's geometrical analogy, the construction also required C and E to be conceiving intelligences.

---

[70] Aristotle, 'Poetics', in *The Complete Works of Aristotle: The Revised Oxford Translation*, ed. Jonathan Barnes, 2 vols (Princeton: Princeton University Press, 1984), II, pp. 2316–40 (pp. 2332–3).

[71] Aristotle, 'Rhetoric', in *The Complete Works of Aristotle: The Revised Oxford Translation*, ed. Jonathan Barnes, 2 vols (Princeton: Princeton University Press, 1984), II, pp. 2152–269 (p. 2244).

[72] See Roger M. White, *Talking about God: The Concept of Analogy and the Problem of Religious Language* (Farnham: Ashgate, 2010).

[73] Sylvester, 'A Plea', p. 238.

[74] Wolfgang Sartorius von Waltershausen, *Gauss: A Memorial*, trans. Helen Worthington Gauss (Colorado Springs, CO: self-published, 1966), pp. 66–7.

This pattern was continued throughout many versions of the dimensional analogy used in the 1870s.

The physicist P.G. Tait, for example, had followed the developments of higher space. Published in 1876, his *Lectures on Some Recent Advances in Physical Science* collected lectures delivered two years previously. In one such lecture Tait synthesized arguments made by Sylvester, Helmholtz, and Clifford:

> consider that in crumpling a leaf of paper, which may be taken as representing a space of two dimensions, we may have some portions of it plane, and other portions more or less cylindrically or conically curved. But an inhabitant of such a sheet, though living in space of two dimensions only, and therefore, we might say beforehand, incapable of appreciating [*sic*] the third dimension would certainly feel some difference of sensations in passing from portions of his space which were less, to other portions which were more, curved. So it is possible that in the rapid march of the solar system through space, we may be gradually passing to regions in which space has not precisely the same properties as we find here—where it may have something in three dimensions analogous to curvature in two dimensions—something, in fact, which will necessarily imply a fourth-dimension change of form in portions of matter in order that they may adapt themselves to their new locality.[75]

Where Helmholtz had speculated nondescript, generalized 'intelligent beings', for other writers the matter could be made clearer by more explicitly anthropomorphizing the beings in question. G.F. Rodwell suggested his own thought experiment in *Nature*:

> Let us now endeavour to realise the condition of a being living in space of two dimensions. If man possessed the eyes and the power of flight of an eagle, superadded to his ordinary intellectual qualities, he would, no doubt, have very enlarged views of space [...] Now, imagine that a man this endowed with our own notions of space of three dimensions, begins to stoop forward and to grow so: his eyes survey less space; he stoops more forward; his body forms angles of 80°, 70°, 60°, 50° in succession, with a horizontal plane. Then he is obliged to go on all-fours, his limbs shorten and are gradually absorbed into the mass of his body; he crawls, he creeps; at length his limbs disappear altogether, and he trails himself along and glides like a serpent, moving in a horizontal plane [...] Now his body begins to diminish in thickness [...] Now he is a mere plane, an infinitely thin surface; he occupies space approximately of two dimensions; his eyes are on a line.[76]

It is notable that Rodwell's man, in his slide down a scale of dimensionality, briefly 'glides like a serpent'. Not only was the two-dimensional form of Sylvester's worm recalled, but the morality play of evolution and advance was stressed, recalling biblical imagery. Bruce Clarke, in an analysis of moral allegories of the period, glosses this: 'Rodwell's imagery of literary metamorphosis [...] is bound up with the Victorian moralization of human evolution as capable of advance or regression.'[77]

---

[75] P.G. Tait, *Lectures on Some Recent Advances in Physical Science* (London: Macmillan, 1876), p. 5.
[76] G.F. Rodwell, 'On Space of Four Dimensions', *Nature*, 8 (1873), 8–9 (p. 9).
[77] Bruce Clarke, *Energy Forms: Allegory and Science in the Era of Classical Thermodynamics* (Ann Arbor: University of Michigan Press, 2001), p. 31.

In W.K. Clifford's *Common Sense of the Exact Sciences*, published posthumously in 1885, a one-dimensional worm similar to Sylvester's was the starting point for a rich development of the dimensional analogy deployed to invoke spaces of different curvature. Clifford imagined his worm inserted into a circular 'tube of exceedingly small bore'. Following Sylvester and Helmholtz he endowed his worm with sophisticated reasoning powers:

> Assuming that the worm is incapable of recognising anything outside its own tube-space, it would still be able to draw certain inferences as to the nature of the space in which it existed were it capable of distinguishing some mark c on the side of its tube. Thus it would notice when it returned to the point c, and it would find that this return would continually recur as it went round in the bore; in other words, the worm would readily postulate the finiteness of space.[78]

Clifford asked his readers to imagine the removal of the identifying marks within the tube to give his imaginary worm no means of measure from which to postulate; he placed it in an elliptical tube, to show how it might determine position from the degree of bend of its own form. Having established the conditions of life in certain spaces for a one-dimensional creature, Clifford then shifted the manifold up a dimension:

> If by analogy to an infinitely thin worm we take an infinitely thin flat-fish, this fish would be incapable of determining position could it leave no landmarks in its plane space […] Now, suppose that instead of taking this homaloidal space of two dimensions we were still to take a perfectly same space but one of finite bend, that is, the surface of a sphere. Then let us so stretch and bend our flat-fish that it would fit on to some part of the sphere. Since the surface of the sphere is everywhere space of the same shape, the fish would then be capable of moving about on the surface without in any way altering the amount of bending and stretching which we had found it necessary to apply to make the fish fit in any one position.[79]

Clifford's flatfish remained an exemplar of bi-dimensionality in work that followed. Karl Pearson, editor of *The Common Sense*, who had compiled the chapter on motion from Clifford's notes, returned to the creature in his own *The Grammar of Science* (1892), while Arthur Eddington extended and expanded the analogy in *Space, Time and Gravitation*:

> A race of flat-fish once lived in an ocean in which there were only two dimensions. It was noticed that in general fishes swam in straight lines, unless there was something obviously interfering with their free courses. This seemed a very natural behaviour. But there was a certain region where all the fish seemed to be bewitched; some passed through the region but changed the direction of their swim, others swam round and round indefinitely.[80]

---

[78] W.K. Clifford, *The Common Sense of the Exact Sciences* (London: Kegan, Paul, Trench & Co., 1886 [1885]), pp. 215–16.

[79] Clifford, *The Common Sense*, pp. 220–1.

[80] Arthur Eddington, *Space, Time and Gravitation* (Cambridge: Cambridge University Press, 1920), p. 95.

Eddington's flatfish were possessed of yet more sophisticated powers of reasoning than Clifford's worms, able to theorize in the manner of William Thomson: 'One fish invented a theory of vortices, and said that there were whirlpools in that region which carried everything round in curves.'

Note that from the first, the beings in the dimensional analogy are intelligent, even when without form. This intelligence was crucial to Helmholtz's argument, concerned with the development of human spatial perception, but for those who simply wished to demonstrate features of a higher-dimensional space the perceiving intelligence was not required. A perfectly scalable geometric analogy would provide information on geometric forms in higher-dimensioned spaces. By inserting biological life forms and the issue of intelligence into the structure of the analogy, authors invited conceptual hybridity and the creatures chosen connected dimensional discourse with the dominant scientific discourse of the period.

DARWIN'S EXAMPLES

On 10 September 1871, C.J. Monro wrote to James Clerk Maxwell. The two had been corresponding about the idea of *n*-dimensional space since Sylvester's address. Monro explicitly borrowed from Darwin to frame his own version of the dimensional analogy:

> 'I can easily believe,' as Darwin would say, that before we were tidal ascidians we were a slimy sheet of cells floating on the surface of the sea. Well in those days the missing dimension, and the two forthcoming ones respectively, kept changing with the rotation of the earth, we now know how, but could not guess then. So now the missing dimension or dimensions, if any, might be determined by circumstances which we could not tell unless we knew all about the said dimension or dimensions.[81]

Monro's description, like Rodwell's, scaled backwards along an evolutionary line, to imagine planar biological intelligence, the mobility of the evolutionary narrative allowing for retrogression as much as progression and mapping onto a dimensionality assumed to be similarly mobile within the structure of the dimensional analogy. The ascidians became a planar agglomeration of single-cellular organisms, a pre-human intelligence: a thinking surface.

Monro's connection of the narrative of the perceiving ascidian or reasoning worm with Darwinian thought was entirely current. The ascidian occupied a particularly privileged position within *The Descent of Man*, published earlier the same year. Here, Darwin had summarized recent works of naturalism noting similarities between a species of fish called the lancelet and the more basic ascidians, described as

> invertebrate, hermaphrodite, marine creatures permanently attached to a support. They hardly appear like animals, and consist of a simple, tough, leathery sack, with two small projecting orifices. They belong to the Molluscoida of Huxley—a lower

---

[81] London, London Metropolitan Archive, MS Monro correspondence, ACC/1063/2109a.

division of the great kingdom of the Mollusca; but they have recently been placed by some naturalists amongst the Vermes or worms.[82]

Observation had shown that ascidians appeared to develop a nervous system similar to that of vertebrates:

> We should thus be justified in believing that at an extremely remote period a group of animals existed, resembling in many respects the larvæ of our present Ascidians, which diverged into two great branches—the one retrograding in development and producing the present class of Ascidians, the other rising to the crown and summit of the animal kingdom by giving birth to the Vertebrata.

Darwin's identification of man's most basic predecessor attracted much attention around the publication of *The Descent of Man*. *The Times* for 8 April 1871 gleefully recounted Darwin's theory with a mocking gloss: 'An Ascidian, it may be necessary to explain—so grossly ignorant are many of us of our blood relations—is an invertebrate, hermaphrodite, marine creature, permanently attached to a support.'[83]

The satirical magazine *Punch* was particularly keen on lampooning the ascidian theory and these obscure creatures cropped up in four different parodies in 1871, the pick of which used them as a punchline in a joke about marriage:

> At present, marriages with near relations are generally considered objectionable, and that in proportion to proximity of kin. Accordingly, therefore, the more remote the relationship between a married pair, the more normal the marriage [...] Any human being, desirous of a perfect mate, would clearly do best of all to marry, if possible, the Larva of a Marine Ascidian.[84]

The ascidian was the most basic form and the most visible in popular print publications and it had been identified variously as a type of worm and a precursor of certain fish species. Among the vast array of life forms discussed by Darwin, the flatfish assumed a similarly privileged position for the more educated scholarly reader with the publication of the sixth edition of *On the Origin of Species* the following year. Considered by Lamarck at the beginning of the century, the asymmetry of flatfish had been the focal point for a specific debate around natural selection and its implications. In the sixth edition, Darwin described the evolutionary oddity:

> The Pleuronectidae, or Flat-fish, are remarkable for their asymmetrical bodies [...] But the eyes offer the most remarkable peculiarity; for they are both placed on the upper side of the head. During early youth, however, they stand opposite to each other, and the whole body is then symmetrical, with both sides equally coloured.[85]

He went on to respond to criticisms of his earlier suggestions of a gradual migration of the eye by Sir George Mivart and to describe further criticisms and

---

[82] Charles Darwin, *The Descent of Man, and Selection in Relation to Sex* (London: John Murray, 1871), p. 205.

[83] Anon., 'Mr. Darwin on the Descent of Man', *The Times*, 8 April 1871, 5.

[84] Anon., 'Most Natural Selection', *Punch, or the London Charivari*, 1 April 1871, 127.

[85] Charles Darwin, *The Origin of Species by Means of Natural Selection, or the Preservation of Favoured Races in the Struggle for Life*, 6th edn (London: John Murray, 1872), p. 186.

controversies centred around the mechanism of that movement, raised by the Swedish naturalist Malm.

As prominent and significant as ascidians and flatfishes were, it was the humble worm that became a Darwinian favourite. By 1881 Darwin's theory of natural selection was more broadly accepted, as reflected in a *Times* review that described the publication of his *The Formation of Vegetable Mould, Through the Action of Worms* as 'a gift to be grateful for'. On this occasion it was Darwin's claims for the intelligence of the creatures about which he was writing that were singled out for attention in *The Times* and in a cartoon in *Punch*.[86] Darwin's anthropomorphic portrayal of the worm as the humble and unsung ploughman of the earth displayed considerable affection. Gillian Beer writes: 'He domesticates the worm and through the delicate and detailed attention he pays to all its proceedings he also celebrates it.'[87]

I have routed into Darwin to show how when specific life forms were also used in versions of the dimensional analogy, evolutionary narratives were fused in the cultural imagination with dimensional expansion. The narrative of evolutionary theory was a cultural dominant, with gradual progression at its core. Darwin's writing endowed life forms thought of as basic with progressive intelligence. When these same life forms were used as exemplars in the dimensional analogy, an analogy whose success was dependent upon a distinct form of progression—between abstract geometric concepts—the dimensional analogy became hybridized with evolutionary narrative. The cultural documents that evidenced this hybridized thought were frequently remarkable and we will encounter examples of these in later chapters.

The scientific analogy—vibrates with generative ambiguity. While it serves to illuminate and suggest, it prompts the imagination in unpredictable ways. Despite its roots in geometrical proportion, as a rhetorical device it becomes unstable. As Gillian Beer writes:

> If allegory is narrative metaphor, analogy is predictive metaphor [...] Darwin's aim is to discover analogies which can move beyond the provisional and metaphorical and prove themselves as 'true affinities' [...] As in hypothesis, the arc of desire seeks to transform the conditional into the actual. And again, as in hypothesis, such a transformation is seen as changing fiction into a truth.[88]

Analogical reasoning is fuzzy, ambivalent, and catalytic, productive of inference. Beer continues: 'The shifty, revelatory quality of analogy aligns it to magic [...] A living, not simply an imputed, relation between unlikes is claimed by such discourse.'[89] In the case I study in Chapter 2, analogy and magic become even more closely aligned.

[86] Linley Sambourne, 'Man is but a Worm', in *Punch's Almanack for 1882* (London: Punch Office, 1882).

[87] Gillian Beer, *Open Fields: Science in Cultural Encounter* (Oxford: Clarendon Press, 1996), p. 241.

[88] Beer, *Darwin's Plots*, p. 80.   [89] Beer, *Darwin's Plots*, p. 84.

Bruce Clarke reads the dimensional analogy as a form of allegory that relies upon the reification of an abstract domain:

> In reification allegory, the kernel of the text is [...] a conceptual or moral abstraction [...] Reification concretizes abstractions into things, often agents of some sort, animated characters or personifications. Reification typically creates daemonic figures—either anthropomorphically, as nonhuman entities are endowed with human forms or traits, or metamorphically, as human beings are transformed into signifiers superseding human or natural functioning.[90]

The German astrophysicist Johann Carl Friedrich Zöllner was arguably the most highly influential literalizer of the dimensional analogy. Zöllner was already prone to literal readings of scientific analogy and metaphor before he encountered the idea of the fourth dimension. In P.G. Tait's review of his *Abhandlungen*, the British scientist noted this habit with exasperation: for Zöllner, vortex rings *were* smoke rings, and the smoking habits of William Thomson were consistently attacked; Maxwell's demon *was* a supernatural creature rather than an illustrative device, and its invocation exposed the stupidity of James Clerk Maxwell and the theories he proposed.[91] Zöllner may have been attempting deliberate provocation or an exaggerated sarcasm with these attacks; he may have been, as was frequently claimed, not in his right mind; but his engagement with higher space displays evidence that he reified elements of the dimensional analogy that were deployed rhetorically to illustrate the conceivability of space. Focusing on the suggested intelligences required by conceivability, and reading across from conclusions about the nature of four-dimensional space made in the context of projective geometry, Zöllner had deduced four-dimensional intelligences to be real.

Zöllner's generative misreading of an analogy was crucial to the cultural development of the idea of higher space. It was a test case of scientific authority in the context of spiritualism and psychic research and it catalysed an occult higher spatial discourse that persists to this day. If reifications his theories were, they were both remarkably broadly read and remarkably resilient.

---

[90] Clarke, *Energy Forms*, p. 30.
[91] See P.G. Tait, 'Zöllner's Scientific Papers', *Nature*, 17 (1878), 420–2.

2

Knots

Topology, Conjuring, and the Spiritualist
Fourth Dimension

In March 1878 Johann Carl Friedrich Zöllner, professor of astrophysics at the University of Leipzig, published an account of an experiment he had undertaken with the spirit medium Henry Slade in the *Quarterly Journal of Science*, edited by the British chemist William Crookes. 'On Space of Four Dimensions' argued that the human conception of space as three-dimensional was based upon experience and that as soon as experience appeared to contradict this conception we would be forced to revise our theories of space.[1] Pausing briefly to appeal to the work of the mathematician Bernhard Riemann, Zöllner went on to describe manipulations of a cord of string.

Confined to a plane, or two dimensions, a knot could be conceived as a simple twisting of the cord. Without access to the third dimension, this twisting or crossing could not be undone except by cutting the cord or passing it back through itself; with access to a third dimension, a simple rotation of the cord would suffice to undo the loop. This loop was the two-dimensional equivalent of a knot. Extending space to three dimensions, and placing a three-dimensional knot in the cord, Zöllner argued that access to a fourth dimension of space would, by analogy, allow the knot to be undone without cutting the cord or passing it back through itself.

Zöllner continued his argument by borrowing highly selectively from the work of the mathematician Carl Friedrich Gauss and the philosopher Immanuel Kant. He quoted from Kant to argue that because a space of four dimensions could be conceived, it would 'probably' exist, and likewise immaterial beings of the spiritual world. His groundwork established, Zöllner proceeded to the core of his argument:

> If a single cord has its ends tied together and sealed, an intelligent being, having the power voluntarily to produce on this cord four-dimensional bendings and movements, must be able, without loosening the seal, to tie one or more knots in this endless cord. Now, this experiment has been successfully made within the space of a few minutes in

[1] Johann Carl Friedrich Zöllner, 'On Space of Four Dimensions', *Quarterly Journal of Science*, 8 (1878), 227–37.

Leipzig, on the 17th December 1877, at 11 o'clock AM, in the presence of Mr. Henry Slade, the American.[2]

Spiritualists in Germany and England trumpeted the triumph through their journals. It was reported uncritically in the *Daily Telegraph*, to which the spiritualist editor William Harrison contributed science stories. Within a week British spiritualists, including T.L. Nichols and the medium William Eglinton, claimed to have reproduced Zöllner's 'splendid manifestation'.[3]

The repercussions of Zöllner's experiment echoed for some time. It became a highly contested episode in the history of psychic research and is to this day cited in spiritualist literature as evidence of scientific plausibility of the spiritual hypothesis. By the early twentieth century, historians of mathematics were noting the spiritualist interest in higher space. Duncan Sommerville wrote of the year 1881 in the introduction to his bibliography of non-Euclidean geometry:

At that time this joint subject was at an interesting stage. HELMHOLTZ was popularising non-Euclidean geometry in England and in Germany, and the interest of philosophers was roused to rescue their favourite example of 'necessary truth' from the general attack of scepticism. Space of *n* dimensions was just emerging as a branch of geometry capable of rigorous treatment and was attaining an unfortunate popularity in the hands of the spiritualists. Both subjects were surrounded by a mysticism which can scarcely yet be said to be entirely dissipated.[4]

Felix Klein, too, remembered this period as unfortunate, describing the 'surprising difficulty from [...] philosophical enthusiasts who [...] inferred the real existence [*Dasein*] of an actual space of four dimensions'. He concluded: 'Here began the great popular mystification, which, in combination with hypnotism, suggestion, religious sectarianism, popular philosophy of nature, etc., soon came to dominate many minds.'[5]

Charles Carleton Massey's translation of Zöllner's collected scientific papers recording the experiment, *Transcendental Physics*, was to prove enormously influential in terms of the popular understanding of higher space well into the twentieth century, particularly in the USA. As Tom Gibbons notes of a 1909 book compiling entries to a competition run in the *Scientific American* for the best popular explanation of the fourth dimension: 'The scientific name most often cited by contributors to *The Fourth Dimension Simply Explained* is that of J.C.F. Zöllner, Professor of Astronomy at the University of Leipzig.'[6]

---

[2] Johann Carl Friedrich Zöllner, *Transcendental Physics*, trans. Charles Carleton Massey, 4th edn (Boston: Banner of Light, 1901), p. 41. The fourth edition contained as its first chapter the entire text of 'On Space of Four Dimensions'. All further references to this edition are given in the body of the text using the abbreviation *TP*.

[3] T.L. Nichols, MD, 'Remarkable Physical Manifestations', *Spiritualist*, 12 (1878), 174–5 (p. 175).

[4] Duncan M.Y. Sommerville, *Bibliography of Non-Euclidean Geometry* (London: Harrison & Sons, 1911), p. vi.

[5] Felix Klein, *Development of Mathematics in the 19th Century*, trans. M. Ackerman (Brookline: Math Sci Press, 1979), pp. 156–8.

[6] Tom H. Gibbons, 'Cubism and "The Fourth Dimension" in the Context of the Late Nineteenth-Century and Early Twentieth-Century Revival of Occult Idealism', *Journal of the Warburg and Courtauld Institutes*, 44 (1981), 130–47 (p. 133).

For Henry P. Manning, the editor of that collection, interest in the fourth dimension had continued 'because of the many curious things about it, and because of attempts to explain mysterious phenomena by means of it'.[7] Zöllner's supernatural hypothesis and subsequent experiments had secured longevity for the popular idea of higher space.

Neither was Zöllner's work of purely cultural or scholarly interest. It had material, political implications. Friedrich Engels responded almost immediately with an essay entitled *Die Naturforschung in der Geisterwelt* (Natural Science in the Spirit World), written in 1878 but unpublished until 1898 in a Hamburg Social Democrat calendar, writing: 'For years, as is well known, Herr Zöllner has been hard at work on the "fourth dimension" of space, and has discovered that many things that are impossible in a space of three dimensions, are a simple matter of course in a space of four dimensions.' Engels called time on the experimental work of spiritualist scientists:

> Enough. Here it becomes palpably evident which is the most certain path from natural science to mysticism. It is not the extravagant theorising of the philosophy of nature, but the shallowest empiricism that spurns all theory and distrusts all thought. It is not a priori necessity that proves the existence of spirits, but the empirical observations of Messrs. Wallace, Crookes, and Co.[8]

While Zöllner was omitted from official twentieth-century accounts of the scientific history of the German Democratic Republic due to Engels's negative response, he resurfaced in a number of influential intellectual and artistic contexts, referenced, for example, in Kandinsky's 1912 essay *On the Spiritual in Art* and cited approvingly by Jung in his juvenile lectures to the Zofingia: 'In 1877 the noble Zöllner published his scientific tracts in Germany, and fought for the spiritualist cause in a series of seven volumes. But his was a "voice in the wilderness".'[9]

The aim of this chapter is to provide a full historical account of Zöllner's experiment, the resources it drew together, the texts it produced and responses to them, in the hope of describing the scope and topography of an event in the cultural history of higher-dimensioned space which, judging by the volume of responses it generated, must be regarded as highly significant.

Unpicking the knot that Zöllner proclaimed as evidence of his success will be the focus of this chapter, a denouement in its literal sense. The knot comes heavily freighted before it was produced by Slade. Folklorists of the period were recording

---

[7] Henry Parker Manning, *Geometry of Four Dimensions* (New York: Dover Publications, 1956; repr. of Macmillan: 1914), 9 n.

[8] Friedrich Engels, 'Natural Science and the Spirit World', Marxists.org, http://www.marxists.org/archive/marx/works/1883/don/ch10.htm [accessed 9 February 2012] (para. 14 of 18) (repr. of *Illustrierter Neue Welt-Kalender für das Jahr 1898*).

[9] See Wassily Kandinsky, *On the Spiritual in Art*, trans. Hilla Rebay (New York: The Solomon R. Guggenheim Foundation, 1946), p. 24; C.G. Jung, *The Zofingia Lectures*, trans. Jan van Huerck (London: Routledge and Kegan Paul, 1983), p. 35; D.B. Hermann, 'Zöllner Studies at Archenhold Observatory 1974–1994', in *Karl Friedrich Zöllner and the Historical Dimension of Astronomical Photometry*, ed. Christiaan Sterken and Klaus Staubermann (Brussels: VUB University Press, 2000), pp. 151–9.

accounts of the use of knot magic in witchcraft. Knots would be tied to curse adulterers, or to control the wind.[10] The knot-work of Celtic artwork indicated an ancient decorative cultural fascination with these structures, and the endless knot—or, to accord its most basic form its proper title, the trefoil knot—was a staple of Victorian funerary ornamentation, a symbol of both the Trinity and infinity.[11] For mathematicians and physicists in the late 1860s the knot was assuming yet greater significance.

In 1867 Peter Guthrie Tait translated Helmholtz's paper on vortex motion in liquids.[12] Helmholtz's '*Wirbelbewegung*' appealed to both William Thomson and James Clerk Maxwell, who dubbed them 'worbles'. When Thomson witnessed a demonstration of smoke rings in a lecture given by Tait that demonstrated similar stability to that observed by Helmholtz in vortices in a perfect liquid, Thomson was moved to propose that atoms were in fact vortices in another perfect medium: the aether. Presenting a paper before the Royal Society of Edinburgh, 'On Vortex Atoms', Thomson demonstrated diagrams and models of various knotted or knitted vortex atoms and wrote:

> two ring atoms linked together or one knotted in any manner with its ends meeting, constitute a system which, however it may be altered in shape, can never deviate from its own peculiarity of multiple continuity, it being impossible for the matter in any line of vortex motion to go through the line of any other matter in such motion or any other part of its own line.[13]

In this theory, movement in the space-filling medium of the ether became matter. Steven Connor describes Thomson's theory as 'a kind of ontological promotion' for the ether. 'No longer merely that which lay between things, the inert and docile soup studded with beads of matter, the ether became primary, an ur-matter, or quasi-matter, out of which all things were made.'[14]

Nearly ten years later, Tait in turn took his lead from Thomson. Clerk Maxwell had introduced his friend to Johann Benedict Listing's *Vorstudien zur Topologie* (1847). Tait would later discuss the significance of this work in his obituary of Listing: 'The consideration of double-threaded screws, twisted bundles of fibres, &c., leads to the general theory of paradromic winding. From this follow the properties of a large class of knots which form "clear coils."'[15] The mathematics of

---

[10] The Folklore Society was founded in London in 1878 and its journal *The Folk Lore Record* included numerous accounts of such magical knots in folklore.

[11] The gravestone of Edwin Abbott Abbott in Hampstead Cemetery is carved with trefoil knots. See Ian Stewart, *The Annotated Flatland* (New York: Perseus Publishing, 2002), p. xxvii.

[12] Hermann von Helmholtz, 'Integrals of the Hydrodynamical Equations, which Express Vortex-motion', trans. P.G. Tait, *Philosophical Magazine*, 33 supplement (1867), 485–511.

[13] Sir William Thomson, 'On Vortex Atoms', in *Mathematical and Physical Papers* (Cambridge: Cambridge University Press, 2011), pp. 1–12 (p. 3) (repr. of 'On Vortex Atoms', *Proceedings of the Royal Society of Edinburgh*, 6 [1867], 94–105).

[14] Steven Connor, *The Matter of Air: Science and Art of the Ethereal* (London: Reaktion Books, 2010), p. 152.

[15] P.G. Tait, 'Johann Benedict Listing', *Nature*, 27 (1883), 316–17 (p. 317).

knots suggested itself as a route of inquiry into Thomson's theory: 'I was led to the consideration of the forms of knots by Sir W. THOMSON'S Theory of Vortex Atoms, and consequently the point of view, which, at least at first, I adopted was that of classifying knots by the number of their crossings.'[16]

Tait's first realization was that some objects appearing to be knots are in fact merely tangles and can be unravelled by pulling. These he termed trivial knots, or unknots. True knots he tabulated according to the number of their crossings, presenting his research to the Royal Society of Edinburgh in 1876, instigating the branch of mathematics now known as knot theory. These true knots were what is described in topological terms as 'closed space curves': the putative cords in which they were tied had no ends. They were unambiguous and unbreakable in their own space and were—possibly—the very bedrock of matter.

By the time Tait's work was published later that year, he had received notice from Felix Christian Klein of Klein's 'very singular discovery that in space of four dimensions there cannot be knots'. In Klein's own description he had demonstrated that 'the presence of a knot can be considered an essential (i.e., invariant under deformations) property of a closed curve only if one is restricted to move in three-dimensional space; in four-dimensional space a closed curve can be unknotted by deformation'.[17] In other words knots, or closed-space curves, could be untied by movements that did not involve cutting the knot or passing the cord through itself, in space of four dimensions. Tait was intrigued, but he was not Klein's only correspondent. In the mid-1870s Klein had also informed Zöllner of his research. According to Klein, 'Zöllner took up my remark with an enthusiasm that was unintelligible to me. He thought he had a means of experimentally proving the "existence of the fourth dimension".'[18] Tait's response to that experiment will be considered in this chapter, in the section 'Zöllner and Slade', but Klein's research demonstrated the essentially topological nature of knots, closed-space curves that retained their characteristics when subject to deformations, twisting, and rotations.

The unbreakability of the knot or the twisted vortex ring emblematized and informed the idea of continuity that had assumed a position of increasing importance in scientific naturalism, buoyed by and inspired by the gradualist implications of the theory of natural selection. In his address to the British Association for the Advancement of Science (BAAS) in 1866, William Grove described continuity as 'a law of nature' inscribed into matter itself:

> Though we may humbly confess our inability to explain why matter is impressed with this tendency to gradual structural formation, we should cease to look for special interventions of creative power in changes which are difficult to understand, because, being removed from us in time, their concomitants are lost; we should endeavour

---

[16] P.G. Tait, 'On Knots', *Transactions of The Royal Society of Edinburgh*, 28 (1877), 145–90 (p. 145).
[17] Klein, *Development of Mathematics*, p. 157.
[18] Klein, *Development of Mathematics*, p. 157.

from the relics to evoke their history, and when we find a gap not try to bridge it over with a miracle.[19]

John Tyndall echoed these sentiments a decade later, declaring that his faith in 'Matter' extended beyond the reach of his ability to observe it: 'Believing as I do in the continuity of nature, I cannot stop abruptly where our microscopes cease to be of use.'[20]

Tait and his colleague Balfour Stewart referred to Grove in declaring a 'Principle of Continuity' in *The Unseen Universe*. For the Scottish physicists, this principle did not imply that temporary discontinuities could not occur, but that such discontinuities would indicate the necessity of discovering, and the route towards, a higher truth. They asked: 'Does not something analogous to the principle of continuity prevent us from supposing that we can ever arrive at the ultimate expression of truth on any, however limited, subject?' They argued that both theologians and 'the extreme scientific school' had violated this principle, the former by ignoring the evidence of matter, and the latter by failing to pursue their principle into the unseen.[21]

The Unseen Universe applied its principle of continuity to the most recent atomic speculation, 'the vortex-atom theory of Sir W Thomson'.[22] For Tait and Stewart, following the Principle of Continuity, vortex atoms were evidence of a creator:

> This definition involves the necessity of a creative act for the production or destruction of the smallest portion of matter, because rotation can only be produced or destroyed by us in a fluid in virtue of its viscosity (or internal friction), and in a perfect fluid there is nothing of the kind.[23]

James Clerk Maxwell held a more nuanced position regarding the position of theology in contemporary physics and disagreed with this claim, expressing his thoughts in the very last poem he wrote before his death. This verse is an intriguing resource through which to consider the knot. It combines an informed overview of spatial theories enjoying currency at the time with a critique of Tait's reading of the idea of continuity. Sent to Tait in 1878, after Tait's review of Zöllner's collected scientific papers and responding to the appearance of the sequel to *The Unseen Universe*, *Paradoxical Philosophy*, to whose hero, 'Hermann Stoffkraft PhD', it was addressed, *A Paradoxical Ode* pastiched the final three stanzas of the second Act of Shelley's *Prometheus Unbound*, a song sung by the ocean nymph, Asia:

---

[19] William Robert Grove, *Address to the British Association for the Advancement of Science* (London: Longmans, Green and Co., 1867), p. 73.

[20] John Tyndall, *Fragments of Science: A Series of Detached Essays, Addresses and Reviews*, 6th edn, 2 vols (London: Longmans, Green and Co., 1879), II, p. 193.

[21] P.G. Tait and Balfour Stewart, *The Unseen Universe or, Physical Speculations on a Future State*, 4th edn (London: Macmillan and Co., 1876), pp. 87, 96.

[22] Tait and Stewart, *The Unseen Universe*, p. 139.

[23] Tait and Stewart, *The Unseen Universe*, p. 140.

A Paradoxical Ode

My soul's an amphicheiral knot
Upon a liquid vortex wrought
By Intellect in the Unseen residing,
While thou dost like a convict sit
With marlinspike untwisting it
Only to find my knottiness abiding,
Since all the tools for my untying
In four-dimensioned space are lying,
Where playful fancy intersperses,
Whole avenues of universes;
Where Klein and Clifford fill the void
With one unbounded, finite homaloid,
Whereby the Infinite is hopelessly
  destroyed.

But when thy Science lifts her pinions
In Speculation's wild dominions,
I treasure every dictum thou emittest;
While down the stream of Evolution
We drift, and look for no solution
But that of survival of the fittest,
Till in that twilight of the gods
When earth and sun are frozen clods,
When, all its matter degraded,
Matter in aether shall have faded,
We, that is, all the work we've done,
As waves in aether, shall for ever run
In swift expanding spheres, through
heavens beyond the sun.

Great Principle of all we see,
Thou endless Continuity!
By thee are all our angles gently rounded,
Our misfits are by thee adjusted,
And as I still in thee have trusted,
So let my methods never be confounded!
O never may direct Creation
Breach in upon my contemplation,
Still may the causal chain ascending,
Appear unbroken and unending,
And where the chain is best to sight
Let viewless fancies guide my darkling
  flight
Through aeon-haunted worlds, in order
  infinite.[24]

Prometheus Unbound: Asia's Song

My soul is an enchanted boat,
Which, like a sleeping swan, doth float
Upon the silver waves of thy sweet singing;
And thine doth like an angel sit
Beside a helm conducting it,
Whilst all the winds with melody are ringing.
It seems to float ever, for ever,
Upon that many-winding river,
Between mountains, woods, abysses,
A paradise of wildernesses!
Till, like one in slumber bound,
Borne to the ocean, I float down, around,
Into a sea profound, of ever-spreading sound:

Meanwhile thy spirit lifts its pinions
In music's most serene dominions;
Catching the winds that fan that happy heaven.
And we sail on, away, afar,
Without a course, without a star,
But, by the instinct of sweet music driven;
Till through Elysian garden islets
By thee, most beautiful of pilots,
Where never mortal pinnace glided,
The boat of my desire is guided:
Realms where the air we breathe is love,
Which in the winds and on the waves doth
  move,
Harmonizing this earth with what we feel above.

We have past Age's icy caves,
And Manhood's dark and tossing waves,
And Youth's smooth ocean, smiling
  to betray:
Beyond the glassy gulfs we flee
Of shadow-peopled Infancy,
Through Death and Birth, to a diviner day;
A paradise of vaulted bowers,
Lit by downward-gazing flowers,
And watery paths that wind between
Wildernesses calm and green,
Peopled by shapes too bright to see,
And rest, having beheld; somewhat like thee;
Which walk upon the sea, and chant
  melodiously!

---

[24] Daniel S. Silver, 'The Last Poem of James Clerk Maxwell', *Notices of the AMS*, 55 (2008), 1266–70 (pp. 1266–7). Silver reinstates the poem from Clerk Maxwell's notebooks. Percy Shelley, *Prometheus Unbound*, ed. Lawrence John Zillman (Seattle: University of Washington Press, 1959), pp. 222–5.

Maxwell's verse is suggestive on many counts. It plays with both cheirality—handedness, a function of standard space dissolved by higher space that Zöllner wanted to test with crystals—and knottedness. As a pastiche it privileges form: we read a close mimicry of the metre of its source, departing at line 6 in the first verse and staying close throughout the second. Images, rhymes, and verbs are repeated between the two, but where Shelley's poem concerns itself with the continuity between the Romantic imagination and the nature of winds, rivers, mountains, woods, and abysses, Maxwell's nature is the speculated, contemporary physical nature of four-dimensioned space, avenues of universes and homaloids, though his imagination is none the less continuous with it.

Replacing Shelley's boat with a knot, Maxwell uses the structure as the vessel through which to explore his physical vision of nature, as he weaves between evolutionary theory and classical thermodynamics. The two poems share a vision of fluidity: Maxwell's 'liquid vortex', drift down the 'stream of evolution', and 'waves of aether' the counterpoints to Shelley's winds, waves, and 'sea profound'. By his third verse, Maxwell's knot has ceded place to the idea of continuity, to which the physicist accedes; but to which he adds the coda 'never may direct Creation / Breach in upon my contemplation', a clear shot across the bows of Tait.

Maxwell welcomes Stoffkraft's words in 'speculation's wild dominions', but is critical of his tolerance of paradox: a belief in a creator and the principle of continuity cannot be held at the same time. Maxwell defends infinity on these grounds—using the fourth dimension to do away with infinity amounts to 'playful fancy'. Infinity is a function of continuity, rigorously applied. 'Speculation'—Tyndall's scientific imagination, perhaps—may be appropriate at the point at which the causal chain 'is lost to sight', but the hand of a creator would end that chain.

The heart of Maxwell's poem contains the keenest paradox: 'When, all its matter degraded, / Matter in aether shall have faded.' We may forgive the amateur poet a lapse for what appears to be a reference to the matter of matter, or we may agree that the very materiality of matter is what was at stake and its ebbing into the space-filling aether precisely what was suggested by vortex atom theory.

Steven Connor describes 'an intensification of the condition known as modernity' in the nineteenth century characterized by thought which identified 'human beings [as] both dislocated from the natural world of matter, and swallowed up in a materialism of their own making'.[25] We might read Clerk Maxwell's poem in this light, as an ode to the physics at the heart of this vortical process in which the human is estranged from the non-human world only to become re-entangled in cultural matter.

ZÖLLNER AND SLADE: SCIENTISTS AND MEDIUMS, ALLIANCES AND RIVALRIES

In the early years of his career, Zöllner's research was centred around the relationship between the observing scientist and the instruments he used, focused on

---

[25] Connor, *The Matter of Air*, p. 147.

light, colour, and optical illusions. In the early 1860s he published a number of papers on optical illusions and invented three instruments for use in astronomy, including a reversion spectroscope still in use today. An interest in after-images on the retina led to correspondence with Helmholtz that revolved around technical aspects of vision.[26] Throughout the 1860s, Zöllner's career was in the ascendant. In 1866 he became the first senior lecturer in astrophysics at a German university and in 1872 the first professor.[27]

By the 1872 publication of *Über die Natur der Cometen* his interests could no longer be described as primarily technical and his once civil relationship with Helmholtz had curdled. In this book Zöllner launched a tirade against the British scientists Tyndall, Thomson, and Tait, and their friend and translator into German, Helmholtz. The reason for this attack would later be identified by Tait, reviewing Zöllner's *Wissenschaftliche Abhandlungen* in *Nature* in 1878: "'the aim of all his scientific efforts has been to contribute, as far as the ability given him permits, to the realisation" of a certain "hopeful project":- viz., the explanation of all molecular actions by means of that Law of Electric Attraction (due to W. Weber)'.[28] Zöllner held the superiority of Weber's theory to be absolute, and saw in the work of these British scientists a threat. He referred several times to Helmholtz's introduction to the German translation of *A Treatise on Natural Philosophy*, lambasting Helmholtz's praise for Thomson and the book Thomson and Tait had written. *Über die Natur* arrived as a shock to both German and British scientists. Rudolf Clausius wrote to Tyndall with suggestions that the astrophysicist was not in his right mind:

> I am absolutely outraged by this book, and furthermore I can tell you that everyone with whom I have spoken about it condemns and decidedly disapproves of the personal attacks contained therein. The manner in which he speaks is not only unheard of in a scientific work but it is also unexplainable and peculiar that one cannot even understand how he came to speak in that way. As a result, there are even rumours about his mental state, which, if they should turn out to be true, will be very sad for him.[29]

The source of these rumours is obscure, but Helmholtz had also heard them, as he too wrote to Tyndall:

> Here too in Germany, Zöllner's book has extremely shocked us. The first impression was that, in general, he may be insane, all the more so as one knew that several cases

---

[26] See Klaus B. Staubermann, 'Tying the Knot: Skill, Judgement and Authority in the 1870s Leipzig Spiritistic Experiments', *British Journal for the History of Science*, 34 (2001), 67–79 (p. 68).

[27] See Klaus. B. Staubermann, *Astronomers at Work: A Study of the Replicability of 19th Century Astronomical Practice* (Frankfurt am Main: Verlag Harri Deutsch, 2007), p. 86.

[28] P.G. Tait, 'Zöllner's Scientific Papers', *Nature*, 17 (1883), 420.

[29] Rudolf Clausius, 'Letter to John Tyndall', 4 April 1872, cited in David Cahan, 'Anti-Helmholtz, Anti-Zöllner, Anti-Duhring: The Freedom of Science in Germany during the 1870s', *Universalgenie Helmholtz: Rückblick nach 100 Jahren*, ed. Lorenz Kruger (Berlin: Akademie Verlag, 1994), pp. 330–44 (p. 334). See also Jed Z. Buchwald, 'Electrodynamics in Context: Object States, Laboratory Practice, and Anti-Romanticism', in *Hermann von Helmholtz and the Foundations of Nineteenth Century Science*, ed. David Cahan (Berkeley, Los Angeles, and London: University of California Press, 1993), p. 371, for Zöllner's disputes.

of mental illness had already occurred in his family, and with his mother's brother the illness began with the publication of a similar book.[30]

The scientific value of Zöllner's work was immediately lost in the climate of shock and outrage. Importantly, *Über die Natur der Cometen* contained Zöllner's first published engagement with non-Euclidean geometry, in which he argued that the universe must be of constant positive curvature, and credited the roll call of innovators in whose footsteps he followed: 'Gauss, Riemann, Helmholtz, Lobaschefsky, Bolyai und in jüngster Zeit Felix Klein in Göttingen'.[31]

Zöllner continued to publish in similar vein. His 1874 paper *Über einen Elektrodynamischen Versuch* focused its attentions on Helmholtz's work on system energies, derived from Tait and Thomson's energetic model of physics, and repeated the criticism of physical research that did not follow with the Weberean theory of charged particles. Helmholtz was roused to respond in the preface to the second part of the translation of Tait and Thomson's *Treatise*, published in the same year. He identified Zöllner's arguments with Schopenhauer's *Metaphysics*, and drew particular attention to the personal nature of his attacks:

> Now, that a man who mentally treads such paths should recognise in the method of Thomson and Tait's book the exact opposite of the right way, or of that which he himself considers such, is natural; that he should seek to ground the contradiction, not where it is really to be found, but in all conceivable personal weaknesses of his opponents, is quite in keeping with the intolerant manner in which the adherents of metaphysical articles of faith are wont to treat their opponents, in order to conceal from themselves and from the world the weakness of their own position.[32]

In the same year Helmholtz also contributed a foreword to the translation of Tyndall's *Fragments of Science*, in which Tyndall published a chapter arguing against the spiritualist movement in England.

In 1875 Zöllner visited William Crookes in London, returning to Germany with new ideas for applications for the radiometer. Staubermann regrets the absence of any record of the encounter between the astrophysicist and the chemist, but argues that this was the seed for the subsequent trajectory of Zöllner's scientific career:

> Zöllner saw a strict analogy between the photometer and the human body. For him the human body could respond to natural forces like an instrument [...] All that had to be done for Zöllner now was to find a person, a medium, who was sensitive enough to respond to such forces as his photometer could.[33]

The dedication to Crookes of his later book *Transcendental Physics*, discussed in detail below, leaves little doubt that the meeting had a significant influence on the subsequent direction of Zöllner's research and writing.

[30] London, Royal Institution, MS John Tyndall, RI MS JT/1/H/48.

[31] Johann C.F. Zöllner, *Über die Natur der Cometen: Beiträge zur Geschichte und Theorie der Erkenntnis* (Leipzig: Engelmann, 1872), p. 306. See Helge Kragh, 'Zöllner's Universe', *Physics in Perspective*, 14 (2012), 392–420 for Zöllner's speculations on the non-Euclidean universe.

[32] Hermann von Helmholtz, 'On the Use and Abuse of the Deductive Method in Physical Science', *Nature*, 11 (1874), 149–51 (p. 150).

[33] Staubermann, 'Tying the Knot', 73.

Zöllner's feud with Helmholtz, according to Jed Z. Buchwald, became immediately significant in encouraging Helmholtz's work with non-Euclidean geometry:

> The extraordinary intensity with which Helmholtz pursued the subject in the mid-1870s at least in part reflected his desire to eliminate the plausibility of Zöllner's Idealistic deductive scheme of Weberean physics by creating situations in the laboratory that Helmholtz, but not Weber, could explain.[34]

Zöllner's arguments, in this account, shape the construction of higher space not only through his own theses, but also through Helmholtz's. Zöllner's work becomes significant not on account of its scientific legitimacy, but rather as a result of its illegitimacy, its negative feedback into scientific knowledge networks.

In Berlin, meanwhile, Helmholtz had taken action over another feud that had taken place closer to home, and had resulted directly from his campaign to establish the empirical and learned nature of space. Eugene Dühring, a critic of non-Euclidean geometry who had in his 1875 book *Cursus der Philosophie* accused Gauss of 'mathematical mysticism', had included Helmholtz in his criticisms in the second edition of his *Kritische Geschichte der allgemeinen Principien der Mechanik*, objecting to Helmholtz's support for the 'piquant absurdity' of non-Euclidean geometry.[35] Helmholtz, by then Rector of the University of Berlin, had in 1877 instigated proceedings against Dühring that had resulted in the latter's expulsion from his post. While these had nominally been concerned with administrative issues, Dühring later claimed that it had been his remarks on 'Helmholtz's involvement in the contagion-sphere of four- and more-dimensional space' that had resulted in his clash with the Rector.[36]

Into this intellectual milieu arrived the medium Henry Slade, who had been itinerant in mainland Europe since the overturning on the basis of a clerical error of his conviction in England in the early months of 1877 for vagrancy. The trial had been a high-profile and occasionally farcical event, in which evidence was given by well-known scientists, popular entertainers, and notorious society figures. It had been reported in detail for its duration as a test case for the legitimacy of spiritualism.[37] Performing seances for society figures and any professors who could be persuaded to attend, Slade had visited France, Holland, Belgium, and Denmark since leaving England, arriving in Berlin in early November.

Over the course of three days in November 1877 and four in December Slade led twelve seances for Zöllner and his colleagues. During their first meeting, over tea at Slade's hotel room in Leipzig on the morning of 15 November, Zöllner had asked Slade if he could deflect the point of a compass. Slade claimed to have

---

[34] Buchwald, 'Electrodynamics in Context', p. 373.

[35] Cited in Wayne H. Stromberg, 'Helmholtz and Zöllner: Nineteenth-century Empiricism, Spiritism, and the Theory of Space Perception', *Journal of the History of the Behavioral Sciences*, 25 (2006), 371–83 (p. 379).

[36] Stromberg, 'Helmholtz and Zöllner', p. 379.

[37] See reports in *The Times* for 11, 23, 28, and 30 October, 1 November 1876, and 30 January 1877.

achieved this before, in a sitting in Berlin on the 11th of that month, and would try to do so again. He came to Zöllner's house that evening, and 'violently agitated' the needle of a compass set in a celestial globe. This was the same phenomenon as witnessed by Zöllner's colleague Gustav Theodor Fechner some twenty years previously in the company of Baron von Reichenbach and his housekeeper, Mrs Ruf, and it confirmed in Zöllner's mind that Slade was a sensitive.[38]

The following evening, in the company of Fechner and Wilhelm Eduard Weber, whose theory of charged particles Zöllner had defended so staunchly, a selection of mediumistic phenomena were observed: the slate-writing, or psychography, for which Slade was best known; the apparently unmediated movement of objects around the room; and knocking on the table at which the four were sat. The same quartet reconvened the following evening, and similar phenomena occurred: a screen was broken and an apology for surprising the professors was written onto a slate.

Encouraged by what he had seen, Zöllner invited three more colleagues, the experimental psychologist Professor Wilhelm Wundt, the surgeon Geheimrath Thiersch, and the physiologist Professor Carl Ludwig, to attend a seance on the evening of 17 November. The three stayed in the room for only half an hour, but Thiersch later reported to Zöllner that they had witnessed a knife flying across the room.

Slade had to leave for Berlin that night, but was invited back to stay with Zöllner's friend Oscar von Hoffmann for another series of sittings. On 11, 13, and 14 December at seances attended either by Zöllner, Weber, and Scheibner, or by Zöllner and Weber alone, slate-writing, bell-ringing, accordion playing, the placing of hand prints in flour, and the appearance of disembodied hands were witnessed. Zöllner wrote: 'I was now sufficiently encouraged gradually to set on foot those experiments I had prepared from the stand-point of my theory of the fourth dimension' (*TP*, 64).

Having observed events consistent with previously reported mediumistic phenomena, Zöllner designed experiments to exploit the properties of higher space suggested by the work of August Möbius and the projective geometer Felix Klein. Möbius's chapter 'On Higher Space' from his 1827 paper on barycentric calculus had speculated on the congruence of geometric figures: specifically, asymmetrical three-dimensional figures might be made to coincide were it possible to transform them in space of four dimensions. In layman's terms, access to higher space would permit an object to be transformed into its mirror image. Zöllner's plan was to have Slade attempt 'the conversion, by a four dimensional diversion of molecules of tartaric acid, which diverts the plane of polarized light to the right, into racemic acid, which diverts it to the left' (*TP*, 64).

The construction of this experiment indicates the currency of Zöllner's research. *Die Lagerung der atome im Räume*, the German translation of the Dutch chemist

---

[38] See Wolfgang G. Bringmann, Norma J. Bringmann, and Norma L. Medway, 'Fechner and Psychical Research', in *G.T. Fechner and Psychology*, ed. Josef Brožek and Horst Gundlach (Passau: Passavia, 1988), pp. 243–56. Bringmann, Bringmann, and Medway refer to Gustav Theodor Fechner, *Erinnerungen an die letzten Tage der Odlehre und ihres Urhebers* [Recollections of the Last Days of Odic Teaching and its Originator] (Leipzig, 1876).

Jacobus Henricus van't Hoff's *La chimie dans l'espace*, the foundation stone of stereochemistry, had been published earlier that same year.[39] Van't Hoff's account of molecular asymmetry described left-handed and right-handed isomers producing opposite effects on polarized light. Zöllner's experiment for Slade would exploit both this newly discovered property of certain molecular structures and Möbius's speculations on fourth-dimensional mirror-image transformations.

As Zöllner explained the experiment to Slade, however, and demonstrated to him the polarizing effect of Nicol prisms, the medium claimed to be able to see light through crossed prisms. This impossible claim distracted Zöllner, and a series of experiments with the Nicol prisms ensued, leaving Zöllner to write: 'The originally intended experiment with the tartaric acid was discontinued in consequence of the above extraordinary observations. I purposed to carry it out at future investigation of Slade's peculiarities' (*TP*, 66).

On 17 December Zöllner conducted the experiment he felt to be conclusive, based on the work of Felix Klein, as published in his 1876 paper 'Bemerkungen über den Zusammenhang der Flächen'. As Klein later remembered:

> I had rather incidentally given Zöllner a purely scientific account of results that I had found on knotted closed space-curves and published in Volume 9 of the Math. Annalen (see Ges. Abh. 2:63). This result was that the presence of a knot can be considered an essential (i.e., invariant under deformations) property of a closed curve only if one is restricted to move in three-dimensional space; in four-dimensional space a closed curve can be unknotted by deformation. Hence knottedness is no longer a property of analysis situs once our considerations have gone beyond the usual space.[40]

The night before the experiment, Zöllner prepared four cords, sealing them with wax and paper. On the morning of the 17th he selected one cord and asked Slade to produce knots in it:

> While the seal always remained in our sight on the table, the unknotted cord was firmly pressed by my two thumbs against the table's surface, and the remainder of the cord hung down in my lap. I had desired the tying of only one knot, yet the four knots [...] were formed, after a few minutes, in the cord. (*TP*, 41)

News of Zöllner's experiments began to reach British readers of the spiritualist press in early February 1878. A correspondent to *Banner of Light* reported that Slade was working with Professors Zöllner, Fechner, and Scheibner in Leipzig, and *The Spiritualist*, which had been providing regular updates of Slade's travels through Europe, reprinted the letter. The following week Aleksander Aksakov, Russian Imperial Councillor of State and founder and editor of the Leipzig-based journal *Psychische Studien*, wrote in *The Spiritualist*, praising Zöllner's research and the publication of his first volume of scientific papers:

> In the first part of this volume, printed last August, Mr. Zöllner shows that, in the course of speculations on the fourth dimension of space, he came to the conclusion of

---

[39] See Harry C. Jones, 'Jacobus Henricus Van't Hoff', *Proceedings of the American Philosophical Society*, 50 (1911), iii–xii.

[40] Klein, *Development of Mathematics*, p. 157.

the possibility of certain medial phenomena, viz., that beings existent in a fourth dimension of space (*Vierdimmensionale Wesen*) could produce knots on a continuous thread by a simple process of a manipulation of matter—a process impossible and incomprehensible to us. (Three dimensional beings) [...] At a seance with Slade on the 17th December, experience confirmed the reality of the fact, the possibility of which had been admitted *a priori*.[41]

Aksakov, a proponent of a scientific approach to spiritualist phenomena, was so heartened by the news that he suspended plans to cease publication of *Psychische Studien*. He continued: 'We have here the first attempt at a scientific hypothesis in explanation of medial phenomena; and more than that, a hypothesis which renders necessary the acceptance of the cardinal dogma of Spiritualism.'[42]

P.G. Tait, the junior victim of Zöllner's polemical attacks on British scientists in *Über die Nature der Cometen*, responded rapidly with a review of the first volume of *Wissenschaftliche Abhandlungen* in *Nature*. An affected lightness of tone barely concealed a seething sarcasm: 'alas, all is not scientific that professes to be science', wrote Tait, arguing that Zöllner was worthy of interest because he was a curiosity, a metaphysician who in the nineteenth century attempted to bring metaphysics into the pure physical sciences. Tait highlighted Zöllner's devotion to Weber's electrodynamic theory and action at a distance, and mocked his literal readings of Maxwell's Demon and the formation of vortex rings by tobacco smokers. His most pronounced contempt was reserved for the knot experiments derived from Klein's discovery: 'He has held the two ends of a cord (sealed together) in his hands while trefoil knots, genuine, IRREDUCIBLE TREFOIL KNOTS, of which he gives us a picture, were developed upon it!'[43]

Of Zöllner's four-dimensional spiritual hypothesis, he was scarcely less contemptuous:

> It is some time since the Astronomer-Royal for Ireland told me his jocular mode of arguing from Klein's discovery:—viz that all the secrets of the spiritualistic 'rope-trick' could be at once explained by supposing that inside the mysterious cabinet (in which the tambourines and the musical boxes fly about) space was of four dimensions—so that the well-corded performers were at once loosed from their bonds on entering it! But Prof. Zöllner (with the assistance of the spiritualists) has tied knots by means of beings in four dimensional space!!![44]

Zöllner's rush to publish his results did not abate: a paper appearing in *Psychische Studien*, exerted from his third volume of scientific papers and later translated for *Light*, was rapidly followed in the April issue of Crookes's *Quarterly Journal of Science* with an article for English-language readers, 'On Space of Four Dimensions'. This publication was picked up by the *Daily Telegraph* of 2 April, the first attention given in a generalist English-language newspaper to higher space: 'Under this title the "Quarterly Journal of Science" publishes a remarkable article giving a curious

---

[41] Aleksander Aksakow, 'A New Manifestation with Dr. Slade at Leipzig University', *Spiritualist*, 12 (1878), 78.

[42] Aksakow, 'A New Manifestation', 78.   [43] Tait, 'Zöllner's Scientific Papers', 421.

[44] Tait, 'Zöllner's Scientific Papers', 421.

illustration of the idea which has recently been developed in Germany that space has another dimension beyond the length, breadth and thickness recognised in geometry.'[45] All the more remarkable for the central role in that illustration of Henry Slade, whose trial only a year previously had occupied many column inches of the same newspaper.

'On Space of Four Dimensions' was canny in its structure. The work of Helmholtz, familiar to British readers, was placed at the forefront of Zöllner's theoretical introduction, and its corrective agenda neatly subsumed into Zöllner's Kantian version of scientific methodology:

> In accordance with Kant, Schopenhauer and Helmholtz, the author regards the application of the law of causality as a function of the human intellect given to man *a priori*, i.e., before all experience. The totality of all human experience is communicated to the intellect by the senses, i.e., by organs which communicate to the mind all the sensual impressions that are received at the *surface* of our bodies.   (*TP*, 32)

Zöllner's theory of spatial perception was briefly sketched on this basis, that we sense only two dimensions and learn through experience to construct a third dimension to allow for perceived changes in the size of objects as we move in relation to them. When we observe facts contradictory to this learned space, he continued, 'our reason would at once be forced to reconcile these contradictions' (*TP*, 33). Arguing by analogy, Zöllner outlined the theory behind the possibility of unknotting three-dimensional knots in higher space, introducing four-dimensional intelligences who could more easily unknot three-dimensional knots just as we three-dimensional intelligences could unknot those of two. Citing Berkeley and Lichtenberg, he suggested that such intelligences might not be aware of their actions, giving the first intimation of the unusual direction in which his argument was proceeding:

> The want of these conceptions [of higher space] would necessarily be felt by us, if in some individuals, and these only occasionally, the will should be capable of producing physical movements for whose geometro-mathematical definition a four-dimensional system of co-ordinates is necessary.   (*TP*, 36–7)

Declaring that the '*real* existence of four-dimensional space can only be decided by experience, i.e., by the observation of facts' (*TP*, 67), he praised Helmholtz's work in describing that the intellect could understand the possibility of higher space without being able to produce a mental image of it, but trumped it with extracts from Kant, using Helmholtz's empirical corrective to Kant as a springboard into metaphysics: 'If it is possible that there be developments of other dimensions in space, it is also very probable that God has somewhere produced them' (*TP*, 67).

Reporters to the spiritualist journals felt that Zöllner's experiments had provided the scientific legitimacy spiritualism had been denied in the wake of

---

[45] Anon. [probably William H. Harrison], 'On Space of Four Dimensions', *Daily Telegraph*, 2 April 1878, 2.

Crookes's experiments and Lankester's prosecution of Slade. Reporting the 'Vindication of Dr Slade' in *Medium and Daybreak*, C. Reimers was in celebratory mood:

> He [Zöllner] found his previously formed hypothesis confirmed by a stupendous experiment with Dr. Slade, and boldly proclaims his result. He anticipated a fourth dimension, hitherto overlooked, to account for strange occurrences, and rested his conclusion on following self-proposed experiment.
>
> Our gallant Mr. Askakow's excellent remarks (or rather congratutions) [*sic*] on this triumph in the latest number of *Psychische Studien* should be translated, if only for the allusion to the gordian knot. How curious! Where the wise German scholars would not listen to the honest assertions of their fellow-students and the endless proofs from all quarters of the globe, they now group round this little perplexing knot and prick up their ears.[46]

The issue of *The Spiritualist* for the same day provided a report of successful replication of Zöllner's experiments by T.L. Nichols, MD:

> Your readers may be glad to know that, on the night of April 7th, we had repeated, in my house, in the presence of six persons, including Mr. AV. Eglinton and Mr. A. Colman, Professor Zöllner's marvel of tying knots in a cord, the ends of which were tied and sealed together. I have the sealed cord, which I prepared myself, with the knotted ends firmly sealed to my card, on which the fingers of every person present rested while five knots were tied, about a foot apart, in the central portion of the cord. I have no doubt that this splendid manifestation can be repeated at any time under like conditions.[47]

Writing to update readers of Slade's travels in Russia, the following week, the man described during his trial as Slade's agent, J. Simmons, summarized the mood of optimism in the spiritualist community, reporting that 'Prof Boutlerof' had published a pamphlet, that Dr Hoffmann had written to the effect that the Leipzig professors were eager for more sittings, and that according to another correspondent, public opinion in Berlin, previously mixed, was beginning to turn in Slade's favour, evidenced by the fact that Dr Wittig had 'obtained a hearing on the subject in an illustrated journal'.[48]

Among mathematicians the response to 'Zöllner's marvel' was rather more muted. Addressing the British Association in Dublin in August 1878, the mathematician William Spottiswoode noted the recent developments in *n*-dimensional work by Klein, Tait, and Simon Newcomb: 'Thus it has recently been shown that in four dimensions a closed material shell could be turned inside out by simple flexure, without either stretching or tearing; and that in such a space it is impossible to tie a knot.'[49]

---

[46] C. Reimers, 'Vindication of Dr Slade', *Medium and Daybreak*, 9 (1878), 232.
[47] Nichols, 'Remarkable Physical Manifestations', 175.
[48] J. Simmons, 'Slade in Europe', *Spiritualist*, 12 (1878), 186.
[49] William Spottiswoode, 'Presidential Address', *Report of the Forty-Eighth Meeting of the BAAS Held at Dublin in August 1878* (London: John Murray, 1879), p. 21.

Whether or not he had in mind news of the Leipzig experiments, Spottiswoode was concerned to curtail misunderstandings about the empirical nature of higher space and its application in analytical geometry:

> This is in fact the whole story and mystery of manifold space. It is not seriously regarded as a reality in the same sense as ordinary space; it is a mode of representation, or a method which, having served its purpose, vanishes from the scene. Like a rainbow, if we try to grasp it, it eludes our very touch; but, like a rainbow, it arises out of real conditions of known and tangible quantities, and if rightly apprehended it is a true and valuable expression of natural laws, and serves a definite purpose in the science of which it forms part.[50]

Searching for analogues in reality, Spottiswoode chose a telling example, blurring the lines between imaginative fantasy and spiritualist theory:

> When space already filled with material substances is mentally peopled with immaterial beings, may not the imagination be regarded as having added a new element to the capacity of space, a fourth dimension of which there is no evidence in experimental fact?[51]

The aftermath of the experiments in Germany was more turbulent. Hermann Ulrici, the co-editor with Fichte of the philosophical journal *Zeitschrift für Philosophie und Philosophische Kritik*, was among those convinced by Zöllner's publications. He published an article in support of Zöllner's fourth-dimensional hypothesis: the astrophysicist had raised valid concerns about the empirical existence of higher space, claimed Ulrici, and, given his position as a respected scientist, notice should be taken.

Wundt, a former assistant of Helmholtz whom Zöllner had called to Leipzig in 1875 to help ground his instrumental astronomical observations in physiology, rebelled at this.[52] Wundt's half-hour session with Slade gave him a position of authority from which to comment, and having been named by Ulrici as among those who attended a seance he felt compelled to respond. He published an open letter to Ulrici in 1879, questioning Zöllner's fitness to conduct investigations into a seance.[53] He acknowledged that he did not have the training to make observations, either, but that he did not trust what he had seen. He suggested that Slade had in fact been the experimenter, and that Zöllner did not have control over the experimental conditions. In a second section he sustained an attack against Ulrici's philosophical position, arguing that it was dangerous for philosophy to allow rapping spirits to overthrow science, allowing freedom to superstition: 'From early times, as you well know, materialism has had two forms; the one denies the spiritual, the other transforms it into matter.'[54]

[50] Spottiswoode, 'Presidential Address', pp. 22–3.
[51] Spottiswoode, 'Presidential Address', p. 23.
[52] See Staubermann, *Astronomers at Work*, p. 104.
[53] Wilhelm Wundt, 'Spiritualism as a Scientific Question: An Open Letter to Prof. Hermann Ulrici of Halle', *Popular Science Monthly*, 15 (1879), 578–93 (repr. and trans. of *Der Spiritismus: Eine Sogenannte Wissenschaftliche Frage* (Leipzig: Engelmann, 1879)).
[54] Wundt, 'Spiritualism as a Scientific Question', p. 593.

Commenting on Wundt's criticisms over the authority of the experiments, Arne Hessenbruch has noted that 'Zöllner's world was peopled with charismatic individuals. Spiritualist research was pursued with individuals who had extraordinary powers and by people who were either trustworthy or untrustworthy.'[55] For Zöllner, friendships and loyalties, personal and national, were more significant than the impersonal assessment and peer review of the broader scientific community. Slade was beyond suspicion because he was a gentleman. His friends were loyal colleagues on the field of battle. Despite his polemical attacks on Berlin professors, Zöllner himself was an elitist, excluding those outside his circle. He expected others to trust his observations and his authority to make them, becoming highly defensive whenever questioned.

Inevitably, Zöllner personalized Wundt's criticism, responding in his fourth volume of *Wissenshaftliche Abhandlungen* with a series of charges against Wundt. The popular press, meanwhile, began to poke fun at the arguing professors.

TRANSCENDENTAL PHYSICS

In December 1880, Charles Carlton Massey, co-founder of the Theosophical Society in Britain and barrister for the defence in Slade's 1876 trial, published a heavily edited translation of sections of Zöllner's work.[56] *Transcendental Physics* reproduced the records of the seances from the third of Zöllner's treatises, *Die transcendentale Physik und die sogennante Philosophie*, from which it took its title, and excerpted theoretical material from each of his two earlier *Wissenschaftliche Abhandlungen*. The translator also contributed a long preface, in which he reassessed Slade's trial. This sanitized edition of the Zöllnerian world view became widely read, going through three English editions, published by W.H. Harrison in 1880, 1882, and 1885, and an American edition published by the spiritualist publisher Banner of Light in 1901, and including the article from the *Quarterly Journal of Science* and various appendices.

The term *Raumgebiete*, used by Zöllner to describe his higher spatial hypothesis, displayed a minor deviation from the interpretation of Riemannian space by Clifford and Helmholtz. Rather than a space of n-dimensions lying within a space of $n+1$ dimensions, in Zöllner's reading Riemann described 'the juxtaposition of different, infinitely extended regions of space' (*TP*, 69). The difference was subtle, but necessary to account for the invisibility of three-dimensional intelligences. Zöllner's fourth dimension was adjacent to our three-dimensional space, not around and through it.

---

[55] Arne Hessenbruch, 'Science as Public Sphere: X-Rays Between Spiritualism and Physics', in *Wissenschaft und Öffentlichkeit in Berlin, 1870–1930*, ed. Constantin Goschler (Stuttgart: Franz Steiner, 2000), pp. 89–126 (p. 122).

[56] For Massey's brief but important involvement with the Theosophical Society see A.P. Sinnett, *The Early Days of Theosophy in Europe* (London: Theosophical Publishing House, 1922) and Chapter 5 of this volume.

If the effects observed by us proceed from intelligent beings occupying (*welche sich befinden*), in the absolute space, places which in the direction of the fourth dimension lie near the places occupied by Mr. Slade and us in the three-dimensional space,* and therefore necessarily invisible to us, for these beings the interior of a figure of three-dimensional space, enclosed on all sides, is just as easily accessible as is to us, three-dimensional beings, the interior of a surface enclosed on all sides by a line—a two-dimensional figure. (*TP*, 70–1)

Any demonstration of the penetration of closed three-dimensional spaces, such as slate-writing on closed slates, would therefore support Zöllner's theory. 'Whatever may be thought of the correctness of my theory with regard to the existence of intelligent beings in three-dimensional space,' he wrote, 'at all events it cannot be said to be useless as a clue to research in the mazes of Spiritualistic phenomena' (*TP*, 71).

Reproducing in their entirety the letters to *The Spiritualist* from T.L. Nichols describing the replication of Zöllner's knot experiments, Zöllner went on to describe a series of further knot experiments conducted with Slade during his third residence in Leipzig, in May 1878.[57] He outlined the possible explanations for his observations, 'according as one supposes a space of three or of four dimensions':

> In the first case, there must have been a so-called passage of matter through matter; or, in other words, the molecules of which the cord consists must have been separated in certain places, and then, after the other portion of cord had been passed through, again united as in the same position as the first. In the second case the manipulation of the flexible cord being, according to my theory, subject to the laws of a four-dimensional region of space, such a separation and re-union of molecules would not be necessary. (*TP*, 85)

The materialist hypothesis was explicitly dismissed in favour of the higher spatial. The disappearance and reappearance of material objects during the sittings was also regarded as supporting the higher spatial hypothesis. A letter to Zöllner from Baron von Hellenbach described such an occurrence and commented:

> I regard as very important a demonstration on your part of a similar disappearance; for if the seen and felt ascent of the slate on my foot proves an unperceived mechanical agency, and the production of knots in the endless cord a four-dimensional agency, so would the entrance and exit of an object prove another space-dimension, as it were in our immediate neighbourhood, in so stupendous a manner that it could not be for a moment doubted in my opinion, which is that our illusion of consciousness is nothing but a three-dimensional world, brought about by a strange organism. (*TP*, 91)

Such demonstrations were obtained, as were the insertion and removal of coins and chalks from closed boxes, which to Zöllner's mind also lent credence to the phenomena of clairvoyance: it was a simpler matter to look inside a closed lower-dimensioned space than it was to remove items from it. In describing this

---

[57] The reproduction of sources in their entirety was a feature of Zöllner's *Wissenschaftliche Abhandlungen*. He would reproduce and comment upon personal letters, reviews, newspaper reports, and even reported conversation without prejudice.

last occurrence, Zöllner gave an intriguing insight into his understanding of the medium's relation to higher space, allowing for the fact that it was sometimes the medium himself who gave answers and sometimes the intelligences he claimed to channel:

> Thus, Slade's soul was, in the first case, so far raised in the fourth dimension that the contents of the box in front of him were visible in particular detail. In the second case, one of those intelligent beings of the fourth dimension looked down upon us from such a height that the contents of the rectangular box were visible to him, and he could describe its contents upon the slate by means of the pencil. (*TP*, 148)

As the thesis was progressed, Zöllner addressed also the need for his empirical fourth dimension to conform to the laws of physics. He referred back to his initial theoretical outline, 'On Action at a Distance', and the arguments contained therein that higher space did not violate the conservation of energy 'if one regards the distance of two atoms and the intensity of their interaction, in our three-dimensional space, as projections of similar magnitudes in space of four dimensions' (*TP*, 95). So, too, did he attempt to describe how a 'luminous appearance' beneath a table that appeared to violate the geometry of shadow projection might have its source in the fourth dimension:

> Since now in the above-mentioned case surprisingly sharp shadows of the feet of the table of perceptibly similar size to the feet themselves were observed, it follows from this that the rays which produced that projection of shadow, must have issued from a light source, first, possessing a very small apparent size, and, secondly, being at a great distance [...] the said phenomenon would thus point to another place as the point of issue, which cannot lie at all in our three-dimensional space. (*TP*, 181)

A largely sympathetic review in the Christmas edition of *The Spectator* voiced an opinion of Zöllner's higher spatial theorizing that would be parroted in correspondence in the spiritualist press: 'We hope our readers understand these last consequences of four-dimensional activity, confessing, for ourselves, our utter inability to follow the Professor into this inconceivable region.' Despite an additional ambivalence towards the theory of intelligent spirits, the reviewer was more comfortable with Zöllner's methodology: 'The important thing to note is that he previously entertained this theory and that the theory was, in his opinion, confirmed by the subsequent experiments, as seems often to happen in similar circumstances.'[58]

Correspondents to the early issues of the new spiritualist journal *Light*, launched in January 1881, responded to *Transcendental Physics* with some confusion. In the inaugural issue, a question entitled 'puzzled' echoed the plea of C.J. Monro to Clerk Maxwell: 'Can any of your readers explain Zöllner's explanation in such simple terms as will make it plain to people of moderate capacity?'[59]

The ensuing discussion featured a muddle of ideas, from correspondents empirically, ideally, or metaphysically inclined. Massey preferred an explanation

[58] Anon., 'Modern Spiritualism', *Spectator*, 2739 (1880), 1661–2.
[59] 'Puzzled', 'Questions and Answers', *Light*, 1 (1881), 7.

based on Zöllner's insistence on the principle of causality: 'as Zöllner shews, and as Bishop Berkeley ("Theory of Vision") had shewn before him [...] we obtain our knowledge of the third dimension not directly from the senses (whose impressions are only of superficial extension, i.e., of space in two dimensions), but by the principle of causality in our minds'.[60] Newton Crosland discussed the 'relativity of space' and E.T.B. suggested 'confining the word "space" to the three dimensions. The "fourth dimension" must be beyond space in the sense of thickness (or solidity), as space (in the same sense) is beyond mere surface (or breadth).'[61] By the fourth issue, Massey was responding to several correspondents at once, Crosland was taking issue with E.T.B. on the grounds that he had described 'powers' and not 'dimensions', and G.W., MD, was arguing that the fourth dimension was merely an 'unknown force' and 'the ultimate sub-stance of matter'. Clarification of the issue seemed the least likely outcome of discussions.[62]

These discussions had petered out by Zöllner's death in 1882. Instead, a report of the appearance of his spirit at a seance appeared in *Medium and Daybreak*.[63] An obituary in the *Astronomical Society Monthly Notices* of February 1883 played down his later career, remarking simply: 'His researches on the philosophy of space, and other subjects, with his theories of light and electricity, appeared in the three volumes of his *Wissenschaftliche Abhandlungen*.'[64]

In 1883 allegations concerning Zöllner's sanity resurfaced noisily. Writing in the *Contemporary Review*, the Russian physiologist Emil de Cyon responded angrily to what he felt was a betrayal by Zöllner in the last book published before his death, an anti-vivisectionist polemic.[65] De Cyon recorded that he had become friendly with Zöllner between 1865 and 1875, meeting him through his colleague Ludwig: 'Indeed it was Mr Zöllner—as he has since reminded me with a certain satisfaction—who initiated me into the study of higher mathematics.' Zöllner had been present at many operations on animals but in the last years of his life had been taken advantage of by anti-vivisectionists. For De Cyon, the explanation was clear:

> In reading M. Zöllner's last book, one follows with a sort of terror the progressive ravages of insanity on so fine and rare intelligence [...] If the conversation of a dozen inmates of Bedlam during twenty-four hours were taken down by stenography, it would not make a more extraordinary collection of nonsense than the ravings of this anti-vivisectionist.[66]

Not for the first time, Massey made the case for the defence, arguing that the accusations of madness were unfounded and that his own last contact with Zöllner had given no indication of madness: 'one must suspect that the whole imputation

[60] C.C. Massey, 'The Fourth Dimension', *Light*, 1 (1881), 15.
[61] Newton Crosland and E.T.B., 'The Fourth Dimension', *Light*, 1 (1881), 23.
[62] Newton Crosland, C.C.M., and G.W., MD, 'The Fourth Dimension', *Light*, 1 (1881), 31.
[63] C.W. Skilin, 'Zöllner', *Medium and Daybreak* (1882), 803–4 (p. 804).
[64] Anon., *Astronomical Society Monthly Notices*, February 1883, 185.
[65] See Johann Carl Friedrich Zöllner, *Ueber den wissenschaftlichen Missbrauch der Vivisectionen* (Leipzig, 1882).
[66] E. de Cyon, 'The Anti-Vivisectionist Agitation', *Contemporary Review*, 43 (1883), 498–510.

of madness rests on no better foundation than the fact that down to the last he held and expressed opinions highly unacceptable to M. de Cyon, and to the scientific world in general'.[67]

Countering various claims made by De Cyon with references from Zöllner's book, Massey finally acknowledged, with no little prescience, the impossibility of his task:

> I have been disappointed in my wish to challenge these imputations on Zöllner before a more general public than the columns of 'LIGHT' will reach; and now, no doubt, every appeal to Zöllner's testimony will be met by the assertion of his insanity, as though that were a proved and admitted fact.[68]

THE SEYBERT COMMISSION REPORT: A TEST CASE IN SCIENTIFIC AUTHORITY

In 1887 The Seybert Commission Report published the results of a two-year-long scientific investigation into 'Modern Spiritualism'. Established at the University of Minnesota at the bequest of the philanthropist Henry Seybert, himself 'an enthusiastic believer' who had funded a philosophy chair at the university on the condition that such a report was undertaken, the Commission had set out to approach the investigation by first-hand observation and interview:

> We decided that, as we shall be held responsible for our conclusions, we must form those conclusions solely on our own observations; without at all imputing untrustworthiness to the testimony of others, we can really vouch only for facts which we have ourselves observed.[69]

To this end it had undertaken its own trials with Slade in America in 1885 and its conclusions had been negative:

> We had a number of sittings, and, however wonderful may have been the manifestations of his Mediumship in the past, or elsewhere, we were forced to the conclusion, that the character of those which passed under our observation was fraudulent throughout. There was really no need of any elaborate method of investigation; close observation was all that was required.[70]

Slade's slate-writing was either pre-prepared or conducted in haste under the table. He was observed using a foot to tip over chairs and tricks such as playing the accordion with one hand beneath the table or tossing pencils onto its surface were deemed 'almost puerile in the simplicity of their legerdemain'.[71]

---

[67] C.C. Massey, 'M.E. de Cyon and the Late Professor Zöllner', *Light*, 3 (1883), 188.

[68] Massey, 'M.E. de Cyon', 189.

[69] George S. Fullerton and others, *Preliminary Report of the Commission Appointed by the University of Pennsylvania to Investigate Modern Spiritualism in Accordance with the Request of the Late Henry Seybert* (Philadelphia: J.B. Lippincott Co., 1887), p. 5.

[70] Fullerton and others, *Preliminary Report*, p. 7.

[71] Fullerton and others, *Preliminary Report*, p. 12.

Of the Zöllner trials the *Report* noted: 'Perhaps no other investigation of Spiritistic phenomena has exercised so strong an influence upon the public mind in America, at least, as that conducted by Professor J. C. F. Zöllner and his colleagues in Leipsic [*sic*] in 1877 and 1878.'[72] Professor George Fullerton, the secretary of the committee, visited Germany on academic business in 1886 and reading that all the witnesses to the Zöllner trials bar Zöllner himself were still alive, made contact with each.[73]

In the summer of 1886, Fechner, Weber, Wundt, and Scheibner all submitted to interview, offering divergent opinions of Zöllner's state of mind and each of his own capacity accurately to observe the trials. Wundt, as he had been in the immediate aftermath of Zöllner's claims, was the most highly critical, arguing that 'the conditions of observation were very unsatisfactory' and doubting the authority of each of the other witnesses. On the question of Zöllner's mental state he told Fullerton: 'That Professor Zöllner himself was at the time decidedly not in his right mind; his abnormal mental condition being clearly indicated in his letters and in his intercourse with his family.'[74]

Theoretically more damaging to the reputation of the experiments were the interviews with Fechner, who admitted to suffering from cataracts and acknowledged Zöllner's 'mental derangement', and Scheibner, who thought Zöllner's later madness was incipient at that time but whose assessment profoundly questioned his friend's methodology:

> Professor Zöllner was, said Professor Scheibner, a man of keen mind, but in his investigations apt to see 'by preference' what lay in the path of his theory. He could 'less easily' see what was against his theory. He was childlike and trustful in character, and might easily have been deceived by an impostor. He expected everyone to be honest and frank as he was. He started with the assumption that Slade meant to be honest with him. He would have thought it wrong to doubt Slade's honesty. Professor Zöllner, said Professor Scheibner, set out to find proof for four-dimentional [*sic*] space, in which he was already inclined to believe. His whole thought was directed to that point.[75]

Following Massey's translation, spiritualist journals such as *Light* had included the names of Zöllner, Fechner, Scheibner, and Weber in their lists of eminent men of science who supported the spiritualist hypothesis. Fullerton's summary of his investigation was unequivocal in its rebuttal to such claims:

> Thus it would appear that of the four eminent men whose names have made famous the investigation, there is reason to believe one, *Zöllner*, was of unsound mind at the time, and anxious for experimental verification of an already accepted hypothesis; another, *Fechner*, was partly blind, and believed because of Zöllner's observations; a

---

[72] Fullerton and others, *Preliminary Report*, p. 104.

[73] Having published an article on higher space, the philosopher Fullerton was an informed observer. See George S. Fullerton, 'On Space of Four Dimensions', *Journal of Speculative Philosophy*, 18 (1884), 113–21.

[74] Fullerton and others, *Preliminary Report*, p. 106.

[75] Fullerton and others, *Preliminary Report*, p. 109.

third, *Scheibner*, was also afflicted with defective vision, and not entirely satisfied in his own mind as to the phenomena; and a fourth, *Weber*, was advanced in age, and did not even recognize the disabilities of his associates. No one of these men had ever had experiences of this sort before, nor was any one of them acquainted with the ordinary possibilities of deception. The experience of our Commission with Dr. Slade would suggest that the lack of such knowledge on their part was unfortunate.[76]

The *Pall Mall Gazette* predicted that the report would receive a 'good deal of attention', but the rearguard action by the spiritualist community was far from overwhelming. In Britain, Massey, who had also been interviewed by Fullerton, published in a supplement to *Light* a lengthy open letter, in which he employed a full range of barristerial rhetoric, quoted from texts to marshal written support for his case, and attacked inaccurate details, to dispute the accusation that Zöllner had been insane, which he claimed 'only to have taken tangible shape at a later date, and in obedience to polemical exigencies'.[77]

He quoted from E. von Hartmann's argument from *Der Spiritismus*, that Zöllner's experiments were 'excellently contrived' and gave 'the best conceivable security against conjuring, show everywhere the skilled hand of an accomplished experimenter and are reported with clearness and precision', and Von L.B. Hellenbach's account of his correspondence with Zöllner from *Geburt und Tod als Wenschel der Anschauungsform, oder der Doppel-Natur des Menschen*:

> But since so many of these gentlemen have not shrunk from declaring that Zöllner was deranged or insane, I declare that I was in frequent correspondence with him, latterly in the subject of my Magic of Numbers—thus a serious and deep topic—on which I received a letter from him a few days before his death, and there was not even the semblance of justification for the above allegation.[78]

One-by-one he dismissed oppositional accounts as defined by 'controversies' in which Zöllner had become embroiled, noting the 'personal insults and contumelies and estrangements' caused by them. Particular focus was given to the subject of observation:

> Only I decidedly object to your (of course accidental) altering of the word 'seeing' in your notes ('He became more and more given to fixing his attention on a few ideas, and incapable of seeing what was against them') into the word 'observe' when you would use your notes argumentatively with reference to Zöllner's capacity as an 'observer'. The note of Scheibner's statement about Zöllner is not very lucid as a whole, but on this point it is unmistakable; and you cannot be allowed to convert a statement of a theorist's inability to 'see' an objection into a statement of an investigator's inability to 'observe' a trick, though you are of course at liberty to argue from one to the other.[79]

---

[76] Fullerton and others, *Preliminary Report*, p. 114.

[77] C.C. Massey, 'Zöllner: An Open Letter to Professor George S. Fullerton', [supplement to] *Light*, 7 (1887), 375–84 (p. 375). Massey had acknowledged in his response to De Cyon that rumours regarding Zöllner's mental health had circulated as early as 1872.

[78] Massey, 'Zöllner', 375, 381.   [79] Massey, 'Zöllner', 381.

Both Massey and the Manchester-based medium and editor Emma Hardinge Britten countered claims published in the *Manchester Guardian*, before Fullerton's response to Massey's letter was printed in *Light*. Acknowledging a slip of which Massey had accused him, Fullerton added: 'two professors [...] to whom I mentioned incidentally this matter, as well as the evidence concerning Zöllner's mental disease and Fechner's defective vision, both spoke of these things as generally known'.[80] Massey naturally disapproved of his opponent's canny admission of additional hearsay evidence, but was left with no choice but to report it.

The subject of Zöllner's experiments re-emerged nine months later with the publication in the German spiritualist journal *Sphinx* of extracts from Fechner's diary. These were translated for readers of *Light* and appeared to show Fechner's agreement with the spiritualist hypothesis:

> Spiritualists attach great importance to the facts of Spiritualism as proving the immortality of the soul; and indeed, they appear to me to justify the views I hold as to the other world, according to which the spirits constantly surround us and influence us without our knowledge.[81]

That these were edited to skew the presentation only became clear in 1901 with the publication of Wundt's biography of Fechner, brought out to coincide with the centenary of his friend's birth. In this he argued that the apparently successful knot experiments in which Zöllner held so much stock were crucial in changing Fechner's opinion.

More complete extracts from these diaries, subsequently discovered among Wundt's papers, reveal that Fechner's 'attitude toward spiritism [...] progressed over a two-month period from ambivalent skepticism to a guarded endorsement'. In his diary entry concerning the first experiment he attended, he wrote:

> What can I say about all this? If I look at the general character of the performance which took place, my overwhelming impression is that of skillfully conducted trickery. One certainly cannot talk of a scientific examination of the phenomena which took place. All the conditions for this were missing.[82]

By December, however, Fechner was reporting:

> Support for spiritistic achievements comes from experiments, which are above suspicion, as was the case with the knot experiment. Opponents of spiritism, however, generally follow the opposite method. They usually base their criticisms on events in which spiritistic activities fail, and they overlook evidence which supports the claims of spiritism.[83]

---

[80] George S. Fullerton, 'A Letter from Professor Fullerton to Mr. C.C. Massey', *Light*, 7 (1887), 451.

[81] Gustav Theodor Fechner, 'Zöllner's Mediumistic Experiments: Extracts from the Diary of Gustav Theodor Fechner, Late Professor in Vienna, Died November 19th, 1887', *Light*, 8 (1888), 256–7 (p. 257).

[82] Bringmann, Bringmann, and Medway, 'Fechner', p. 248.

[83] Bringmann, Bringmann, and Medway, 'Fechner', p. 252.

Fechner was now privileging the results of the experiments, rather than the a priori deductions that had prompted them.

In surveying the ambivalent reception of higher spatial ideas in the British spiritualist press, K.G. Valente has warned against readings that stress the unanimous acceptance of Zöllner's theories, arguing that an analysis of the coverage of these ideas in such journals reveals that 'British Spiritualists espoused a range of positions that typically included guarded optimism, apprehension, and skepticism in relation to the potentiality of a fourth spatial dimension'.[84] This assessment should be applauded for maintaining the standards of symmetry called for by Latour: Valente assesses the spiritualist press as one would assess scientific journals. It also serves to highlight that the fourth dimension, despite being the supposed object of Zöllner's experiment, was marginalized in the Seybert Commission Report and in discussions which focused on issues of scientific legitimacy. In these networks, shadowing the original event, discourse and speech acts dominated; researchers were undertaking the work of purification. Here, the fourth dimension is like the vacuum in Boyle's air pump, the object of the experiment that is nevertheless sidelined.

It seems as though symmetry remains elusive, though. These textual translations are not the only non-humans participating in the Zöllner event. There are also the assorted instruments he used: radiometers, slates, crystals, lenses, and ropes. The knots procured in these ropes, knots around which so much of the public representation of the event clustered, need to be granted equal standing to the human participants. In this case, just as much attention as has been given to Zöllner's career, and significantly more than has been allotted to Slade's legal biography, should be devoted to the knot's earlier life. The knot is like a 'blackbox' in the Latourian sense, in that we take for granted its interior, ignoring the considerable volume of information that is bound up in its coils. The physical content of this pregnant form has been sketched at the beginning of this chapter, but given its central role in the Zöllner event we should make explicit the other resources it binds together. Precisely what kind of knots are we dealing with?

THE CONJUROR'S KNOT CONSIDERED

American spirit mediums like Slade had been the objects of attention for professional stage conjurors and magicians—professed entertainers—since their arrival in Europe in the 1860s. The spiritualist Davenport Brothers, who exhibited to their audience purported spiritual manifestations occurring inside a cabinet, such as the playing of instruments, were bound by ropes before being shut into their box. Their bindings were proof to their audience that these manifestations could not be

---

[84] K.G. Valente, '"Who Will Explain the Explanation?": The Ambivalent Reception of Higher Dimensional Space in the British Spiritualist Press, 1875–1900', *Victorian Periodicals Review*, 41 (2008), 124–49 (p. 126).

performed by the brothers themselves but must be created by the spirits they summoned. When the Davenports performed in Cheltenham in 1864, a young man called John Nevil Maskelyne interrupted their performance to announce that he knew how it was done. Two months later Maskelyne and his partner Cooke gave a performance with their own spirit cabinet. Special attention was given to their bindings, knotted by a sailor, with additional ropes tied and sealed with wax by other members of the audience. Maskelyne and Cooke nevertheless reproduced the manifestations the Davenports had earlier claimed to be due to spirits, launching a stage conjuring career that continued well into the twentieth century.

Maskelyne was not alone in mastering the knot tricks of spiritualist performers. In France, Jean Robert-Houdin, who had already successfully reproduced various aspects of the performances of the medium Daniel Home, announced upon witnessing the Davenports' act in 1865 that 'the article wherein lay all the deception was the rope'. Houdin described the act, barely changed since Maskelyne had copied it:

> The two Americans are seen securely tied; their legs, arms, and bodies are alike covered with a network of cords binding them to the chairs on which they are seated [...] Spectators gather round, they examine the various knots, and are constrained to admit that they are honestly tied.[85]

Again and again audience attention was focused on the knots and precisely the techniques used in the Zöllner event were deployed as means of ensuring the material honesty of the bindings:

> In order to afford the company absolute certainty that the ligatures are not unfastened, one of the spectators who happens to be nearest is asked to apply some melted sealing-wax to the knots which bind the wrists, and to impress a seal thereon.[86]

Alas, Robert-Houdin declared, all was misdirection: 'Vain precaution! Every knot, every form of ligature, is necessarily capable of being again untied.'[87] Knots that could be tied—even by sailors—could also be untied: they were trivial. This was no secret and its connection to Slade was direct. Maskelyne had been a witness for the prosecution at the trial of Slade in London in 1876. Maskelyne and Slade had encountered each other across a courtroom before Zöllner and Slade had met.[88]

---

[85] Jean Eugene Robert-Houdin, *The Secrets of Stage Conjuring*, ed. and trans. Professor Hoffmann [pseud. Angelo John Lewis] (London: G. Routledge & Sons, 1881), p. 192.

[86] Robert-Houdin, *Secrets*, pp. 193–4.

[87] Robert-Houdin, *Secrets*, p. 200. Papers published in English translation posthumously in 1880 give illustrations of the type of slip-knot, or cat's paw, Robert-Houdin believed to have been used. The first book-length guide to magical knots written by Harry Houdini in 1923 was seeded in this sketch.

[88] Roger Luckhurst has written of the Lankester seance and trial of Slade that 'by stepping into a non-scientific terrain, Lankester invested the standards of evidence in legal discourse with greater authority than those of scientific naturalism'. See *The Invention of Telepathy* (Oxford: Oxford University Press, 2002), p. 47. The legacy of this can be seen in Massey's extraordinary defence of Zöllner in *Light* and in a preface to the fourth edition of *Transcendental Physics* in which he criticized the trial of Slade employing the same legal rhetoric. The legal witness was given precedence over the scientific witness by the culmination of discussions, as Fullerton and Massey mobilized texts and hearsay against each other's claims.

Little surprise, then, that the knot had focused the investigative attention of British psychical researchers before the Seybert Commission. The Society for Psychical Research had conducted its own text-based research into Zöllner's experiment. Eleanor Sidgwick had trawled through the more than 4,000 pages in the original German of Zöllner's *Abhandlungen*, and had found, in volume 2, on page 1191, a report of an experiment that had been omitted from Massey's translation:

> That Professor Zöllner did not always perceive and avoid important sources of possible error may, I think, be inferred from his writings. For instance, in describing the séance on December 17th, 1877, wherein he obtained four knots in a string of which the ends were tied and sealed together, he omits to mention that the experiment had been tried and failed before. We learn that this was so, accidentally, as it were, from his mentioning it in another place and in another connection, where he tells us that it was a long time before the spirits understood what kind of knot was required of them, and that before they did so he obtained knots, but not such as he wanted—knots, I infer, which could be made by ordinary beings without undoing the string.[89]

Indeed they could. In Listing's obituary Tait revealed that he had not forgotten his engagement with Zöllner. He observed that one of Listing's notable discoveries had concerned certain features of twisted closed-space curves:

> A special example of these, given by Listing for threads, is the well-known juggler's trick of slitting a ring-formed band up the middle, through its whole length, so that instead of separating into two parts, it remains in a continuous ring. For this purpose it is only necessary to give a strip of paper one half-twist before pasting the ends together. If three half-twists be given, the paper still remains a continuous band after slitting, but it cannot be opened into a ring, it is in fact a trefoil knot. This remark of Listing's forms the sole basis of a work which recently had a large sale in Vienna:— showing how, in emulation of the celebrated Slade, to tie an irreducible knot on an endless string![90]

Slade was, in the final analysis, perhaps as much mathematician as magician.

The conjuring strand bound up in Zöllner's knot is therefore particularly important and often obscured by its shadow brother, spiritualism. Zöllner's fourth dimension is magical but a crucial element of this magic is the performed, dramatized magic of the stage conjuror. This is intermingled with the mathematics of knots in a way that is difficult to separate: our denouement remains elusive. Zöllner's knots are truly quasi-objects, drawing together geometry, scientific rivalries, national politics, belief systems, and conjuring techniques such as sleight of hand. They permit Zöllner to publish by giving him the evidence he reads as supporting his claims. They also enable the mobility, or portability, of the experiment, the replication of the laboratory: just as a contributory factor to Pasteur's success, in Latour's account, was his ability to replicate his laboratory in the outside world, so was the replication of Zöllner's experiment by Eglington a success for

[89] Mrs. Henry Sidgwick, 'Results of a Personal Investigation into the Physical Phenomena of Spiritualism with Some Critical Remarks on the Evidence for the Genuineness of Such Phenomena', *Proceedings of the Society for Psychical Research*, 4 (1887), 45–74 (p. 66n).
[90] Tait, 'Johann Benedict Listing', 317.

the network within spiritualism. It carried little weight beyond this narrow field, however, and failed in a final test of strength with the network mobilized by British physicists.

TRANSLATION: 'THE MIDDLE TERM IN A SYLLOGISM'

The final node in the network of the event: in 1878 G.S. Hall, pupil and correspondent of William James, was in Leipzig. He knew many of Zöllner's circle, including Fechner, Wundt, and Weber, and was also acquainted with Helmholtz. In his 1912 book *The Founders of Modern Psychology*, Hall recalled the scandal surrounding Zöllner's experiments, and gave his own opinion, with the benefit of some considerable hindsight, that:

> We cannot however entirely acquit Professor Helmholtz from the charge of making an indiscreet appeal to the scientific imagination. His vivid characterization of intelligent beings of two dimensions [...] tickles the popular consciousness to which it is expressly addressed by the cunning of its pseudo-conceivability to fancy tridimensional space as the middle term of a syllogism.[91]

This criticism of Helmholtz's popularization of non-Euclidean and higher spatial ideas echoed the criticisms levelled at Helmholtz by both Dühring and, indeed, Zöllner himself. What was unique in the period was a critique of Helmholtz's rhetoric that singled out his 'syllogism'. Hall identified Helmholtz's use of analogy as 'indiscreet', an 'appeal to the scientific imagination' and as 'pseudo-conceivab[le]'. It may be useful at this stage to develop Chapter 1's groundwork on analogy. We might note that the model Helmholtz had constructed, with language, did not match up to reality and that, as Hall pointed out, it addressed itself specifically to the imagination.

The knot in the Zöllner experiment shows us that our thought is prone towards analogy, demonstrates that models work by giving our thought of the abstract, or the unseen, structure. It mediates not just between space, or the ether, and matter, but between thought and matter. The vortex atom is, after all, an idea; Klein's closed-space curves are mathematical entities; Tait's knots only ever drawings. In moving them onto string, Zöllner made material models of the knots. Had he thought them only models, he would have remained on safe ground: the location of the fourth dimension in relation to the prior three is certainly twisting and knot-like. Where is the fourth dimension? It is through, around, beyond, and beside, all at the same time: we have not the direction to describe it. We might well picture this relationship as a vortex and note that computer animations of rotating projections of tesseracts have a dizzying sense of continually falling in on themselves.

---

[91] G.S. Hall, *The Founders of Modern Psychology* (New York and London: D. Appleton and Company, 1924), p. 266.

When August Kekulé recounted the way the vision of another closed-space curve had come to him in his half-sleep in Ghent, the idea of the model was very much to the fore. At the Benzolfest of 1890, a celebration of the twenty-fifth anniversary of Kekulé's discovery of the ring structure of benzene, Kekulé famously described how the structure had come to him in a dream:

> My mind's eye, sharpened by repeated visions of a similar kind, now distinguished larger forms [*Gebilde*] in a variety of combinations. Long lines [*Reihen*], often fitted together more densely [*dichter zusammengefügt*]; everything in motion, twisting and turning like snakes. But look, what was that? One of the snakes had seized its own tail, and the figure whirled mockingly before my eyes. I awoke as by a stroke of lightning, and this time, too, I spent the rest of the night working out the consequences of the hypothesis.[92]

The truth of Kekulé's ouroboric narrative has been debated but the fact that he contributed this account to the 1890 publication of the papers given at the Benzolfest cannot be disputed. Whether or not he was thinking of such strikingly vortical torsion in 1865, he surely was in 1890. For Alan Rocke what is more interesting than the truth of Kekulé's dream account is the model itself. Convenor of the Benzolfest and former student of Kekulé, Adolf Baeyer, presented to the audience a rhetorical question: 'Is Kekulé's benzene theory a true depiction of the molecule, or is it simply a heuristically useful fiction?' Baeyer noted Heinrich Hertz's famous remark that Maxwell's equations for the electromagnetic field seemed to live independently of their creator. 'To be sure, Baeyer said, what we are talking about are pictures or representations (*Bilder*), which must never be confused with real things themselves, but Hertz's comment applies whenever our theoretical pictures approach the unseen reality.'[93] Another former student of Kekulé, Hermann Wichelhaus, reiterated the same idea: '"What the bodily eye of man has never seen and never will see," Wichelhaus intoned, "has appeared to your searching mind as a picture whose features speak to us as things that are real and alive."'[94]

The scientific model—the 'heuristically useful fiction'—is alive with ambiguity. It mediates for us the abstracted, performing translations between mind and matter, fleshing out the unseen but theorized, but it can be symbolically profligate, urging the very audience it is constructed to enlighten towards greater entanglements in its contusions.

Over the past decade, led by research into artificial intelligence, cognitive science has explored the ways in which scientific thinking uses model-based reasoning to generate creative change in theories and concepts. Dedre Gentner argues that 'analogical reasoning can lead to change of knowledge—not only to enrichment of existing representations but also to true conceptual change'.[95] Gentner describes the conditions required for successful analogy as a threefold alignment of

---

[92] Alan J. Rocke, *Image and Reality* (Chicago: University of Chicago Press, 2010), p. 194.

[93] Rocke, *Image and Reality*, p. 296.   [94] Rocke, *Image and Reality*, p. 297.

[95] Dedre Gentner, 'Analogy in Scientific Discovery: The Case of Johannes Kepler', in *Model-based Reasoning: Science, Technology, Values*, ed. Lorenzo Magnani and Nancy J. Nersessian (New York: Kluwer Academic/Plenum, 2002), pp. 21–40 (p. 21).

relational structure, such that the base domain, or the original concept to be described, and target domain, or the structure at which the reader is to arrive, observe 'parallel connectivity and one-to-one correspondence'; that they have common relations, rather than common objects; and that they 'tend to match connected systems of relations'. This cognitive account builds productively on Aristotle's germinal analyses. At least six different ways in which knowledge can be changed by analogy are noted:

> highlighting and schema abstraction [...] projection of candidate inferences from one domain to another [...] noticing alignable differences [...] re-representation—altering one or both representations so as to improve the match [...] incremental analogizing: extending the mapping by returning to the base domain for more material to add to the analogy [...] re-structuring—altering the domain structure of one domain in terms of the other.

What emerges is a picture of analogical reasoning as fuzzy, ambivalent, and catalytic. Indeed, we might note that in the Zöllner event we witness an inferential catalysis prompted by a peripheral element of the original analogy and the construction of a deviant concept. Specifically, Zöllner was drawn to the intelligence of the proposed higher- or lower-dimensional creatures, and inferred connections with the supernaturalist hypothesis of unseen 'intelligence' behind mediumistic phenomena. A series of further connections between aspects of higher space theorized in geometry—amphicheirality and the impossibility of knottedness—prompted him to fail to observe a translation between a model—geometry—and the domain it modelled—space. In short, the Zöllner event exemplifies a Latourian translation that the scientist fails to recognize, a switch between word and world that was repeatedly ignored in the case of higher space.

3

A Square

Flatland, Play, and Tradition

Flatland, purportedly written by one A Square, was published by J.R. Seeley and Co. at the end of October 1884.[1] The inside title plate bore an illustration by the author, a map of kinds, depicting a nebulous mass of clouds, or fog, and lines of text curtailed as they scrolled behind and through these obscuring fronts: 'Ten Dim'; 'Five Dimen'; 'Eight D' (see Figure 1). Beneath the mass were a point, a line, a square, and a projection drawing of a cube, depicting Pointland, Lineland, Flatland, and Spaceland. The subtitle, 'A Romance of Many Dimensions', ran between these lower-dimensional illustrations. At the top of the page ran a line from *Hamlet*, Act I Scene v, uttered by Horatio on witnessing Hamlet's vision of the ghost of his slain father: 'Oh day and night, but this is wondrous strange.'

A smaller illustration on the inside back page repeated the Shakespearean theme and clarified the nebulous illustration (see Figure 2). Lines from Prospero's speech in *The Tempest*, Act IV Scene i, ran through the clouds: 'the baseless fabric of vision'; 'melted into air, into thin air'; 'such stuff as dreams [are] made on'. *Flatland* announced itself as strange, a vision, a dream, and a romance. What played out within its pages was certainly 'such stuff as dreams are made on': a story, part satire, part parable, in which abstract geometric figures were the characters, and whose setting was a world as flat as the page.

The text was divided into two parts—'This World' and 'Other Worlds'—and illustrations were dotted throughout, hand-drawn by the author. 'This World' recounted the history, geography, architecture, biology, social structure, and belief system of *Flatland*. It described a rigid gender and social hierarchy, with circles (priests) occupying the most powerful position in society, and lines (females) the least. Between these poles were polygonal geometric figures whose standing decreased with the number of their sides and their 'regularity'.

Flatland was an immediate success, its first edition selling out within the month. A second was printed in December 1884, containing a new foreword by A Square in which he answered some of his critics' objections. By this time already a handful of notices were giving clues to the identity of the likely author hiding behind the pseudonym A Square: Edwin Abbott Abbott, the headmaster of City of

---

[1] Edwin A. Abbott, *Flatland: A Romance of Many Dimensions,* 2nd edn (London: Seeley and Co., 1884; repr. New York: Dover Publications, 1992). All further references to this edition are given in the body of the text using the abbreviation *F*.

Figure 1 Edwin A. Abbott, *Flatland* (1884), inside front

London School, a noted theological writer and influential, progressive educationalist.[2] In the brief period between these two editions, meanwhile, the publisher Swan Sonnenschein printed a pamphlet containing the first of a planned series of *Scientific Romances* by Charles Howard Hinton, science master at Uppingham College.

---

[2] 'That curious little book Flatland, which we noticed last week is said to be the production of the head master of a well-known school.' Anon., 'Literary Gossip', *Athenaeum*, 2978 (1884), 660.

Prometheus up in Spaceland was bound for bringing down fire for mortals, but I—poor Flatland Prometheus—lie here in prison for bringing down nothing to my countrymen. Yet I exist in the hope that these memoirs, in some manner, I know not how, may find their way to the minds of humanity in Some Dimension, and may stir up a race of rebels who shall refuse to be confined to limited Dimensionality.

That is the hope of my brighter moments. Alas, it is not always so. Heavily weighs on me at times the burdensome reflection that I cannot honestly say I am confident as to the exact shape of the once-seen, oft-regretted Cube ; and in my nightly visions the mysterious precept, " Upward, not Northward," haunts me like a soul-devouring Sphinx. It is part of the martyrdom which I endure for the cause of the Truth that there are seasons of mental weakness, when Cubes and Spheres flit away into the background of scarce-possible existences ; when the Land of Three Dimensions seems almost as visionary as the Land of One or None ; nay, when even this hard wall that bars me from my freedom, these very tablets on which I am writing, and all the substantial realities of Flatland itself, appear no better than the offspring of a diseased imagination, or the baseless fabric of a dream.

LONDON : R. CLAY, SONS, AND TAYLOR, PRINTERS.

Figure 2 Edwin A. Abbott, *Flatland* (1884), inside back

Flatland's subtitle, including the term 'romance', shared by both Abbott's book and Hinton's early collection of short stories and essays for Swan Sonnenschein, is of particular interest. While a far from unusual description for narrative fiction, the romance was very current in 1884 and I will examine the text's moment in some detail. I argue that the text was a specific and considered intervention into contemporary literary debates. This argument rests upon a development of the text's modal status and its relationship to analogy.

Critical responses to the text have been highly ambivalent. I have summarized these in sketch: a first wave of reviews on publication; mid-century reassessment, followed by incorporation into the early scholarly canons of SF; and, most recently, approaches from nineteenth-century studies that have aimed to locate the text in Victorian culture. I sketch these in part to stress the mobility of the text over time in terms of its reception, a mobility that evidences the multivalence I read as central to *Flatland*'s epistemology, but also in order that the most useful insights of each critical moment might be maintained.

Flatland has been well served by nineteenth-century studies and two recent and very detailed annotated editions. This excellent work provides solid foundation upon which to build. I extend genealogies of the text that have been established by the annotators Iain Stewart and Thomas Banchoff and supplemented by K.G. Valente's rediscovery in *The City of London School Magazine* of 'A New Philosophy'.[3] I am keen to demonstrate a longer heritage for geometrical satire than tends to be stated in order to look more closely at what *Flatland* maintains from its ancestors and how it moves its specific satirical tradition forward. To the same end, I want to expand what I think is the most important discovery of the nineteenth-century studies by Rosemary Jann and Jonathan Smith—*Flatland*'s concern with reasoning by analogy—and bring this to bear on the text's self-awareness.

Readings of *Flatland*'s position vis-à-vis analogy have tended to be polarized; the text is either for or against model-based reasoning. I want to suggest a nuanced response that differentiates between uses of analogy: that *Flatland* is aware of the strengths and weaknesses of reasoning by analogy but, more significantly in terms of higher-dimensional thought, that it is interested in our tendency to use models in our language and thought.

I argue that we need to expand these arguments about analogy to consider the position of fiction, the form of analogy that might be a more or less mimetic narrative art. *Flatland* is interested in analogy, in model-based thinking, and it is interested in its own status as a kind of model. The text both uses analogy and comments on it. As a satire it is a kind of analogy itself: it models one society through another; it projects. I want to think about what kind of model *Flatland* is, how its form reflects and reproduces its content. To enable this consideration I want to expand the picture of the text's moment: both its position in a lineage of geometric satire and its responsiveness to contemporaneous debates about fiction. Andrea Henderson argues that 'the features that make *Flatland* hard to assimilate to the canonical novel tradition, its flouting of the protocols of realist characterization and plotting, are the same features that make it uniquely useful to the critic seeking to understand late-nineteenth-century theories of representation'.[4] Following Henderson, I want to examine the unassimilability of the text.

---

[3] Edwin A. Abbott, *The Annotated Flatland: A Romance of Many Dimensions*, ed. Ian Stewart (Reading, MA and Oxford: Perseus, 2002); *Flatland: An Edition with Notes and Commentary*, ed. William F. Lindgren and Thomas F. Banchoff (Cambridge: Cambridge University Press; Washington, DC: Mathematical Association of America, 2010); K.G. Valente, 'Transgression and Transcendence: *Flatland* as a Response to "A New Philosophy"', *Nineteenth Century Contexts*, 26 (2004), 61–77.

[4] Andrea Henderson, 'Math for Math's Sake': Non-Euclidean Geometry, Aestheticism, and "Flatland"', *PMLA*, 124 (2009), 457.

Studies that have focused on Abbott's quotidian writing as a theologian have tended to underemphasize *Flatland*'s formal playfulness. This is entirely understandable. This playfulness is continuous with its modal status as a satire but entirely discontinuous with everything else Abbott wrote. Its playfulness is also of relevance to its engagement with analogy; it approaches the idea of satirical modelling with a ludic determination. My interest has been drawn particularly to the second half of the text, 'Other Worlds', not only because it is the part of the text concerned with higher- and lower-dimensioned spaces, but also because it is less obviously traditionally satirical than the first half, 'This World'. 'Other Worlds' does not work as a refracted representation of reality but as a representation of the refraction of reality and I argue that its exceeding of traditional satirical modes is continuous with its subject matter.

FLATLAND CRITICISM

In November 1884, *The Literary World* published a detailed and engaged review of *Flatland*, quoting large sections of the text and predicting an elite readership, 'the favoured few'. It warned that the 'subject is too abstruse [...] to appeal to the multitude', and noted that 'interpretation [is] not always easy', also criticizing 'a certain likeness in it here and there to the precise and formal lessons demonstrated on a blackboard to a class of schoolboys'. The reviewer was confident of the breadth of *Flatland*'s interests:

> The writer has aimed his shafts sometimes at the world of society, sometimes at that of politics, sometimes at that of religion. His references are now plain and palpable, now recondite and obscure. But about the broad drift of his parable, there can be no mistake whatever. His allegory is in the chiefest of its aspects a magnificent protest against self-sufficiency and dogmatism; against cherishing the idea that we have, in reference to any matter of experience whatever, seen the end of all perfection; against all narrowness, bigotry, and intolerance in any region of supposed knowledge, whether that of scientific self-assurance on the one hand, or religious fanaticism on the other.[5]

The Athenaeum was, by comparison, altogether less certain, doubting even that there were satiric elements to the text:

> At first it read as if it were intended to teach young people the elementary principles of geometry. Next it seemed to have been written in support of the more transcendental branches of the same science. Lastly we fancied we could see indications that it was meant to enforce spiritualistic doctrines, with perhaps an admixture of covert satire on various social and political theories.[6]

The Athenaeum continued with a legitimate mathematical gripe: 'Of course, if our friend the Square and his polygonal relations could see each other edgewise, they

---

[5] Anon., 'Flatland', *Literary World*, 15 (1884), 389–90; reproduced on Flatweb [http://library.brown.edu/cds/flatweb/1884litworld.html] [17 January 2013] (para. 1).

[6] Anon., 'Flatland', *Athenaeum*, 2977 (15 November 1884), 622.

must have had some thickness, and need not, therefore, have been so distressed at the doctrine of a third dimension.'

The reviewer for *The Spectator*, meanwhile, provided an interesting insight into the status of higher spatial theories at the time of *Flatland*'s publication: 'These comfortable doctrines have long been a speculation and a pious opinion of the few, and a stumbling block and foolishness to the uninitiated.' The review was positive about the material appearance of the text—'It is very pleasantly got-up in paper, print, and cover'—and anticipated a mixed reception offering a different type of text to different readers: 'Much of it will also be read with amusement, as satire, by those who do not appreciate its scientific bearing, or as pure nonsense by those who are not searching for satire.'[7]

It seems that reviewers were non-committal about the book, uncertain about its audience or what it would mean to different readers. A large audience was certainly not anticipated and this was echoed in American reviews when the US edition was published in February 1885. The *New York Times* described it as 'a very puzzling book and a very distressing one, and to be enjoyed by about six, or at the outside seven, persons in the whole of the United States and Canada'.[8] Specialist literary journals were less sceptical, the *Literary News* declaring it 'an effective satire on social differences, and on the assumption of absolute knowledge', and the *Literary World* proclaiming a 'brilliant *jeu d'esprit*'.[9]

This situation persists in scholarship on *Flatland* to this day, although for different reasons. Writing in 1996, Jonathan Smith summarized:

> When not treated as a joke, *Flatland* has tended to be approached in ways that divorce it from its cultural position in the debate over non-Euclideanism and its implications. Historically, literary critics have treated it as an early example of science fiction and fantasy, while scientists and mathematicians have used it as a clever way to introduce their students to concepts of dimensionality and non-Euclideanism. It has only been recently that the book has been brought back to the center of the study of Victorian culture, and it will be to further that movement that I approach the novel here.[10]

Smith's work beds *Flatland* into non-Euclidean contexts thoroughly, contexts that have already been discussed in relation to higher space in Chapter 1 of this book. His summary is accurate but would perhaps benefit from some expansion. Despite the fact that it remained in print throughout the second half of the twentieth century, writing on *Flatland* in the middle of the century is scarce. Banesh Hoffmann's 1952 introduction to the Dover edition, which has been reproduced several times since, may be the only critical account of the text for a generation. In laudatory mode, Hoffmann situated the text in scientific history, as pre-Einsteinian, a reading both accurate and reflective of the post-war popular scientific orthodoxy.[11]

---

[7] Anon., 'Flatland: A Romance of Many Dimensions', *Spectator*, 2944 (1884), 1583–4.

[8] Anon., 'Flatland', *New York Times*, 23 February 1885, 3.

[9] Anon., 'Humor and Satire', *Literary News*, 6 (1885), 85; Anon., 'Flatland', *Literary World*, 16 (1885), 93.

[10] Jonathan Smith, *Fact and Feeling: Baconian Science and the Nineteenth-century Literary Imagination* (Madison: University of Wisconsin Press, 1994), p. 191.

[11] Banesh Hoffmann, 'Introduction', in *Flatland* (New York: Dover Publications, 1952), pp. iii–iv.

Interest began to increase in the 1980s as the centenary of publication approached, led by SF writers, publishers, and the critics who had begun to found a scholarly response to the genre.[12] In 1983 Darko Suvin published a scholarly genre prehistory of science fiction. The progressive strain in *Flatland* was central to Suvin's elevation of the text to lofty status within his pantheon of Victorian Science Fiction. For Suvin, *Flatland* is categorized as a 'sophisticated alternative history' and he also observes: 'Cleverly adapting Carroll's and Verne's strategy of subsuming but transcending the juvenile reader, Abbott's is in truth "A Romance of Many Dimensions"; in its thoroughgoing democratism, it is addressed to the best minds in the new reading public, issuing from the newly introduced obligatory primary schooling.'[13] Suvin's argument that *Flatland* works for both younger and older readers has often been lost by Victorian studies that have worked closely with the full extent of Abbott's literary output, an output otherwise dedicated in its entirety towards adult readers and teachers. Even if we are to disagree with other aspects of Suvin's approach, this valuable point is worth maintaining. It chimes with reviewers of the period who noted Abbott's 'headmasterly' tones, and frequently compared *Flatland* to the work of the two other authors Suvin names, Carroll and Verne. Suvin's observation is surely correct, and criticism since has tended to sideline this younger readership.

Such criticism has been more concerned, like Smith, with contextualizing *Flatland* in the broader cultural field of the late nineteenth century. Most influentially, Rosemary Jann has argued that 'as part of Abbott's wider commentary on the role of imagination in cognition, *Flatland* alludes to contemporary debate over the role of hypothesis in scientific discovery and the relationship between material proof and religious faith'.[14] For Jann, *Flatland* is a paean to 'the progressive force of the imagination', and negotiates a middle way through inductive science, responding to debates over the unseen in the natural world, and dogmatic faith, allowing for a less absolute faith in the literal truth of the written scriptures.[15]

The trajectory of current criticism on *Flatland*, then, does indeed follow Smith's aim to bring the text back to the 'center of the study of Victorian culture'. While this shift in emphasis has produced some inspiring work and has, perhaps, rescued *Flatland* from ghettoization as an SF precursor text, it would benefit from a more direct engagement with the ghetto. SF criticism has recognized to a greater extent the juvenile reader and the text's sense of fun.

It is important, particularly when attempting to recreate the 'cultural position' of the text, to hold in mind the description of the book as a 'jeu d'esprit', repeated by Abbott's pupil William Garnett in the preface to the third edition. There are certainly consistencies with Abbott's theological writings, as one would expect, but it is unhelpful to narrow the matrix of concerns informing the text. I want to

---

[12] See Walter J. Kaplan and Alison Chaiken, 'Flatland Fans', *Science News*, 126 (1984), 355 for interesting centenary responses from an optometrist and a graduate researcher of 2D physics.

[13] Darko Suvin, *Victorian Science Fiction in the UK: The Discourses of Knowledge and Power* (Boston: G.K. Hall & Co., 1983), p. 373.

[14] Rosemary Jann, 'Abbott's *Flatland*: Scientific Imagination and "Natural Christianity"', *Victorian Studies*, 28 (1985), 473–90 (p. 473).

[15] Jann, 'Abbott's *Flatland*', 486.

restress *Flatland*'s sense of fun, to 'treat it is a joke', and in so doing to highlight areas of confusion encountered when reading the text. Significant among these is the text's modal status. For the *Literary World*'s reviewer it was an allegory and a parable while the critic at *The Athenaeum* detected 'an admixture of covert satire'. It is today almost unanimously considered, as the jacket to the Oxford's World Classics edition describes it, 'part geometry lesson, part social satire'. Satire is a notoriously difficult idea to pin down and *Flatland* is a complex example of the mode. Satirical texts can be highly normative, assuming that their readers share their world view, but so too can they be disorientating, particularly, perhaps, when their allegory is either very closely matched to the world of the reader or radically estranged. It is not simply that there are some readers who don't get the joke, though there often are: it is more accurate to claim that different readers get different jokes. Before considering this modality in greater detail it would be useful to attend to *Flatland*'s literary forebears.

FLATLAND PRECURSORS

Lindgren and Banchoff include in the introduction to their annotated edition of the text a section on precursors to *Flatland* that names Hermann von Helmholtz's essays in *Nature* and *The Academy*, Charles Howard Hinton's essays on the fourth dimension, Gustav Theodor Fechner's collection of satirical essays written as Dr Mises and published as *Kleine Schriften* in Germany in 1875, Lewis Carroll's 'Dynamics of a Parti-cle', and the anonymous essay 'A New Philosophy', published in the edition of the *City of London School Magazine* for October 1877. Helmholtz's essays have already been discussed in Chapters 1 and 2 here; Hinton's work will be discussed in Chapter 4.

J.N.P. Land's article in *Mind*, also discussed in Chapter 1, had sketched Dr Mises's story 'Space has Four Dimensions', an early fictional versioning of the dimensional analogy. Lindgren and Banchoff draw attention to two other stories from the same Mises collection: 'The Shadow is Alive', and 'Why Should the Sausage be Sliced Slantways?'. I want to add to this survey a further story in the same collection: 'The Comparative Anatomy of the Angels', a favoured source for Alfred Jarry. 'The Comparative Anatomy' is of particular interest because of its rich array of familiar sources and its clear influence on *Flatland*. It fuses a sly version of the Platonism of the *Timaeus* with evolutionary biology and Fechner's own psycho-physics. Mises writes:

> In recent years, certain people have achieved high credit for their industrious collection and dissemination of knowledge about human anatomy based on the comparative anatomy of the lower creatures. Apparently, it has occurred to no one yet to direct attention with the same purpose to the anatomy of higher creatures, though equally astounding progress may be anticipated using this approach.[16]

---

[16] Gustav Theodor Fechner, 'The Comparative Anatomy of Angels', trans. Hildegard Corbet and Marilyn E. Marshall, *Journal of the History of the Behavioral Sciences*, 5 (1869), 135–51 (p. 135).

Using analogy to suggest projection up the scale of comparative anatomy in the same way that science has projected down, Mises takes his readers to a perspective very similar to that from which A Square will find himself viewing *Flatland*: 'Now we find ourselves standing high above the earth, viewing it and the other planets objectively, and comparing their creatures.' The piece then argues that to perfect the human form it is necessary to pare away 'all the uneven parts and asymmetrical outcroppings of the human'.[17] What remains is a sphere.

A sphere is presented as the ultimate form, angelic, a form to which mankind will eventually evolve. In its giddying ascent the sphere is made analogous to the eye, and then the planets. Mises's spheres, like the geometric figures of *Flatland*, can communicate through chromatic alteration: 'normally angels are transparent, but they can assume different colours if they wish. Whatever one angel wishes to tell another, he paints on his surface; the recipient sees the picture and knows the soul of the communicating angel.'[18] Mises's collection would offer much to inform *Flatland*—not least its playfully erudite humour—but this perfect sphere, derived, as Mises notes, from Xenophanes, seems to hop out of one text and into the other.[19]

The same authors offer a translation of 'Why Should the Sausage be Sliced Slantwise?' that relates a conversation between Mises (Fechner) and August Möbius in which they discuss the inhabitants of a one-dimensional world and address the problem of how these Linelanders would pass one another. In the first model they discuss, they imagine the line as an ellipse 'with divine monads at its foci'. The inhabitants would reverse their direction and meet halfway on the other side.

> In the second model, not subject to this limitation, one had to imagine the people as linear waves. As is well known, waves can pass through one another without interference, and since our thoughts are already attached to aether waves in the brain, in this setting one being could exchange places with another in reality simultaneously with thinking of doing so.[20]

The interest of this section lies in materialization of thought in ether waves, an idea that would be rehearsed in a less ambivalent mode by Hinton, and the topological influence of Möbius. We are in a world of switches and inversions that will be replicated in *Flatland*.

In the English language we might note earlier forebears that personify geometric figures for satirical effect. 'The Loves of the Triangles', a parody of Erasmus Darwin's 'The Loves of the Plants', was published in three parts in *The Anti-Jacobin* in 1798. *The Anti-Jacobin*, founded by George Canning, the Foreign Secretary of the time, satirized and parodied a number of influential poets and public sympathizers with the French Revolution. 'The Loves of the Triangles' was reportedly sent to the paper by Mr Higgins of St Mary Axe, a caricature of William Godwin. The form

---

[17] Fechner, 'The Comparative Anatomy', p. 136.
[18] Fechner, 'The Comparative Anatomy', p. 143.
[19] 'The substance of God is spherical, in no way resembling man.' Diogenes Laertius, 'Xenophanes', in *The Lives of Eminent Philosophers*, trans. R.D. Hicks, 2 vols (London: William Heinemann, 1950), II, pp. 425–9 (p. 427).
[20] Abbott, *Flatland: An Edition with Notes and Commentary*, p. 121.

of the didactic poem and Darwin's popular productions was continually referenced and Euclidean expansion of geometric dimensions described:

> But chief, thou Nurse of the Didactic Muse,
> Divine NONSENSIA all thy sense infuse;
> The charms of secants and of tangents tell,
> How Loves and Graces in an Angle dwell;
> How slow progressive points protract the line,
> As pendant spiders spin the filmy twine;
> How lengthen'd lines impetuous sweeping round,
> Spread the wide plane
> How Planes, their substance with their motion grown,
> Form the huge Cube, the Cylinder, the Cone.[21]

The poetry of *The Anti-Jacobin* contributed so much to its popularity that it was collected in an edition that was reprinted five times in eight years. The parody replaces botanic terms with geometric terms, reproducing Darwin's anthropomorphization in a new setting to render it absurd. What, after all, could be more ridiculous than the 'charms of secants and of tangents' to anyone who had sat through a lesson on Euclidean geometry?

Geometry provides a formal structure whose rigidity lends itself to parody and in such examples we see also a residual antagonism towards the books of Euclid, a site of communion between schoolboy and adult readers. Soon after its launch in 1841, *Punch* used geometry as the grounding for a political satire, focusing exclusively on Euclid:

BOOK I.—DEFINITIONS.   A point in politics is that which always has place (in view,) but no particular party. A line in politics is interest without principle. The extremities of a line are loaves and fishes. A right line is that which lies evenly between the Ministerial and Opposition benches. A superficies is that which professes to have principle, but has no consistency. The extremities of a superficies are expediencies. A plain superficies is that of which two opposite speeches being taken, the line between them evidently lies wholly in the direction of Downing-street.[22]

In this piece the strictures of Euclid are a framework into which political terms are transposed. The transfer of concepts from one domain to another is playful. The domain of geometry provides no specific traction with the political concepts themselves but provides a fresh setting, structure, and vocabulary through which to highlight absurdities or make puns. 'The Political Euclid''s witty juxtaposition also reveals an overlap between *Punch*'s educated readership and *Flatland*'s.

While a number of critics responding to *Flatland* in 1884 compared it to Lewis Carroll's *Alice* books, an altogether different text by the same author, writing

---

[21] G. Canning, John Hookham Frere, and G. Ellis, 'The Loves of the Triangles', in *Poetry of The Anti-Jacobin,* ed. Charles Edmonds, 3rd edn (London: Sampson Low, Marston, Searle & Rivington, 1890), pp. 151–64 (pp. 153–4).
[22] Anon., 'The Political Euclid', *Punch*, 1 (1841), 149.

under his own name, has also been noted as a possible forebear. Dodgson's squib *The Dynamics of a Part-icle*, first published in 1865 when the author printed it at his own cost and distributed it among his colleagues at Oxford, had been reprinted as one of a series of pamphlets entitled *Notes by an Oxford Chiel* in 1874. Its introduction reads like a lost abstract for *Flatland*:

> It was a lovely Autumn evening, and the glorious effects of chromatic aberration were beginning to show themselves in the atmosphere as the earth revolved away from the great western luminary, when two lines might have been observed wending their weary way across a plane superficies. The elder of the two had by long practice acquired the art, so painful to young and impulsive loci, of lying evenly between his extreme points; but the younger, in her girlish impetuosity, was ever longing to diverge and become an hyperbola or some such romantic and boundless curve [...] We have commenced with the above quotation as a striking illustration of the advantage of introducing the human element into the hitherto barren region of Mathematics. Who shall say what germs of romance, hitherto unobserved, may not underlie the subject?[23]

Like 'The Loves of the Triangles', *The Dynamics* anthropomorphizes geometric figures using the geometric as a domain into which to perform satiric transposition.

Another possible, though by no means certain, forebear for *Flatland* foregrounds the idea of character. A Square is not the first fictional A Square. Rev. G.G. Oliver published *The Revelations of A Square* in 1855, a fictionalized history of Freemasonry narrated by a talking Masonic set square, a square that is therefore a triangle. The set square's narrative history of Freemasonry is nested within the account of its owner:

> A dull and dreamy sensation came over me and I saw, or fancied I saw, the Square, which had just been reposing before me, raise itself up, with great solemnity, on the exterior points of its two limbs, which seemed to assume the form of legs. Body it had none, but the heart which was delineated at the angle, put forth two eyes, a snub nose, and a mouth—a sort of amplification of the letter J.[24]

Whether or not we read in *Flatland* a satirical glance at Freemasonry, described by Oliver in the introduction to *The Revelations* as 'like all sciences [...] a system of progression', we might note parallels to *Flatland*'s own parody of a 'system of progression', its exclusion of women, its rigid hierarchies, all of which could be read as satires of late nineteenth-century Freemasonry.[25]

K.G. Valente has recently identified a text even closer to home for Abbott that works with higher geometry specifically—the essay 'A New Philosophy' that had appeared in the *City of London School Magazine* in November 1877. It proceeds by seemingly impeccable reasoning to argue for a philosophy of belief 'based on a branch of science which is in its essence unchangeable': mathematics.[26] Engaging with contemporary discussions of non-Euclidean geometry—'it has been said by

---

[23] Charles Lutwidge Dodgson, *The Dynamics of a Parti-cle* (Oxford: James Parker and Co., 1874), pp. v–vi.

[24] Rev. George Oliver, *The Revelations of A Square* (London: Richard Spencer, 1855), p. 2.

[25] Oliver, *The Revelations*, p. v.

[26] Anon., 'A New Philosophy', *City of London School Magazine*, 1 (1877), 277–81 (p. 277).

an eminent writer that it is possible that animals of two dimensions, or of four, may exist in the universe'—it reveals its schoolboyish humour slyly and gently at the beginning, through understatement and sarcasm, arguing, for instance, that algebraical geometry would not become 'harder' in four dimensions, because 'our mathematical capacity would increase proportionately', and referring to the dark patch on the ground beside a person on 'one of the few sunny days England can furnish' as 'popularly called his shadow'.[27]

From the idea that a space of n-dimensions is a shadow of a space of $n+1$ dimensions, the author extrapolates to imagine the logical conclusion of such a line of thought, 'the Function of Infinite Positive Dimensions or the Infinite Plus', and, likewise, its negative, 'the very darkness of darkness, the boundless negation, the Infinite Minus'.[28] This *reductio ad absurdum* of strictly analogical reasoning becomes the springboard for further parodic speculations, some of which are prescient in their similarity to later, entirely serious, uses of the idea of higher space. 'Surely the development of science will, by an analogical process enable us to lift ourselves into the world of four dimensions, and rejoice in the sense of new abilities, new strength, new enjoyments,' writes the author. The Infinite Plus is described as 'an idea imperfectly indicated in the Buddhist's Nirvana' and 'the seeming waste of evolution can be explained by the new and nobler form of the doctrine of transmigration'.[29] The author proposes determining 'the equation of man' and, in a passage in which the satirical mode of the piece most clearly breaks the surface, asserts that

> the saints of the new philosophy will commence with Pythagoras and Euclid and will include all eminent mathematicians. The New Pope will be the annually elected Senior Wrangler, assisted by a College of Wranglers—Cambridge will be the new Rome.[30]

Underwriting 'A New Philosophy' is a critique of analogical reasoning evident from the earliest passages, in which 'the defective state of our geometry' prevents the author from pursuing a demonstration, but allows him to conclude that 'it is at least probable' that his conclusions hold good for higher- or lower-dimensioned space.[31] This critique finds its summary in the final paragraph's parodic euphoria: 'Progress is the watchword of the age—progress the only possible true life of the human race.'[32] 'A New Philosophy''s satire is multiple: not only is it a broadside against the pitfalls of reasoning by analogy, implicit in the idea of the Infinite Plus is a critique of the theory of continuity as adapted by Peter Guthrie Tait and Balfour Stewart in *The Unseen Universe*.

The fourth edition of this popular exposition of a scientific accommodation of religion, published in April 1876, had addressed the idea of higher-dimensioned space with an account that bears close comparison to 'A New Philosophy'. The basis of the arguments of *The Unseen Universe* on the theory of continuity, for the authors the idea that 'the Great Whole is infinite in energy, and will last from

---

[27] Anon., 'A New Philosophy', p. 278.
[28] Anon., 'A New Philosophy', p. 279.
[29] Anon., 'A New Philosophy', p. 280.
[30] Anon., 'A New Philosophy', p. 281.
[31] Anon., 'A New Philosophy', p. 278.
[32] Anon., 'A New Philosophy', p. 281.

eternity to eternity', was extended to account for the higher-dimensioned space emerging in discussions of non-Euclidean geometry by Helmholtz and Clifford. The principle of continuity was founded on the idea of the uniformity of nature and argued against supernatural interventions in the natural order of cause and effect. It was supported by the conservation of energy, and supported Darwinian evolution. Tait and Stewart adapted this principle to argue for continuity between a visible and an invisible, allowing divine intervention to be accounted for by energy transfer between the two. In a later addition to their original account, incorporating modish geometrical speculations, higher-dimensioned space extended the invisible, ad infinitum:

> Just as points are the terminations of lines, lines the boundaries of surfaces, and sur-faces the boundaries of portions of space of three dimensions:—so we may suppose our (essentially three-dimensional) matter to be the mere skin or boundary of an Unseen whose matter has four dimensions. And, just as there is a peculiar molecular difference between the surface-film and the rest of a mass of liquid—wherever such a surface film exists, even in the smallest air-bubble—so the matter of our present uni-verse may be regarded as produced by mere rents or cracks in that of the Unseen. But this may itself consist of four-dimension boundaries of the five-dimensional matter of a higher Unseen, and so on [...] reflection leads us to the ultimate conception of an infinite series of Universes, each depending on another, and possessing of course among them an infinite store of energy.[33]

The Infinite Plus seems to parody this idea of an abyssal series of universes, present-ing the discovery with a giddy sense of revelation, echoed in both style and content by A Square's insistence upon the 'Argument from Analogy of Figures':

> And once there, shall we stay our upward course? In that blessed region of Four Dimensions, shall we linger on the threshold of the Fifth, and not enter therein? Ah, no! Let us rather resolve that our ambition shall soar with our corporal ascent. Then, yielding to our intellectual onset, the gates of the Sixth Dimension shall fly open; after that a Seventh, and then an Eighth.   (*F*, 74)

Flatland's heritage indicates some of its themes. The contemporary physics that informs 'A New Philosophy' is closest to home.[34] Lindgren and Banchoff make the essential observation that *Flatland* is a society modelled on that of ancient Greece. We read late Victorian society refracted through this lens, bringing into focus both similarities between nineteenth-century imperial Britain and ancient Greece, and, indeed, the hypocrisy of British thinkers' tendency to compare Britain with classical civilization:

> He heightened his satirical commentary by making prominent in this imaginary civ-ilization some of the very aspects of classical Greece that its Victorian apologists had

---

[33] P.G. Tait and Balfour Stewart, *The Unseen Universe or, Physical Speculations on a Future State*, 4th edn (London: Macmillan and Co., 1876), p. 220.

[34] William Garnett, who was headboy at the City of London School in Abbott's first few years as headmaster, studied at Cambridge and went on to become James Clerk Maxwell's assistant. Garnett was one of Abbott's obituarists.

rationalized away—for example, slavery, a rigid class system, misogyny, and ancient forms of social Darwinism.[35]

The range of classical reference in *Flatland* is rich. We hardly need restate identification with Euclid and I will explore Platonism in Chapter 5 but we see also, through Mises, Neoplatonism's favoured Pythagoreanism. A Square refers on several occasions to his being 'initiated into mysteries'. Lindgren and Banchoff make clear the resonances here with the mystery cults of ancient Greece that practised initiation.[36] A Square's induction into Spaceland replicates an induction into Pythagoreanism.

FLATLAND AS SATIRE

Darko Suvin's account of *Flatland* stressed that 'satire—however vehement—will also be playful—i.e., more concerned with the multiplicity than with the focusing of its references (only the masters, such as Swift or Wells at his best, manage to fuse both)'.[37] For Suvin, *Flatland* was such a 'masterpiece':

> The consistent and radical novum of such a *Flatland* is inseparable from its witty parable: the book is not an allegory about England but its (and therefore not only its) hidden truth, arrived at by an interaction of science, political philosophy, and satire.[38]

Suvin was following Mikhail Bakhtin in this identification of satire as multiple and ambivalent and these descriptions seem particularly appropriate for *Flatland*, a text that fuses 'many dimensions' of thought and perspective in an extremely playful fashion.[39]

Perhaps rather than describing *Flatland* as ambivalent we might borrow a term from mathematics and term it 'non-orientable'. A two-dimensional object moving along a non-orientable surface in a three-dimensional space, such as a Möbius strip, will be transformed into its mirror image by the end of its journey. One can also view orientability as a choice between handedness at every stage of the journey: if this handedness remains consistent throughout the journey along the surface, the surface is orientable. If left-handedness becomes right-handedness, or vice versa, it is non-orientable.

Why might this be a better term for *Flatland*? Not only because the text is a surface, and about a surface, but because, as we have read above, it is a structurally complex satire. In its first section, 'This World', the reader encounters a version of

---

[35] Abbott, *Flatland: An Edition with Notes and Commentary*, pp. 2–3.

[36] Abbott, *Flatland: An Edition with Notes and Commentary*, pp. 2, 4.

[37] Darko Suvin, 'Victorian Science Fiction, 1871–85: The Rise of the Alternative History Sub-Genre', *Science Fiction Studies*, 10 (1983), 148–69 (p. 150).

[38] Suvin, 'Victorian Science Fiction', 163.

[39] See Mikhail Bakhtin, *Rabelais and his World*, trans. Helene Iswolsky (Bloomington: Indiana University Press, 1984), pp. 11–12. It is perhaps worth diluting Suvin's assessment of the radical nature of *Flatland*'s 'novum' in light of the precursors assembled above, which fell outside the scope of Suvin's methodology, limited by a generic categorization of SF.

life in late nineteenth-century England refracted through classical Greece and transposed into a geometric plane. This shifting of real-world character and event analogues into a freshly modelled domain conforms to what the reader might traditionally expect of satire, and operates in the tradition of 'The Lives of the Triangles' or 'The Political Euclid'. We recognize this satire in the terms of Samuel Johnson who described a satirical poem as 'a poem in which wickedness or folly is censured'.[40] We can identify specific targets—Frances Galton and his development of eugenics, 'the science of improving the stock', for example—and the text assumes that we agree that the behaviour of these targets should be censured. This type of satire is highly normative, and as Joe Brooker has noted:

> Normative satire requires that laws of right conduct be understood, not merely by the lone satirist, but by the work's audience. It implies consensus around shared values, and implicit agreement that transgression of those values should be pointed out and punished at the level of representation [...] The satirist seems to be on the side of change, of progress, or at least of correction.[41]

Critical assessments of the satirical function of the text tend to work from this ground. Take Rosemary Jann's account:

> The Square's unquestioning acceptance of angularity as destiny and his conflation of natural with moral order afford Abbott a means of gently satirizing his own society, with its ethos of social climbing, its fetish of conformity, and its refusal to acknowledge the sociological roots of 'immoral' behavior.[42]

In the second section, 'Other Worlds', however, the text becomes more complicated, both structurally and modally. We are shuttled between these other worlds and between A Square's dreams and his experiences: between Lineland, Flatland, Spaceland, and Pointland (and A Square's speculations on the existence of Thoughtland, the entire array of spatial manifolds of ascending dimensionality). We might seek to compare each of these worlds with the assumptions of norms that guided us in the first section, but we would struggle to do so. In Lineland the King communicates with his subjects through the emitting of harmonious tones from each of his points. What wicked behaviour might this allude to? In Spaceland A Square experiences Flatland as a Spacelander. Neither Flatland nor Spaceland is held up for mockery in this episode. Indeed, we learn little about Spaceland beyond its spaciousness, little about Flatland that is not its flatness: we learn, instead, about shifts of perspective.

The dream of Pointland indicates how we might most usefully figure these sections. The King of Pointland 'has no cognizance even of the number Two; nor has he a thought of Plurality; for he is himself his One and All, being really Nothing' (*F*, 75). He 'plumes himself upon the variety of "Its Thought" as an instance of

---

[40] *Johnson's Dictionary*, quoted in J.A. Cuddon, *A Dictionary of Literary Terms* (Harmondsworth: Penguin, 1982), p. 598.

[41] Joseph Brooker, 'Satire Bust: The Wagers of *Money*', *Law and Literature*, 17 (2005), 321–44 (p. 327).

[42] Jann, 'Abbott's *Flatland*', 475–6.

creative Power' (*F*, 76). As was remarked in a contemporary review in the *Literary World*, this is a mockery of solipsism or of a misreading of idealism in which thought becomes all and the world nothing; of 'self-sufficiency and dogmatism'. It usefully demonstrates what the second part of *Flatland* is in fact concerned with: modes of thought, rather than modes of behaviour. 'Other Worlds' does not assume a static, or normal, position for the reader but indicates that the assumption of a single position is the folly to be censured.

It is in this sense, then, that the text operates in a non-orientable fashion, leading the reader first on a path through a world that provides a familiar literary model for his own, before twisting her into consideration of the kind of thought that produces 'other worlds'. We are presented with a series of these other worlds, but we are reading about the thought processes that lead to their mental construction as we pass from one to the next. The text leads us in a thinking game, instructs us in the playfulness of thought as it flips this way and that. As readers, we are exploring the relativities of space. We are moving, as A Square remarks to the King of Lineland, 'Out of your World. Out of your Space. For your Space is not the true Space' (*F*, 50).

The text leads the reader into a representation of thought of space, of higher spatial thinking. What sort of thinking is this? Late in the text, A Square gives a definition of dimension, or rather, what it implies: 'Dimension implies direction and measurement' (*F*, 58). Instinctively, we agree with this assertion, although we might note that it is not Euclidean. Euclid did not define dimensionality but enumerated the three dimensions of Euclidean space when he defined a solid as 'that which has length, breadth, and depth'.[43] *The Elements* gives no guidance on direction—beyond definitions that work with perpendicularity or adjacency. Measurement is not a concern of Euclid's internally coherent system. Direction and measurement are spatial, rather than purely geometrical, notions. Yet in 'Other Worlds', geometric figures inhabit and discuss space. The distinction between Euclidean geometry and lived spaces is already present in, and violated by, the formal conceit of a story narrated by a geometric figure.

It cannot be overstated that the text is concerned throughout with this distinction. Indeed, the point in the text at which this distinction is most explicitly discussed is precisely the point at which A Square's world is expanded. A Square's grandson has extrapolated from his geometry lesson on area to inquire about volume and to ask what might be the meaning of three to the power of three. A Square is perturbed:

> 'The boy is a fool, I say; 3^3 can have no meaning in Geometry.' At once there came a distinctly audible reply, 'The boy is not a fool; and 3^3 has an obvious Geometrical meaning.' (*F*, 54)

---

[43] Euclid, *The Elements*, trans. Sir Thomas Heath, 3 vols (New York: Dover Publications, 1956), III, p. 260. Victorian editions of Euclid glossed the definitions to include dimensionality: see Euclid, *Euclid's Elements of Geometry: The First Six Books and the Portions of the Eleventh and Twelfth Books Read at Cambridge*, ed. Robert Potts (London and New York: Longmans, Green and Co., 1895), p. 2.

A Square was wrong in the first instance—in Spaceland terms, at any rate—because by 1884 Cayley and Salmon had been developing *n*-dimensional geometries for some thirty years: but the advent of the Sphere, for whom 3^3 is a geometric commonplace, has made this most certainly 'obvious'.

The Sphere, for example, attempts to demonstrate to A Square the existence of a third dimension by reason: 'If a Line were mere length without "height," it would cease to occupy Space and would become invisible' (*F*, 57). Here, the Sphere is also already a non-Euclidean: Euclid's line is by definition 'a breadthless length', never mind a length whose breadth is inferred and whose 'infinitesimal' height is also seen. The Sphere, despite being a Platonic solid, in this statement compounds two types of space: the abstract mathematical model of space that is geometry and the sensual space of the lived world.

Let us make no mistake that despite its abstract conceit, *Flatland* is a text that remains concerned with the senses: and so it should be in its depiction of spatial thought, thought that is, for the reader, impinged by sense information. A Square addresses himself to 'Our Methods of Recognizing One Another' in 'This World', and details processes of hearing, feeling, and seeing in *Flatland*. As we Spaceland readers might distinguish between the distance senses of sight, sound, and sometimes smell and the proximity senses of touch and taste, *Flatland* has its own disputed sensual hierarchy. Male Flatlanders privilege sight recognition while female lines privilege touch. A Square's wife's use of two proverbs popular with female Flatlanders illustrates this dispute: 'Feeling is believing' and 'A Straight Line to the touch is worth a Circle to the sight' (*F*, 54).

Compare this to Lineland, where given the limitation of spatial conditions, proximity—or 'approximation'—and therefore touch, is taboo:

> 'If you mean by feeling,' said the King, 'approaching so close as to leave no space between two individuals, know, Stranger, that this offence is punishable in my dominions by death.' (*F*, 48)

In Lineland, the sense of hearing has overcome even the challenges of distance:

> So exquisite is the adaptation of Bass to Treble, or Tenor to Contralto, that oftentimes the Loved Ones, though twenty thousand leagues away, recognize at once the responsive note of their destined Lover; and, penetrating the paltry obstacles of distance, Love unites the three. (*F*, 46)

Love as the conqueror of space is a poetic notion, but so is this an imaginative trick that aligns one-dimensional space with four-dimensional space. Without Lineland's quirky swerve towards audition, a one-dimensional creature would experience extension as all of space. The experience of the inhabitants of Lineland, their mode of communication overcoming extension, is in itself a form of raised dimensionality: sense qualia beyond the dimensions of space are registered as new dimensions in a sensual manifold.

This is analogous to our own experience of space: we fill out visual space—projected onto the two-dimensional interior surface of the eyeball—with assorted sensual qualia in order to construct a three-dimensional spatial field that matches external reality. In *Flatland*, as A Square has noted, there is both sight recognition and touch. Sight recognition plots a different qualium to apprehend two-dimensional depth: 'We see length and brightness' (*F*, 58). This is sufficient to give A Square depth-recognition in Flatland, but in Spaceland his sensory apparatus is, at first, not up to the job. Shown a cube by the Sphere, he can only distinguish its face: 'It appears to you a Plane, because you are not accustomed to light and shade and perspective' (*F*, 69).

What we might characterize as a fear of excessive proximity in Lineland anticipates concerns regarding the fourth dimension that recur. A being in a higher-dimensional space is able to see and touch the interior of a being in a lower-dimensional space: extension into the higher dimension being unavailable to the lower being, this interaction is one-way and intrusive. This is not an experience that is enjoyed by lower beings. A Square is poked in his insides during his first encounter with the Sphere, who has explained that every interior in *Flatland* is 'lying open and exposed to my view': 'even your insides and your stomachs' (*F*, 56). We should note that the domestic space of *Flatland* is violated—its houses and their internal organization are laid bare—and that at this stage of the narrative the intruder has not yet been given a geometric title and is identified as a 'Stranger'. The text also anticipates the elisions between mind and world that become increasingly confused in higher-dimensional thought, when it grants the Sphere access to the interior of A Square's dreams, as well as his material insides: 'I, who see all things, discerned last night the phantasmal vision of Lineland written upon your brain' (*F*, 58).

Having extrapolated higher dimensions from his experience of Spaceland, A Square wishes to reciprocate the visual intrusion into the Sphere. He repeatedly asks to see 'thy stomach, thy intestines' (*F*, 70). The Sphere finds this 'impertinent' and demonstrates in his response the fact that he finds the idea of a fourth dimension 'utterly inconceivable' by failing to understand that the experience of higher-dimensional intrusion is passive: 'Would you have me turn my stomach inside out to oblige you?' (*F*, 71).

The spatial thought that *Flatland* represents in 'Other Worlds' is extensive. Through its models of spaces of varying dimensionality, 'Other Worlds' begins to work through anxieties over co-presence and the possibilities of intrusion into the person and the mind. It explores the limits of sense information in the realm of philosophical space, and works at the problem of the discontinuity between what can be conceived and what can be perceived, a problem tested almost continually in higher-dimensional thought in this period. 'Other Worlds' continually refuses its reader a static viewpoint through its non-orientability: one moment we are with A Square, staring down on Flatland; the next we are in his dream of Pointland; yet later we are opening our minds to the possibility of an endless array of dimensionalities. It similarly refuses a realism/fantasy binary, oscillating between forms of mimesis and imaginative speculation. It occupies multiple topoi figuratively and rationally.

How, then, to orientate a text that we have identified as non-orientable? To answer this question we should return to consider analogy, the rhetorical structure around which *Flatland*'s story is built and to which higher-dimensional thought owed much of its life: a hinging point between language and world.

ANALOGICAL *FLATLAND*

In its concern with the thought of space, and its relationship with geometry, *Flatland* is inevitably concerned with the idea of analogy, both as a mode of rhetoric and more broadly as a form of thinking that employs models. We should recall analogy's roots in geometrical ratio—ἀναλογία—and its development by Aristotle as the structure behind metaphor and its use in legal argument. In *Flatland* we first encounter the dimensional analogy in a form that was particularly current in the 1880s, having been used by mathematicians to explain to a general audience why they were working with *n*-dimensional geometry.

In 'Other Worlds', the Sphere founders in his first attempt to explain to A Square that he has come from a higher-dimensional space. The Sphere considers how he can make A Square understand: 'One resource alone remains, if I am not to resort to action. I must try the method of Analogy.' The Sphere resumes a dialogue with A Square:

> But I will describe it to you. Or rather not I, but Analogy. We began with a single Point, which of course—being itself a Point—has only one terminal Point. One Point produces a Line with two terminal Points. One Line produces a Square with four terminal Points. Now you can give yourself the answer to your own question: 1, 2, 4, are evidently in Geometrical Progression. What is the next number?
>
> I. Eight.
> SPHERE. Exactly. The one Square produces a Something-which-you-do-not-as-yet-know-a-name-for-But-which-we-call-a-Cube with eight terminal Points. Now are you convinced?
> I. And has this Creature sides, as well as angles or what you call 'terminal Points'?
> SPHERE. Of course; and all according to Analogy. (*F*, 61)

The Sphere's argument fails to convince A Square, forcing him to resort to action and to lift A Square out of the plane. Despite the failure of reasoning by analogy in this instance, A Square has picked up on the technique and uses it himself in his attempt to persuade the Sphere of Thoughtland:

> In Three Dimensions, did not a moving square produce—did not this eye of mine behold it—that blessed Being, a Cube, with *eight* terminal points?
>
> And in Four Dimensions shall not a moving Cube—alas for Analogy, and alas for the Progress of Truth, if it be not so—shall not, I say, the motion of a divine Cube result in a still more divine Organization with *sixteen* terminal points?
>
> Behold the infallible confirmation of the Series, 2, 4, 8, 16: is not this Geometrical Progression? Is not this—if I might quote my Lord's own words—'strictly according to Analogy'? (*F*, 72)

A Square continues in this vein, asking 'in that blessed region of Four Dimensions, shall we linger on the threshold of the Fifth and not enter therein?' (*F*, 74). His reasoning is too much for the Sphere to take, however, and he is cast back into *Flatland* never to see higher space again.

This engagement with the idea of reasoning by analogy has focused some of the more recent critical attention of Victorianists. Rosemary Jann has described the process as 'Abbott's prescribed journey through illusion to truth', the necessary leap of faith required for interpreting the material world that leads the mind on to the higher truth of religious revelation.[44] Baker, Berkove, and Smith contest this, arguing that A Square's attempt to reason by analogy the existence of the higher dimensions of Thoughtland is presented as a failure:

> By embodying analogy's limitations, *Flatland* forces its readers to acknowledge the consequences of inappropriate analogical thinking. It becomes itself a model of how analogy should and should not be employed.[45]

In making this argument, the authors are informed by definitions in a textbook Abbott co-wrote with his friend J.R. Seeley in 1871. *English Lessons for English People* gives some guidelines on rhetoric, and in so doing deals with 'induction [...] from enumeration of instances to a general statement about a class', arguing that it is 'evidently an insecure method of proof [...] It is based upon the principle of uniformity in nature, "what has been is and will be".'[46] Because we cannot know the future, 'induction is always incomplete'.[47] Baker, Berkove, and Smith argue that 'this is precisely the error that A Square makes'.[48]

I want to complicate this reading. A Square is not simply enumerating instances, although his 'Progression' is certainly the enumeration of a mathematical series: he is also describing structural similarities in form, for which analogy is a perfectly appropriate tool. Where A Square seems to differ from his creator is in his making continuous the discontinuity between geometry and space, between analogue and source. In *The Kernel and the Husk* Abbott acknowledged that geometry was a model, and one which we could not do without, using it as an example of the imagination leading thought to truth—Jann's argument—but he recognized that the model was distinct from the source: 'If you step from your ideal triangle in Dreamland into your material triangle in chalk-land, you step from absolute truth into statements that are not absolutely true.'[49]

Flatland's relationship with analogy is, as with its relationship with satire, multiple and ambivalent. It not only comments upon analogy but enacts it. Analogy is used structurally: scenes are repeated with variations. For example, A Square's dream of Lineland anticipates his own experience when visited in *Flatland* by the Sphere. A Square swaps roles with the King: in his dream, he is

[44] Jann, 'Abbott's *Flatland*', 487.

[45] Jonathan Smith, Lawrence I. Berkove, and Gerald A. Baker, 'A Grammar of Dissent: *Flatland*, Newman, and the Theology of Probability', *Victorian Studies*, 39 (1996), 129–50 (p. 137).

[46] Edwin A. Abbott and J.R. Seeley, *English Lessons for English People* (London: Seeley, Jackson and Halliday, 1871), pp. 262–3.

[47] Abbott and Seeley, *English Lessons*, p. 263.

[48] Smith, Berkove, and Baker, 'A Grammar of Dissent', 140.

[49] Edwin A. Abbott, *The Kernel and the Husk: Letters on Spiritual Christianity* (London: Macmillan, 1886), p. 30.

the visitor from higher space; in the second episode, he is visited. The structural analogy of these two episodes is underlined by descriptive terms that are almost identical between them: the King of Lineland refers to A Square as a 'Monstrosity' (*F*, 46), just as A Square calls the Sphere a 'monster' (*F*, 52); the King accuses A Square of exercising 'some magic art of vanishing and returning to sight' (*F*, 51) as A Square assumes the Sphere to be 'some extremely clever juggler; or else the old wives tales were true, and that after all there were such people as Enchanters and Magicians' (*F*, 59); each expresses astonishment and concern over his 'insides'. The first scene models the second; A Square dreams the experience of higher spatial visitation before he experiences it, but when he does experience it the roles are reversed.

The modelling of episodes or scenes, this structural form of analogy, is appropriate for the text. If we define structural analogy in its broadest sense, we might note that satire is frequently a structural form, reliant upon processes of transposition, substitution, and projection and that in this it has aspects in common with analogy. Where allegory might broadly be read as a directly analogical form—structure A:B represented as structure C:D—as we read in Aristotle, metaphor makes use of substitutions between the pairs: A's D, or B's C. Satire similarly operates on the basis of switches between structural domains on a larger scale. The relationship of A with its setting B must be structurally similar enough with that of C with D such that C can be transposed into B. Once transposed, C looks absurd: the structure A in B matches with the structure C in D, but placing C in B illuminates the aspects of C that readers will find ridiculous.

The structural analogy of 'This World' is complicated by the fact that we read Victorian England through classical Greece in *Flatland*. We might represent this as A (Victorian society) has the same relationship with B (Victorian England) as D (2D geometric figures) have with E (a plane), the relationship being that B and E are the spaces in which A and D are located. But A is also related to C (classical Greek society) through an interest in, and idolization of, that society, while D also has a relation to C, having originated within that culture. Relationships between A and C and D and C are not analogous, but C is used as a mediating idea. In 'Other Worlds' the structural analogies are more direct, more like a chain—or perhaps a knot, crossing backwards and forwards—a non-orientable surface. A's visitation by B is like B's visitation by C. B proposes to C an experience that would introduce him to D.

As the position of A Square was inverted between his dream of Lineland and his first encounter with the Sphere so is his position in the argument by analogy also inverted in chapter 19. He asks the Sphere to take him 'into the blessed region of the Fourth Dimension'. The Sphere replies that such a land is utterly inconceivable, but A Square has learnt from the argument by analogy:

> But, just as there was the realm of *Flatland*, though that poor puny Lineland Monarch could neither turn to left nor right to discern it, and just as there was close at hand, and touching my frame, the land of Three Dimensions, though I, blind senseless wretch, had no power to touch it, no eye in my interior to discern it, so of a surety there is a Fourth Dimension, which my Lord perceives with the inner eye of thought.

And that it must exist my Lord himself has taught me. Or can he have forgotten what he himself imparted to his servant? In One Dimension, did not a moving Point produce a Line with two terminal points? In Two Dimensions, did not a moving Line produce a Square with four terminal points? In Three Dimensions, did not a moving Square produce—did not this eye of mine behold it—that blessed Being, a Cube, with eight terminal points? And in Four Dimensions shall not a moving Cube—alas, for Analogy, and alas for the Progress of Truth, if it be not so—shall not, I say, the motion of a divine Cube result in a still more divine Organization with sixteen terminal points? Behold the infallible confirmation of the Series, 2, 4, 8, 16: is not this a Geometrical Progression? Is not this—if I might quote my Lord's own words—'strictly according to Analogy'? (*F*, 72)

The reader is likewise spun around by these inversions of position. The analogies remain stable, but if the reader moves with A Square his perspective shifts. The reader must think for herself as the balance of power in the argument shuttles between A Square and the Sphere. This persistently switching dialogical structure means the text enacts inversions we might read as topological. It is as if we are Flatlanders, not on a plane but a Möbius strip: as we pass from one section to the next we find ourselves on the opposite side of the strip to that we once inhabited, without ever leaving its single surface.

Flatland is persistently dialectical. Its two halves tell different stories and perform different functions, but they address each other. The first half, in its traditionally satirical frame, urges the reader to think on the world in which he lives; its second part tells the story of leaving that world in allegorical terms that urge the reader to examine her thought. It is about ancient Greece and Victorian England, each reflecting the other. Throughout, it pits illustrations into conversation with text, disrupting the temporal flow of reading and engaging a different mode of textual practice, a visual hermeneutics. It reproduces entire sections in the form of dramatic dialogue—sections which recall Socratic dialogue, as one participant in the dialogue attempts to lead another to understanding, perhaps nowhere better indicated by the Sphere's remark to A Square: 'Then you see you have answered your own question' (*F*, 61). This perhaps leads us to recall specifically the passage in the *Meno* in which Socrates claims to demonstrate that Meno's Thracian slave is capable of doubling the area of a square without knowledge of geometry. This echo of the *Meno* alerts us to a concern with the idea of innatism in *Flatland*. In the preface to the second edition, A Square responded to some of his critics in a way that suggested that he, at least, did not subscribe to the view of knowledge as innate:

It is true that we have really in Flatland a Third unrecognized Dimension called 'height,' just as it is also true that you have really in Spaceland a Fourth unrecognized Dimension, called by no name at present, but which I will call 'extra-height'. But we can no more take cognizance of our 'height' than you can of your 'extra-height'. Even I—who have been in Spaceland, and have had the privilege of understanding for twenty-four hours the meaning of 'height'—even I cannot now comprehend it, nor realize it by the sense of sight or by any process of reason; I can but apprehend it by faith. (*F*, vii–viii)

Pulling back yet further and taking a yet more elevated view of structure, much like A Square being raised into Spaceland, we might observe that the novel itself might be a form of model that uses language or, more specifically in the case of *Flatland*, forms of representation, including illustrations, dramatic dialogue (as a representation of direct speech), first-person narrative, reported speech, and narrations of dreams. For J. Hillis Miller, resistance to being reduced to a 'monological voice' is what distinguishes the written text, the fact that we cannot quite line up author and character, or author and voice. The images he chooses to illustrate his point are particularly familiar and appropriate to higher-dimensional thought:

> This forbids imputing the language back to a single mind, imagined or real. In one way or another, the monological becomes dialogical, the unitary thread of language something like a Möbius strip, with two sides and yet only one side. An alternative metaphor would be that of a complex knot of many crossings.[50]

Flatland continued its game beyond publication in a manner that highlighted this schizophrenic literary problematic and in the midst of a debate over the soul of literature that inclined towards issues of modelling and representation.

BEYOND THE TEXT

A month after publication A Square wrote to *The Athenaeum* to clarify a point 'metaphysical or psychological' regarding the difference between an ideal line in the ideal plane and a 'visible' line in the material plane. A Square's response to a 'not unfriendly, but, as I venture to think, too hasty critic' is exemplary of *Flatland*'s mode of operation. A playful extension of the reach of the text beyond the pages of the book, it made its narrator act in the material world to query his own nature. The critic, said A Square, had argued that:

> any visible line must really have thickness as well as length; and therefore all our so-called plane figures, besides having length and breadth, must really have some degree of thickness, or height—in other words a Third Dimension; and of this, he implies, we ought not to be ignorant.[51]

A Square admitted these facts but, echoing his remarks in the text, denied the conclusion:

> Dimension implies measurement. Now, our lines are so thin that they cannot be measured. Measurement implies degrees, the more and the less; but all our lines are equally and infinitesimally thin, or thick, whichever you please to call it; so that we in Flatland can neither measure their thinness, nor even take cognizance of it. Where you speak of a line as being long and thick (or thin), we speak of it as being long and bright; 'thickness' (or 'thinness') never enters our heads, and we do not know what you mean by it.[52]

[50] J. Hillis Miller, *Ariadne's Thread: Story Lines* (New Haven: Yale University Press, 1992), p. 22.
[51] A Square, 'The Metaphysics of Flatland', *Athenaeum*, 2980 (1884), 733.
[52] A Square, 'The Metaphysics of Flatland', 733.

Most of A Square's points could already be found in the detail of the text. In many ways, his letter of clarification was entirely unnecessary. Lindgren and Banchoff report that he also wrote to R.A. Proctor, the editor of *Science*, but that Proctor did not print A Square's missive. However, *The Athenaeum* did print his letter, and the editor's response was in turn appended to the letter and this discussion revisited in A Square's preface to the second edition of *Flatland*.

This meta-textual correspondence is an entirely natural blossoming of the ludic character of the text but we should also note its potency. A Square's letter extended Abbott's satirical practice, deploying a technique that destabilized the distinction between the imaginative text and the real world. It refused the closure of the text and so encouraged proliferation that further blurred the distinction between text and world. We may read later respondents to *Flatland* as participants in a game inaugurated by A Square's letter. This is an important aspect of *Flatland*'s literary practice and it indicates an interesting relationship with character.

We should consider A Square. He is a geometric figure. He is literally, and classically, a character. J. Hillis Miller observes that the word character derives from the Greek verb 'kharassein', to scratch, or inscribe, with a *kharax*, a pointed stick. A Square, like any literary character, is scratched or inscribed on the page, but whereas an author would need to do considerable descriptive work to aid us to imagine a character more closely resembling a human form in appearance, A Square's name alone performs this for us. We picture him immediately. We know A Square. Miller continues:

> The word character, like the word lineaments and the word person (from the Latin word for mask), involves the presumption that external signs correspond to and reveal an otherwise hidden inner nature. The visible design made by the features of a face—nose, mouth, eyes, lines of forehead, cheek, or chin—are taken as a hieroglyphic sign telling accurately what that person is like inwardly. What was originally a synecdoche, part for whole, or a metonymy, contingent visible element as sign for a secret adjacent element, in a complex which must be assumed to be homogeneous for the figure to work, gradually comes to be the 'literal' name for what it used only to figure. What is problematic about the figure is obvious. Does a man's face really correspond to his inner nature? Does he really 'have' an inner nature, separate from the signs that point to it? This problem is likely to be forgotten when the figure is literalized.[53]

A Square presents few of these problems at first glance. We have him all. He has no *in*sides, but is all sides. But it is not quite so simple because A Square carries with him his own symbolic heritage, his own mythology. A Square, a unit on a geometric grid, is more than he might at first seem. He activates our recollection of classical philosophy. We might recollect the square Socrates inscribes in the earth when instructing Meno's slave in geometry. This square, a basic unit in geometry, becomes a tool for demonstrating arguments over innatism, as Socrates guides the slave to the geometric knowledge to double the area of a square through construction. The text is aware that the square is the perfect structure for dimensional

---

[53] Miller, *Ariadne's Thread*, p. 32.

demonstration: 'I mean that every point in you—for you are a Square and will serve the purpose of my illustration—every Point in you [...] is to pass upwards through space' (*F*, 60).

A Square *is* a character, fundamentally, a symbol scratched on a surface carrying within his apparently plain form the ideas of innate knowledge of geometry, of Platonic philosophy as well as the human characteristics and emotions he displays throughout the narrative of *Flatland*. When A Square leaves his fixed text to write letters to the editors of journals, he further exploits and makes ambivalent distinctions between the ideal and the material that are coded into his form. A Square's apparent flatness, emptiness, and blankness encourage the reader to imagine him more fully.

Rosemary Jann has written that Abbott made the 'argument that imagination was the basis of all knowledge [...] sanctioning [...] a religion independent of material proof'.[54] It was a career-long concern, and it is interesting to note that he used the concept to illustrate the analogical structural design of metaphorical language. In 1870, in his *A Shakespearean Grammar*, written for use in the English classroom, he glossed Hamlet's phrase 'in my mind's eye' to demonstrate the expansion of metaphor:

> As the body (known subject) is enlightened by the eye (known predicate), so the mind (subject whose predicate is unknown) is enlightened by a certain perceptive faculty (unknown predicate).[55]

I am struck when reading this passage by the recurrent use of eyes in *Flatland*'s illustrations. In these we see the eye as geometric instrument, organ of perception, and, as a representation of a material structure in an illustration describing an ideal process, a mediator. Abbott's imagination was the faculty with which the ideal was brought into mind, and he made the distinction between the ideal of geometry and the material figures with which we describe its processes:

> The whole of what we call 'Euclid' is based upon a most aerial effort of the imagination [...] Obviously these things have no existence except in the dreams of Imagination; yet Euclid's severe reasoning applies to none but these things. If you step from your ideal triangle in Dreamland into your material triangle in chalk-land, you step from absolute truth into statements that are not absolutely true.[56]

It is worth asking where *Flatland* might lie in relationship to both 'Dreamland' and 'chalk-land', particularly given the narrative twists that send Flatland into its own dreams and the fact that the text of *Flatland* was inscribed and then printed upon a plane sheet: its compositor can hardly have failed to recognize the passage of his text from his imaginative dreamland into the chalk-land of the material book. For Andrea Henderson, 'Flatland celebrates the fact that in the late nineteenth century,

---

[54] Jann, 'Abbott's *Flatland*', 484.
[55] Edwin A. Abbott, *A Shakespearean Grammar* (London and New York: Macmillan and Company, 1870), p. 521.
[56] Abbott, *Kernel*, p. 30.

geometry relinquished its connection with realism in favor of the aesthetic.'[57] Geometry, in becoming speculative, became art, and *Flatland* was a response to this development. We might develop this idea further. Abbott's book was a romance, certainly, a product of and stimulus to the imagination, but explicitly also a romance of 'many dimensions'.

It is also a tale of revelation, though A Square decides against communicating his experience for fear of persecution; he is to appeal to the reason rather than the emotions of his audience: 'It would be better to avoid it [the danger of being executed] by omitting all mention of my Revelation, and by proceeding on the path of Demonstration' (*F*, 77). He decides to write a book explaining Spaceland but in this endeavour he encounters obstacles encountered by all higher spatial thinkers of the late nineteenth century; obstacles of illustration and cognition:

> In writing this book I found myself sadly hampered by the impossibility of drawing such diagrams as were necessary for my purpose [...] I tried to see a Cube with my eye closed, but failed [...] This made me more melancholy than before. (*F*, 80)

As the book in our own space represents objects in text and illustrations printed onto a two-dimensional plane, so in *Flatland* the text is marked onto a line. A Square does not happen upon the representational breakthroughs of Charles Howard Hinton that will be discussed in Chapter 4: the use of a medium of $n-1$ rather than $n-2$ dimensions, through which to plot the cross-sections of higher spatial objects; the development of a system with which to enable the repeated and practised imagination of such cross-sections. A Square instead presents us with the artefact we have in our hands—he does not explain how it was translated from his *Flatland* linear script into the traditionally planar text we have before us, and despite the protestations of critics at the time, this is the greatest continuity error in his game. He does, though, explain that he began his work as fiction before it became the non-fictional account we are now reading: 'At first, indeed, I pretended that I was describing the imaginary experiences of a fictitious person' (*F*, 81). We find ourselves once again on a non-orientable surface, but this time with regards to literary representation, fictionalization, and mimesis. Our narrator, an ideal object, insists that the book we are now reading, that he could not possibly produce, began life as a work of fiction before becoming the non-fiction we are now reading.

Abbott was a literary scholar of considerable repute, who had published books on Shakespeare and Bacon, contributed to journals such as the *Contemporary Review*, the *Modern Review*, *The Academy*, and *The Athenaeum*, as well as conducting correspondences with *The Spectator* and *The Times*. He was a close friend and co-author of his publisher's brother, J.R. Seeley. A Square's correspondence with *The Athenaeum* suggests that the author of *Flatland* took an interest in the literary pages of the journals to which he contributed: he was not only reading journals for his own reviews but was replying to these reviews in character. The publication of his novel and A Square's correspondence arrived at a very specific moment in literary

---

[57] Henderson, 'Math for Math's Sake', p. 470.

critical history during which the very soul of the novel was under debate in the pages of some of the journals to which Abbott contributed.

What would be termed the 'romantic revival' by the critic Andrew Lang only a few years later, and has recently been described by Nicholas Daly as 'the literary current that began to overwhelm the domestic novel in the 1880s', had emerged with Robert Louis Stevenson's *Treasure Island*, published as a novel in November 1883, and the essay by the same author, 'A Gossip on Romance', published in *Longman's Magazine* in 1882.[58] In this essay Stevenson advocated a fiction that was 'absorbing and voluptuous', capable of filling the mind of the reader with 'a kaleidoscopic dance of images'. He privileged 'fit and striking incident' over 'the passionate slips and hesitations of conscience', with his dagger drawn against realism and his banner flown for a narrative that would please both young and old.[59] In private, he was less guarded about his intended audience, writing: 'If this don't fetch the kids, why, they have gone rotten since my day.'[60]

While the term romance was neither new nor exclusive to the 1880s, when *Flatland* appeared with a map on its inside cover, like *Treasure Island*, and, like many other texts, declaring itself a romance, was it seeking to reference this approach to fiction and to address itself to the same readership? There is no evidence that Abbott was as interested in the market as Stevenson, no expression of a desire to 'fetch' an audience, but he was certainly interested in writing for a juvenile reader. In *Hints on Home Teaching*, a manual Abbott had published the previous year, he wrote approvingly of literature for children:

> Fairy stories encourage the imaginative faculty because they present things old, in combinations so new, as to take the child altogether out of the range of things which he sees, and stimulate him by pleasurable associations to realise visions utterly unlike his own experiences.[61]

The resonances with *Flatland* are apparent. The construction of Euclidean figures on a plane was something old for the juvenile reader; endowing them with intelligence and having them narrate a story, something new. His stressing of the 'imaginative faculty' was consistent with a philosophy that elevated the crucially generative power of this aspect of thought, as described by Jann and Elliot L. Gilbert.[62]

[58] Nicholas Daly, *Modernism, Romance and the Fin de Siècle: Popular Fiction and British Culture, 1880–1914* (Cambridge: Cambridge University Press, 1999), p. 7.

[59] R.L. Stevenson, 'A Gossip on Romance', *Longman's Magazine*, 1 (1882), 69–79 (pp. 69, 72, 70).

[60] R.L. Stevenson, *The Letters of Robert Louis Stevenson*, ed. Sidney Colvin, 3 vols, 10th edn (London: Methuen and Co., 1911), I, p. 220.

[61] Edwin A. Abbott, *Hints on Home Teaching* (London: Seeley, Jackson, and Halliday, 1883), p. 25. An interesting tension is emphasized by *Hints on Home Teaching*, a text addressed to the parent, or governess, who taught female children excluded from the state education system. Nicholas Daly writes: 'From the beginning romance is a gendered genre. Pervasive in the critical accounts is the assumption that the romance is a more healthily masculine form than the realist novel.' *Modernism, Romance and the Fin de Siècle*, p. 18. Was *Flatland*, despite its toying with the romantic genre, addressed to the female reader, too?

[62] See Elliott L. Gilbert, ' "Upward, not Northward": *Flatland* and the Quest for the New', *English Literature in Transition*, 34 (1991), 391–404.

The other pole of the engagement was taking shape in the months before *Flatland*'s publication. In 1884 Walter Besant gave a lecture at the Royal Institution that sparked discussion over what the novel could and should be, with Andrew Lang responding in the *Pall Mall Gazette* in April, R.H. Hutton in *The Spectator* in May, and Henry James in *Longman's Magazine* in September. Besant's lecture was vague and imprecise but his cheerleading gave the opportunity for James to hone his own thoughts. The English novel had not been 'discutable', he argued: 'It had no air of having a theory, a conviction, a consciousness of itself behind it—of being the expression of an artistic faith, the result of choice and comparison.' James defended the mimetic against the imaginative terms of the romance: 'The only reason for the existence of the novel is that it does attempt to represent life.' He believed it 'a terrible crime' to make clear the artifice of efforts to achieve this mimesis. He described the novel broadly as the author's 'personal impression of life'.[63]

Nicholas Daly summarizes the continuing debate and response to James: 'Realism/romance seems to have provided the principal axis of difference [...] Realism was often represented as essentially a noxious weed of foreign growth.'[64] The editors of an edition of James's collected critical writing read an increasing tension over the status of the novel in these debates: 'What criticism in 1884 also manifests is that an increased self-consciousness about fiction is part of novelists' larger awareness that their relations with orthodoxy are growing tense.'[65]

Contemporary criticism has located Abbott's text in relation to non-Euclideanism and religious debates; Suvin located it in relation to what he categorized as SF: but where does it lie in the broader literary landscape of its time? It was surely, at some level, an intervention into James's 'era of discussion'. As a novel it declared itself on the side of romance, but it was not quite a romance in the Stevensonian sense: this was no boy's own adventure, unless the boy was unusually minded towards geometry; 'fit and striking incident' was certainly subsidiary to the play of the imagination. What it did do was fill the mind with a 'kaleidoscopic dance of images', perhaps more literally than any pirate story could: from the 'chromatic sedition' of 'This World', to the telescopic sensual rearrangements of 'Other Worlds', *Flatland* was an imaginative romance that put into practice its author's belief about the power of the fairy story. It resisted the Jamesian insistence on realism most clearly by committing the 'terrible crime' of making clear its own artifice. In fact, it went some way further than simply making this artifice clear, it revelled in its own artifice, played with it, extended it beyond the page and invited others to join it in artifice; it centred itself on a problem that illuminated the fundamental nature of artifice in thought, perhaps even the artifice *of* thought: that thought models world.

[63] Henry James, 'The Art of Fiction', in *The Art of Criticism: Henry James on the Theory and Practice of Fiction*, ed. William Veeder and Susan M. Griffin (Chicago and London: University of Chicago Press, 1986), pp. 165–83 (pp. 165, 166, 167, 170); repr. of *Longman's Magazine*, 4 (1884), 502–21.

[64] Daly, *Modernism, Romance and the Fin de Siècle*, p. 17.

[65] William Veeder and Susan M. Griffin, 'Commentary on The Art of Fiction', in *The Art of Criticism*, pp. 184–8 (p. 185).

As an intervention in literary debates of the period, contested over realism and romantic adventure, it unsettles the polarization. It not only says, but demonstrates, that the literary imagination is an active and creative force in the world and need not limit itself to attempted mimesis. A Square resists, despite his flatness, the idea of flat characterization before that becomes a critical commonplace. A Square is far fuller than most literary characters *because* he acts in the material world.

FLATLAND RESPONSES

By 1888 Swan Sonnenschein had become quite the specialist in fourth-dimensional literature. *Another World*, originally entitled *The Fourth Dimension*, was written by A.T. Schofield, a medical doctor who had previously published short-story pamphlets with Swan Sonnenschein. Corresponding with Schofield, Sonnenschein noted that 'it is a subject for a very limited public, thus involving comparatively small sales', but went ahead with publication after agreeing with the author the emendation of its original title, 'so as not to interfere with Hinton's Fourth Dimension'.[66] Schofield was aware of Hinton's work, then, but *Another World* wore its debt to *Flatland* on its sleeve:

> I would here take the opportunity of acknowledging my deep indebtedness to the anonymous author of a small book, called 'Flatland,' which I have used extensively throughout, and without which I am quite sure the public would never have been troubled with these remarks; my object being to carry on the line of argument there brought forward, to what seems to me its true and necessary conclusion.[67]

Schofield's 'true and necessary conclusion' to *Flatland*'s argument advanced 'the spiritual claims of the Christian religion' with single-minded zeal. This agenda is stated from the outset, in the formulation that the most popularly understood 'higher world' is to be found in the pages of the Bible. The likely opponents of such a reading are identified:

> Materialists will, we know, have nothing of this. To them, if true to their creed, there is, and can be, nothing beyond the material. Mind, morals, feelings, passions, are to them only protoplasmic changes of ganglion nerve cells, producing carbonic acid gas and water.[68]

Schofield painstakingly reconstructs the dimensional analogy from *Flatland* in the form of a conversation between A Square and a 'Lord' instructing him in the mysteries of higher space. His language becomes increasingly scriptural as A Square rejoices in the vision granted him by his higher father—'I would go forth, methought, and evangelize the whole of *Flatland*'—before the contrivance of a

---

[66] Reading, Reading University Library, MS Swan Sonnenschein and Co., 3282, Swan Sonnenschein to Arthur Schofield 9 March 1888, 16 March 1888.

[67] A.T. Schofield, *Another World* (London: Swan Sonnenschein, 1888), p. 3.

[68] Schofield, *Another World*, pp. 2–3.

biblical geometry is vividly demonstrated in an attempt to demonstrate life after death through algebra:

> Let, for example, the body, material and solid, be represented fairly enough by x^3, and the spirit, higher and possessing an unknown power, by x^4. Then $(x^3 + x^4)$ represents the man in life, while $(x^3 + x^4) - x^4$ represents the departure of the spirit (x^4) at death, which returns to its own dimension, while the body (x^3), which is left, returns to the earth to which it belongs.[69]

Schofield's aggressive fusion of characterization appropriated from *Flatland*, geometry, and popular Christian theology does not make for elegant prose, but it demonstrates succinctly that the fourth dimension was open to ingress from diverse quarters. Steven Connor notes that Schofield's 'bold renderings' represent 'one of the earliest spiritualised, if not precisely Spiritualistic, appropriations of the fourth dimension in England'.[70]

Schofield's claim that he was following the line of thought of *Flatland* was not supported by its author. Abbott had pre-emptively warned against such literal readings of the fourth dimension in *The Kernel and the Husk*, writing: 'It seems to me rather a moral than an intellectual process, to approximate to the conception of a spirit: and towards this no knowledge of Quadridimensional space can guide us.'[71] Nevertheless, as the producer of a text, by contributing to *Flatland*'s written legacy, Schofield does more than simply misread Abbott. Such a respondent's act of cacography, by introducing noise into the signal of a text, creates further ambiguity, and fuels further production. Schofield's intervention into the idea of the fourth dimension nourished millennial accounts that appeared in the 1890s, and to which I will turn my attention in Chapter 5.

Flatland's cultural reverberations, if they were not always clear, resounded far and wide. Karl Pearson, the most positivist of hard scientists, referred his readers to Abbott's book in his *Grammar of Science*, using a version of the dimensional analogy based on Clifford's flounder to illustrate his idea of ether squirts, before insisting that he only provided the illustration 'for those minds which, strive as they will, cannot wholly repress their metaphysical tendencies, which must project their conceptions into realities beyond perception'.[72]

A report of a staged production of *Flatland* at Haberdashers' Aske's Girls' School in Acton in 1913 provides an interesting illustration of the cultural position of the text a generation after its publication. The audience, members of the Mathematical Association, were introduced to an expanded cast of characters 'represented by cardboard models of circles, pentagons, etc., carried horizontally on the girls'

---

[69] Schofield, *Another World*, pp. 68–9.

[70] Steven Connor, 'Afterword', in *The Victorian Supernatural*, ed. Nicola Brown, Carolyn Burdett, and Pamela Thurschwell (Cambridge: Cambridge University Press, 2004), p. 268.

[71] Abbott, *Kernel*, p. 259.

[72] Karl Pearson, *The Grammar of Science*, 2nd edn (London: Adam and Charles Black, 1900), p. 269. Pearson's disapproval of higher-dimensional metaphysicians was coloured by his involvement with Olive Schreiner and sternly moralistic response to Charles Howard Hinton's conviction for bigamy. See Theodore M. Porter, *Karl Pearson: The Scientific Life in a Statistical Age* (Princeton: Princeton University Press, 2004), p. 193 and Chapter 4 of this book.

heads, thus showing the flat surfaces to the spectators seated in the gallery above'. The plot was adapted to include greater incident, including murder, and less exposition. The adaptation stressed the dramatic elements of Abbott's book, conceptualizing from its geometry this ingenious mode of staging. The teachers responsible for the production were keen to stress its educational function:

> Miss Brown and Miss Griffiths (both teachers at the school) stated how much the interest of the girls had been roused by the preparations involved. The discussions on the nature of Flatland, the making of the cardboard models, the search for inconsistencies in the representation of Flatland, the enquiries about the meanings of the names of the characters, all these were of educational value, not only to the performers, but also to other girls in the school.[73]

The producers of this staged version exploited the dramatic structure already built into *Flatland*: its two acts and the sections of dialogue already written for the director. It also responded to its pedagogical impetus, an aspect of the text that has more recently frequently defined it. Time and again, forewords and introductions remark upon its use in the geometry classroom as an instructional tool. Two films made in 2007 are sold on DVD with educational licences.[74] This is a natural extension of Abbott's text's performance of instruction, its instantiation of pedagogy betraying its author's profound theoretical interest in teaching. It is connected with the subject matter that was taught in the classroom—Plato, Euclid—and what Abbott wanted to see taught—Shakespeare. It mimicked and questioned the Socratic mode of instruction. It created a space for its reader to question. We might usefully consider it an adjunct to Abbott's pedagogical practice.

From these briefly sketched examples we can see that readers of the first editions valued the text for wildly varying reasons, including its applicability as an educational text, ensuring an enduring juvenile audience, but also for its adaptability, even to projects explicitly alien to its author's stated concerns. These respondents are significant readers and reproducers of the text and care should be taken to include them in an account of the text's cultural contribution which is multidimensional, multivalent, and mobile. It is a text about many dimensions that acts on many dimensions: intra-textual and meta-textual; in the world and about the world; about ideas and matter and particularly about the passages to and from, and discontinuities between, the one and the other.

A key respondent to Abbott, Charles Howard Hinton, had already published work on higher space. His project included fictions but, centrally, was developed beyond the page in a system that sought to materialize abstract space into more solid form. Hinton's project began, and ended, in conversation with Abbott's *Flatland*.

---

[73] Anon., 'Summer Meeting of the London Branch: Flatland', *Mathematical Gazette*, 7 (1914), 228–31 (p. 231).
[74] *Flatland: The Film*, dir. Ladd Ehlinger Jr (F.X. Vitolo, 2007); *Flatland: The Movie*, dir. Dano Johnson and Jeffrey Travis (Flat World Productions, 2007).

4

Cubes

Hintonian Higher Space and its Thinking Subject

Charles Howard Hinton's essay 'What is the Fourth Dimension?' was first published in the final issue of the ailing *University Magazine* in 1880.[1] Arriving shortly after Massey's translation of Zöllner's *Transcendental Physics*, it was read in this context: writing anonymously in *The Nonconformist and Independent*, the physicist William Barrett, a founding member of the Society for Psychical Research, noted Hinton's essay and used it as a springboard for speculations on beings 'superior to ourselves, but unknown to us, because living in space extended in the fourth dimension'.[2] Hinton's essay lent itself to such readings. Grounded in geometry, working by analogy from two dimensions to three and then four, and referencing contemporary science, it announced itself clearly as a speculative piece:

> It is the object of these pages to show that, by supposing away certain limitations of the fundamental conditions of existence as we know it, a state of being can be conceived with powers far transcending our own.[3]

The powers a four-dimensional being would possess would have been familiar to readers of Zöllner: 'such a being would be able to make but a part of himself visible to us [...] would suddenly appear as a complete and finite body and as suddenly disappear' (*SR*, 25); it would have the ability to 'get out of a closed box without going through the sides' (*SR*, 27).

---

[1] C.H. Hinton, 'What is the Fourth Dimension?', *The University Magazine*, 96 (1880), 15–34. Hinton's mother-in-law, Mary Everest Boole, had been a frequent contributor to the magazine in its earlier guise as the *Dublin University Magazine*. For ease of navigation, page references to 'What is the Fourth Dimension?' will be given from the 1886 edition of the essay collected in *Scientific Romances* (see footnote 3 below): this final version expanded on the two previous versions of the essay that had appeared in journals, but was identical to the content published in a stand-alone pamphlet of 1884. All further references to this essay will therefore appear in the body of the text using the abbreviation *SR*. While retaining page referencing to this edition, I will refer to the pamphlet edition of the essay as *What is the Fourth Dimension?* in accordance with style for stand-alone publications.

[2] William Barrett, 'Invisible Beings', *Nonconformist and Independent*, 4 (1881), 16–17. Barrett can be identified as the author through the following quotation: 'one of our leading English scientific men has said, in a letter to the present writer, "I am not aware of any law of nature (except the most obvious, such as are seen by common observers) which is sustained by so many assertions, so well attested as far as respectability of evidence goes"' (p. 17). The same letter, written by Dr R. Angus Smith, FRS, is cited by Barrett in his book *On the Threshold of the Unseen* (London: Kegan Paul, 1918).

[3] C.H. Hinton, 'What is the Fourth Dimension?', in *Scientific Romances*, 2 vols (London: Swan Sonnenschein, 1886), I, pp. 3–32 (p. 4).

While Hinton employed the full analogical toolkit for imagining beings confined to lower-dimensioned spaces, a toolkit now tried and tested through repeated use in scientific journals, he offered considerable original insight on the subject. A striking metaphor of threads passing through a sheet of wax was developed to pose the question 'is it possible to suppose that the movements and changes of material objects are the intersections with a three-dimensional space of a four-dimensional existence?' (*SR*, 23). Hinton was concerned to bed the fourth dimension of space into physical models and drawn towards materialization. The ultimate focus of his thought experiments, though, was the human consciousness: 'Why then, should not the four-dimensional beings be ourselves, and our successive states the passing of them through the three-dimensional space to which our consciousness is confined?' (*SR*, 18).

'What is the Fourth Dimension?' was reprinted with an extra paragraph of material in 1883 in the magazine of the Cheltenham Ladies College and, despite its author's social connections, it seems likely that this second iteration would have been the piece's final printing were it not for the publication of the first edition of *Flatland* at the end of October 1884.[4] Within weeks the publisher William Swan Sonnenschein was corresponding with Hinton, now science master at Uppingham College, and a pamphlet containing the first of a planned series of *Scientific Romances* was printed. In the hands of the new publisher *What is the Fourth Dimension?* was expanded and subtitled 'Ghosts Explained'.

The text that expanded the first Sonnenschein edition of Hinton's essay solidified his thesis, developing the idea that 'the matter we know extending in three dimensions has also a small thickness in the fourth dimension'. This speculation was founded upon another: that gases might behave in four dimensions as liquids do in three, might 'have a centre of attraction off in the fourth dimension' (*SR*, 28). The conclusions drawn from such speculations were either that physical space was four-dimensional, but that humans were three-dimensional beings and appeared to four-dimensional beings as nothing but abstractions; or that we have an 'infinitely minute' four-dimensional existence (*SR*, 31).

The encroaching materiality of Hinton's higher spatial thought is the arc around which this chapter orbits. Hinton repeatedly mediated higher space through matter, both in his theories, which constantly accounted for the physical implications of a fourth dimension or speculated a physical source, and in his practice, based on the manipulation and contemplation of a set of colour-coded (or named) cubes. Hinton was sufficiently conversant with the ethereal physics of his time, and the conceptual dissolution of matter into its atomic components, that he was able to weave his higher spatial theories into the electromagnetic spectrum. Over the course of his career he honed a key set of theories about the fourth dimension: that it was evidenced by the phenomena of electricity; that it existed at the macro and micro scales; that thinking it brought about material changes in the brain.

---

[4] C.H. Hinton, 'What is the Fourth Dimension?', *Cheltenham Ladies College Magazine*, 8 (1883), 31–52.

His body of work emerges as an immense force in shaping the idea of higher-dimensioned space. Where the Zöllner event nourished spiritualist appropriations of the fourth dimension, Hinton's work prised open new channels. Linda Dalrymple Henderson treats Hinton as the chief popularizer of hyperspace philosophy.[5] Henderson's description is particularly useful because Hinton has often been mis-characterized as a writer of SF. This chapter outlines the scope of his project, a scope that Hinton himself described in *A New Era of Thought* (1888):

> I propose a complete system of work, of which the volume on four space is the first installment. I shall bring forward a complete system of four-dimensional thought-mechanics, science, and art.[6]

The ambition of this claim is continuous with the visionary zeal that peaked in Hinton's writing around the publication of *A New Era* but, as an overview of his thought shows, he did indeed attempt such an all-encompassing programme, even if the artistic aspect, beyond his own fiction, was left to the modern artists about whom Henderson writes. His interest was philosophical, but so too was it practical and informed by his approach to pedagogy.

It is telling that it took the work of an art historian to recuperate Hinton. Henderson reads Hinton as a 'pioneer', a popularizer whose work became a source, through mathematicians such as Jouffret and Theosophists such as C. W. Leadbeater, for Futurism, Cubism, Suprematism, and more. For Elizabeth Throesch it is a form of '"protomodernism" because [...] his project is a transitional one'.[7] Precisely the reasons Hinton's work does not fit into the literary canon—its hybridity and materiality—are the reasons it is more amenable to art history, a discipline more used to dealing with objects, even if they lurk within texts. Henderson also treats occultists with an equanimity that has not always been afforded them in literary studies. The influence of Hinton's work today can be felt most keenly in the slightly embarrassing esoteric sections of bookshops: as an important source for both Rudolph Steiner and P.D. Uspensky, Hinton has been assured a legacy in this section of the publishing market.[8]

After outlining Hinton's work and its salient characteristics, this chapter focuses on the practical system at its heart which it recreates through archival and textual sources. Reconstituting the cubes he designed and described, the very things with which he thought the fourth dimension, we can ascertain how the author conceived them, how they were used, how *read*, and how we might think them. This work will involve the voicing of key sources for Hinton's approach to these objects before recounting their cultural trajectory. In closing I will consider again

---

[5] Linda Dalrymple Henderson, *The Fourth Dimension and Non-Euclidean Geometry in Modern Art* (Princeton: Princeton University Press, 1983), p. 25.

[6] C.H. Hinton, *A New Era of Thought* (London: Swan Sonnenschein, 1888), p. 86. All further references to this edition will be given in the body of the text using the abbreviation *ANE*.

[7] Elizabeth Lea Throesch, 'The *Scientific Romances*' of Charles Howard Hinton: The Fourth Dimension as Hyperspace, Hyperrealism and Protomodernism', doctoral thesis, University of Leeds, 2007.

[8] See Rudolf Steiner, *The Fourth Dimension* (Great Barrington, MA: Anthroposophic Press, 2001); P.D. Uspensky, *Tertium Organum*, trans. Nicholas Bessaraboff and Claude Bragdon (London: Kegan Paul & Co., 1923).

the mediations between space and matter that loop through this book and which in the case of Charles Hinton were informed by his father's published philosophy.

Also informed by the work of James Hinton was the idea of 'casting out the self', a voiding of spatial subjectivity that was an essential element in the use of the cubes. James Hinton inspired the altruistic currents that swirled beneath the more visionary aspects of Charles Hinton's project and provided a unique intellectual and social platform for such work. Before turning to the son, what of the father?

JAMES HINTON: NATURE PHILOSOPHER

James Hinton split his life between practice as an aural surgeon and writing as a philosopher. From 1858 to 1862 he gave up practice to write, publishing *Man and his Dwelling Place* (1859), writing *The Mystery of Pain* (1866), and contributing numerous articles to *Cornhill Magazine* which became the basis for *Life in Nature* (1862) and *Thoughts on Health* (1871).

Man and his Dwelling Place outlined a perceived lack in man's knowledge of the universe: 'The defective state of man causes our feelings not to correspond with the truth of things; so that we can only understand aright either ourselves or the world by remembering that man is wanting in life.'[9] *Life in Nature* clarified these ideas:

> Not only are the organic and inorganic worlds, which seem to be so different, truly one, exhibiting the same forces, powers, and laws; but life itself, or that which we have called so, appears as a mere result of chemical and mechanical agencies, into the effects of which its most distinct phenomena are resolved. We find no special power which we can call by that name.[10]

Process and systems of relations became of central importance, the flux and continuity between the organic and inorganic, man and the world. Shadworth Hodgson, writing after his friend's death, cast this in terms of mind and matter:

> The operations which have mind and those which have matter for their field are parts of one system of operations; and just because they are parts of a single whole do they recall and seem to repeat each other when each kind is separately examined.[11]

Hodgson read Hinton's view of nature as an extension of the Romantic philosophy of Coleridge and identified later encounters with German idealism as the source for an appreciation of the phenomenal and an alignment of the noumenal with the spiritual. Hinton wrote: 'This physical world, known to be an appearance (or phenomenon,) is the appearance of that spiritual world which we also know.'[12] Hinton argued that instinct and the emotions, more attuned to the noumenal

[9] James Hinton, *Man and his Dwelling Place* (London: John W. Parker and Son, 1859), p. 309.
[10] James Hinton, *Life in Nature* (London: Smith, Elder and Co., 1862), p. 155.
[11] Shadworth Hodgson, 'Introduction', in *Chapters on the Art of Thinking*, ed. C.H. Hinton (London: C. Kegan Paul and Co., 1879), pp. 1–14 (p. 11).
[12] Hinton, *Life in Nature*, p. 166.

world, should be allowed to guide thought: 'Our heart, in a word, asserts the true; science reveals to us the apparent.'[13]

Also writing after Hinton's death, Henry Havelock Ellis recorded that Hinton referred to his philosophical position as 'actualism' and emphasized the moral tone that inflected his ideas with the emotional and instinctive in his later writing: 'Hinton, when he began philosophising, firmly believed in an absolute which might be known; not indeed known intellectually but through the moral sense.'[14]

This moral philosophy, incipient in his first work, dominated his later thought. It began to surface in *The Mystery of Pain*, which set out to offer consolation to all those who had suffered. Hinton argued that pain was its own reward if viewed as sacrifice:

> These facts are evident in human life even as it is: that man is framed for joy in sacrifice; that until it can be made his joy, sacrifice must be his torment, for it can never be banished; that without the willing acceptance of sacrifice, no end is really answered in human life, no satisfaction that is worthy of humanity achieved.[15]

As Thomas Dixon has noted, Hinton claimed that the causal relationships of the physical world illustrated the inevitability of self-sacrifice. Ellis quoted a passage from Hinton's unpublished manuscripts, written around 1858:

> How that idea of self-sacrifice (as the source of all life) is involved in the correlation of forces! 'Each force merging itself as the force it produces becomes developed,' says Grove. This is the very fact of creation, the exact statement of that self limit which is creative action. And this is the phenomenal, the 'instinctive' view of Nature.[16]

Ellis explained that the notion of 'self-sacrifice' was later supplanted by the idea of 'service' to account also for the role of pleasure. He glossed this modification: 'By sacrifice he had meant the willing acceptance of pain, all thought of self being cast out; by service he now meant the acceptance of pleasure also, the thought being still not on the self; that is to say the acceptance of all things, either pleasure or pain, that served.'[17]

James Hinton's work is gradually being recovered by Victorian scholarship. Thomas Dixon has highlighted the significance of his contribution to the naturalization of altruism in the second half of the nineteenth century. Hinton wrote to a friend: 'The word altruistic I borrow from Comte. Is it not a capital word? I am resolved to naturalise it. We want it. It is the antithesis to "self"; self-being = deadness; altruistic being = life; and so on.'[18] Dixon observes how altruism fitted in to his broader system: 'He used altruistic to describe not just philanthropy, nor just the benevolent instincts so named by Comte, but the connectedness and relatedness

---

[13] Hinton, *Life in Nature*, p. 170.
[14] H. Havelock Ellis, 'Hinton's Later Thought', *Mind*, 9 (1884), 384–405 (p. 386).
[15] James Hinton, *The Mystery of Pain* (London: Smith, Elder and Co., 1866), p. 66.
[16] Ellis, 'Hinton's Later Thought', 392.   [17] Ellis, 'Hinton's Later Thought', 394.
[18] Ellice Hopkins, *Life and Letters of James Hinton* (London: C. Kegan Paul, 1878), p. 194.

of all natural phenomena, and the possession of "a consciousness co-extensive with humanity".'[19]

Seth Koven, another scholar who has worked to recover James Hinton, blames the 'quicksand of sexual scandal, based wholly on unsubstantiated rumour' that came in the wake of Charles's 1886 conviction for bigamy as a reason for the almost complete erasure of James Hinton's influence on his contemporaries from the cultural record of the period.[20] This reputational damage is certainly significant for both generations of Hinton; for Charles it had geographical repercussions, sending him into exile at a time when his work was gaining a foothold in Britain.

A significant inheritance from his father that had bolstered this burgeoning public life was a network of intellectual and social connections. Old family connections between the Hinton family and the Nettleship family from Nonconformist circles in Oxford had been maintained, providing a significant philosophical and educational matrix for Charles. Richard Lewis Nettleship was a tutor of philosophy at Balliol, Charles's Oxford college, and literary executor of the idealist T.H. Green; he was also formerly head-boy at Uppingham, where Charles later found employment, and a correspondent and friend of the headmaster Edward Thring; Richard's brother, John Trivett Nettleship, the landscape painter and friend of the Pre-Raphaelites, who cited James Hinton in his study of Robert Browning, later married Charles's sister Adaline.

Beyond the family were those who were, like Havelock Ellis, drawn to James Hinton's philosophy. Seth Koven writes of the 'dense networks of discipleship and affiliation surrounding [James] Hinton' that included key figures in philanthropic and social purity movements and drove discussions of sexuality and sexual reform. Well documented in scholarly studies, and significant with reference to Charles, is the membership of the Men and Women's Club. Henry Havelock Ellis, who had made contact with the Hintons immediately on his arrival in England and became friend and confidant to Charles, was the conduit. The South African novelist Olive Schreiner records frequent visits to the Hintons in her correspondence with Ellis. Caroline Haddon, James's sister-in-law, and Margaret, his wife, were also attendees at these meetings, as they were at the Fellowship of the New Life, the group that became the socialist Fabian Society.[21] James Hinton's biographer, the social reformer Ellice Hopkins, Ellis's lesbian wife Edith Ellis Lees, Arnold Toynbee, Roden Noel, and Charles Ashbee all produced work directly inspired by James Hinton's writing.[22]

While James Hinton's work as a 'philanthropic hedonist', as Koven describes him, drew sexual reformers and progressives to his work, his books also met with

---

[19] Thomas Dixon, *The Invention of Altruism: Making Moral Meanings in Victorian Britain* (Oxford: Oxford University Press, 2008), pp. 85–6.

[20] Seth Koven, *Slumming: Sexual and Social Politics in Victorian London* (Princeton: Princeton University Press, 2004), p. 17.

[21] See Yaffa Claire Draznin, *My Other Self: The Letters of Olive Schreiner and Havelock Ellis, 1884–1920* (New York: Peter Lang, 1992) for detail of the scandal and its fall-out among members of the Men and Women's Club.

[22] See Koven, *Slumming*, pp. 16–17.

the appreciation of figures more central to late Victorian culture. Having impressed Tennyson, James Hinton was invited to be a founder member of the Metaphysical Society, alongside William Gladstone, Thomas Huxley, John Tyndall, Henry Sidgwick, Walter Bagehot, W.K. Clifford, and, later, John Ruskin, with whom he became friends.[23]

In his history of the Society, Alan Willard Brown argues that the foundation of the Society for Psychical Research in 1882 by Sidgwick 'reflects the failure of the Metaphysical Society to bridge the gulf between the institutionist and empiricist positions and the admitted inability of the scientists to relate "the facts of consciousness" to their material hypotheses'.[24] Charles Howard Hinton's work clearly continues aspects of this endeavour independently of the SPR. Brown also comments on the establishment in 1876 by Metaphysical Society member George Croom Robertson of the journal *Mind*:

> And it is true that at the very time when the great popular reviews were beginning to exclude the more profound philosophical and scientific papers and turn to more popularly based articles, *Mind* appeared in answer to a need. As the Metaphysical Society declined, its most philosophically minded members turned increasingly to *Mind* rather than to the *Nineteenth Century*; and this, too, was a sign of the times.[25]

Mind is perhaps the most important context: even though James died in 1876 before he could contribute, he was obituarized in the journal and all of Howard's *Scientific Romances* were subsequently reviewed in its pages, with doffs of the cap to their author's lineage. Its serious treatment of 'mental philosophy', a nexus between emergent psychology and philosophies of thought, mirrors a similar nexus in Charles's higher space philosophy and his development of neurological ideas in *A New Era of Thought*. It also provides further connections, between Charles and Shadworth Hodgson, James Hinton's friend and a contributor from the first issue, and later William James, a contributor from 1879 with whom Charles corresponded when he was living in the USA.

The tradition Charles Hinton inherited from his father was resistant to the processes characterized by Bruno Latour as purifications of nature from the social. James Hinton argued for a society and culture permeated by nature and natural processes. Under the guises of philosopher and surgeon he worked to complicate late Victorian social and scientific orthodoxy.

CHARLES HOWARD HINTON:
SPATIAL PHILOSOPHER

Charles's interest in higher geometry began as he was building a public career as a young schoolmaster and scholar. In 1878 his paper 'On the Co-ordination

---

[23] See Alan Willard Brown, *The Metaphysical Society: Victorian Minds in Crisis, 1869–1880* (New York: Columbia University Press, 1947).
[24] Brown, *The Metaphysical Society*, p. 245.   [25] Brown, *The Metaphysical Society*, p. 199.

of Space' was read at the Physical Society. William Crookes's *Chemical News* summarized:

> If a cubical space be divided into 27 numbered cubes, and each of these be again subdivided in the same way, and so on, the position of any point within the initial cube can be expressed by a reference to the numbers of the several cubes in which it is placed, and the more this series of numbers is extended, the more accurately is its position defined.[26]

No copy of his first monograph, *Science Notebook* (1884), is known to survive, but reviews describe a well-received engagement in debates over non-Euclideanism that seems derived from the same system:

> The author does not presuppose continuous elements as has been generally done, but only sets of points equally distributed in two dimensions, which, merely for the sake of convenience, are connected by straight lines [...] The practical advantages of this new method in the form in which it is now published are purely educational, though it is wholly based on the principles just mentioned.[27]

Later that year, the success of *Flatland* provided the impetus for Swan Sonnenschein's interest in republishing Hinton's twice-round-the-block essay 'What is the Fourth Dimension?'. Correspondence from the publisher to Hinton began in November 1884. A specific publishing context in which the success of *Flatland* made pieces of fringe scholarly or pedagogical interest commercially viable contributed to the expansion of higher-dimensional thought.

Sonnenschein was canny and innovative, publishing an eclectic list that later included the first English translations of both Marx and Freud.[28] The son of a German mathematics teacher, he built his list in the early years (*c*.1878–82) around books for children, educational texts, and translations of German-language books, such as *Grimm's Teutonic Myths*. Although Sonnenschein described himself as a liberal, he was closely connected socially to a number of Fabians and socialists, publishing, as well as *Capital*, George Bernard Shaw's *Unsocial Socialist* in 1887. Stepniak, exiled Russian revolutionary, was apparently often to be encountered taking tea *chez* Sonnenschein.

The Swan Sonnenschein list also always included philosophy, and the publisher was a member of the first Ethical Society in the late 1880s. Commissioned to write a history of the firm's predecessors by George Allen and Unwin in the 1950s, F.A. Mumby wrote: 'Throughout his life Swan Sonnenschein was a remarkable blend of other-worldliness and business acumen; a man of wide erudition whose interests were quickly roused by the simplest human problems.'[29] This business acumen was developed in the turbulent book market of this period. Alexis Weedon writes in her analysis of Victorian book publishing:

[26] Anon., 'Proceedings of Societies', *Chemical News and Journal of Industrial Science*, 37 (1878), 271–2 (p. 272).

[27] Karl Heun, 'Science Note-Book', *Nature*, 31 (1884), 51–2 (p. 51).

[28] See F.A. Mumby and Frances H.S. Stallybrass, *From Swan Sonnenschein to George Allen and Unwin Ltd.* (London: George Allen and Unwin, 1955).

[29] Mumby and Stallybrass, *From Swan Sonnenschein*, pp. 17–18.

When the economic interdependence between novel publishers and the libraries began to fail in Britain in the 1880s, a newly competitive market-place arose. Shrewd publishers looked to their strengths and developed innovative publishing strategies to exploit them [...] careful price structuring and timing of the release of each edition was crucial for them to sustain revenue and reap the full economic potential of the work [...] publishers were able to capitalize on the cost savings of the more efficient printing technologies and cheaper raw materials by marketing the text in a range of formats.[30]

One such format, the part issue, became a viable tool for publishers looking to bring a work to market rapidly to capitalize on favourable conditions. As the entry for 'Serials and the Nineteenth Century Publishing Industry' in the *Dictionary of Nineteenth-century Journalism* notes: 'The principal motivations underlying the rise of serial publications were speed and economy.'[31] The timing, pamphlet format, and re-editing of Hinton's essay for publication by Swan Sonnenschein in November 1884 suggest a rapid commercial response to *Flatland.*

Swan Sonnenschein had a clear view of its readers, advertising in popular scientific journals such as R.A. Proctor's *Knowledge*, and had the commercial nous to engage as many audiences as possible for its output.[32] The new subtitle to Hinton's essay, 'Ghosts Explained', was the work of the publisher and piggybacked on the currency of higher spatial ideas in spiritualist and occult circles. Like this subtitle, the term *Scientific Romance* had not been used by Hinton before his essay was brought out by Swan Sonnenschein and this phrase is yet more indicative of commercial expediency.

The reviewer of Hinton's pamphlet for *Nature* had already read *Flatland*, and discussed Hinton's ideas in relation to Abbott's fictional narrative, noting that 'these ideas are coming to the front again'.[33] It was not only readers who noted the relationship: the work of the two writers was also mutually aware. Hinton praised Abbott but stressed the difference between the two in the introduction to his third romance, *A Plane World*, first published in the summer of 1886:

> And I should have wished to be able to refer the reader altogether to that ingenious work, 'Flatland.' But on turning over its pages again I find that the author has used his rare talent for a purpose foreign to the intent of our work. For evidently the physical conditions of life on the plane have not been his main object. He has used them as a setting wherein to place his satire and his lessons.   (*SR*, 129)

Abbott returned the compliment in *The Kernel and the Husk*, a collection of theological essays published in 1887:

> You know—or might know if you would read a little book recently published called Flatland, and still better, if you would study a very able and original work by

[30] Alexis Weedon, *Victorian Publishing: Book Publishing for the Mass Market 1836–1916* (Aldershot: Ashgate, 2003), p. 141.
[31] 'Serials and the Nineteenth Century Publishing Industry', in *Dictionary of Nineteenth-century Journalism*, ed. Laurel Brake and Marysa Demoor (Ghent: Academia Press; London: British Library, 2009), p. 567.
[32] See James Mussell, *Science, Time and Space in the Periodical Press: Movable Types* (Aldershot and Burlington, VT: Ashgate, 2007), p. 42.
[33] R. Tucker, 'Flatland: A Romance of Many Dimensions', *Nature* (1884), 76.

Mr C.H. Hinton—that a being of Four Dimensions, if such there were, could come into our closed rooms without opening door or window, nay, could penetrate into, and inhabit, our bodies.[34]

A degree of social contact between the two writers has been noted. Specifically, Hinton's colleague at Uppingham, Howard Candler, was a close friend of Abbott and, indeed, the dedicatee of *Flatland*. There is a slight echo of the extra-textual Hinton in the text of *Flatland* that adds support to the idea of the cubes as the central element to Charles's work and sits comfortably within *Flatland's* playful satire. To see it we also have to have already pictured Hinton's preferred method of visualizing space, which he had not published by 1884, but which he had certainly been using in the classroom at Uppingham: that of working with sets of nine one-inch cubes. At the beginning of Section 15 of *Flatland*, A Square describes giving a domestic geometry lesson to his grandson, a hexagon:

> Taking nine Squares, each an inch every way, I had put them together so as to make one large Square, with a side of three inches, and I had hence proved to my little Grandson that—though it was impossible for us to see the inside of the Square—yet we might ascertain the number of square inches in a Square by simply squaring the number of inches in the side: 'and thus,' said I, 'we know that three-to-the-second, or nine, represents the number of square inches in a Square whose side is three inches long.'   (*F*, 52–3)

The hexagon is a bright student and extrapolates by analogy from this planar system to inquire about three to the third, much as Hinton hoped students of his cubic system would: 'It must be that a Square of three inches every way, moving somehow parallel to itself (but I don't see how) must make Something else (but I don't see what) of three inches every way—and this must be represented by three-to-the-third.' The passage is brief, as A Square behaves in an un-Hintonian fashion and dismisses his grandson's speculations, but the echo of this central aspect of Hintonian practice—flattened—is there. Abbott's pencil sketch grasped the core of Hinton's project: material objects deployed in a pedagogical context and mediating between geometry and space.

At the beginning of 1885 Hinton presented a refined version of his system of co-ordinate relations, now called 'poiographs' after Sir William Hamilton's 'hodographs', to the Physical Society. *Nature's* review suggests either a different approach or a different perception of Hinton: 'As a result of a process of metaphysical reasoning, Mr. Hinton has come to the conclusion that relations holding about "number" should be extended to space.'[35] In 1886 he was commissioned to contribute an essay on the fourth dimension that ran to double the length of the entry on 'Psychical Research, and the Society for', to Hazell's *Cyclopaedia* [*sic*], a publication that purported to provide 'up-to-date information on such subjects as are now, or are likely to be, in the mind of the public'.[36]

---

[34] Edwin A. Abbott, *The Kernel and the Husk: Letters on Spiritual Christianity* (London: Macmillan, 1886), p. 259.

[35] Anon., 'Societies and Academies', *Nature*, 31 (1885), 328–32 (p. 329).

[36] C.H. Hinton, 'Fourth Dimension', in *Hazell's Annual Cyclopaedia, 1886* (London: Hazell, Watson and Viney, 1886), p. iii, pp. 183–5.

Meanwhile, Swan Sonnenschein began to publish a series of essays in pamphlet form as *Scientific Romances*. The second *Romance, The Persian King*, was a physical parable describing the system of governance in an isolated Persian valley, in which pain and pleasure assumed by a ruler and distributed among citizens was laid out on a thermodynamic basis, with accompanying calculations provided in a second part:

> When the king wished to start a being on the train of activity he divided its apathy into pleasure and pain. The pleasure be connected with one act which we will call A. The pain he associated with another act which we will call B.
>
> [...]
>
> The sensation in the first A was 1000, in the first B it was 998, giving a disappearance of 2. In the second A it was 980, and in C, which starts concurrently with the second A, it was not 20 as might have been expected, but 16, giving a loss of 4.[37]

Mind, as sympathetic a reviewer to Hinton's concerns as existed, described it as 'somewhat less effective' than its predecessor, and as a piece of narrative fiction it was hamstrung by its didactic aims.[38] Considered as part of Hinton's larger programme, however, it was both consistent with his teaching and an expansion of his higher spatial project, inserting his work into thermodynamic discourse and hybridizing it with moral elements from his father's late philosophy of 'service'. Bruce Clarke argues that *The Persian King* 'seeks moral asylum from the materialism of the second law in the conceptual haven of the spatial fourth dimension'.[39] *The Persian King* illustrates, too, Hinton's tendency towards allegory, and his debt to one allegory in particular: the character of Demiourgos, a creator controlling the valley from a distinct, if not higher, space, signalling the Platonic Demiurge of the *Timaeus*.

The physical concerns of *The Persian King* were extended in 'A Plane World', a post-*Flatland* meditation on congruence in two dimensions, and 'A Picture of Our Universe', which outlined two theories extended in his later work. The first considered congruence with regard to spiral twists, the mutual cancellation of opposite twists, and suggested a fourth-dimensional rather than ethereal explanation for the phenomena of electricity:

> Thus if we suppose that in the minute motions which go on about us there is the possibility of moving in a four-dimensional way, then it is perfectly legitimate to assume that in a medium which cannot be twisted, but which is elastic, a twist calls up a real image twist. And thus the assumptions which we have made as the basis of an electrical theory are justified on the assumption of a four-dimensional space, are untenable except on that supposition.[40]

---

[37] C.H. Hinton, 'The Persian King', in *Scientific Romances*, 2 vols (London: Swan Sonnenschein, 1886), I, pp. 33–128 (pp. 55, 60).

[38] Anon., 'Scientific Romances. No. II', *Mind*, 10 (1885), 613.

[39] Bruce Clarke, *Energy Forms: Allegory and Science in the Era of Classical Thermodynamics* (Ann Arbor: University of Michigan Press, 2001), p. 5.

[40] C.H. Hinton, 'A Picture of Our Universe', in *Scientific Romances*, 2 vols (London: Swan Sonnenschein, 1886), I, pp. 161–204 (p. 180).

The second theory was worked out in an appendix:

> For suppose the aether, instead of being perfectly smooth, to be corrugated, and to
> have all manner of definite marks and furrows. Then the earth, coming in its course
> round the sun on this corrugated surface, would behave exactly like the phonograph
> behaves [...] Corresponding to each of the marks in the aether there would be a move-
> ment of matter, and the consistency and laws of the movements of matter would
> depend on the predetermined disposition of the furrows and indentations of the solid
> surface along which it slips [...] Thus matter may be entirely passive, and the history
> of nations, stories of kings, down to the smallest details in the life of individuals, be
> phonographed out according to predetermined marks in the aether. In that case a man
> would, as to his material body, correspond to certain portions of matter; as to his
> actions and thoughts he would be a complicated set of furrows in the aether.[41]

Bruce Clarke routes his investigations of 'Hinton's groovy phonographic ether'
through Friedrich Kittler's theoretical writing, arguing: 'In the transformation of
the ether medium into the spatial and temporal fourth dimensions, mediated reality
is metamorphosed into art forms—imaginary realms and symbolic structures.'[42]
We might adapt this to recognize the grooved ether as another physical model, a
form into which the immaterial is compressed.

The grooved ether model was reprised in the full-length book *A New Era of
Thought* (1888) to argue for the existence of both a material and an eternal ethereal
body for any organism, and an 'essential unity of the race':

> We find an organism which is not so absolutely separated from the surrounding
> organisms—an organism which is part of the aether, and which is linked to other
> aethereal organisms by its very substance—an organism between which and others
> there exists a unity incapable of being broken, and a common life which is rather
> marked than revealed by the matter which passes over it.   (*ANE*, 64)

A New Era not only drew together the theoretical work of Hinton's *Romances*, but
was the book in which he explicitly advanced the ethical, and therefore social,
theories implicitly stated in those works. His writing found a new tone, less speculative, more visionary, as he approached the content that dominated the book:
'And then those instincts which humanity feels with a secret impulse to be sacred
and higher than any temporary good will be justified—or fulfilled' (*ANE*, 94).

Charles paid direct, but anonymous, tribute to his father in *A New Era of
Thought*, referring to 'one with whose thought I have been very familiar, and to
which I return again, after having abandoned it for the purely materialistic views
which seem forced on us by the facts of science'. He summarized what he saw as
the most significant element of his father's thought:

> He looked for a time when, driven from all thoughts of our own pain or pleasure,
> good or evil, we should say, in view of the miseries of our fellow-creatures, Let me be
> anyhow, use my body and my mind in any way, so that I serve.   (*ANE*, 72)

---

[41] Hinton, 'A Picture of Our Universe', pp. 196–7.
[42] Clarke, *Energy Forms*, pp. 183, 184.

A New Era created a hinge between the ethical and the spatial by equating the absenting of the egoistic self from altruistic activity with the need to remove 'self-elements', the impositions of the corporeal self, from thought of absolute space: 'Thus altruism, or the sacrifice of egoism to others, is followed by a truer egoism, or assertion of self' (*ANE*, 27). The altruism and egoism binary might be equated with space in terms of the space occupied by the self: altruism is the voiding of the self from space; egoism the occupation of all of space to the exclusion of others.

Also extended from James Hinton's work on the continuity of the organic and inorganic, mind and matter, was a striking theory of the action of higher-dimensional thought on the brain. Charles argued that 'it is by a structure in the brain that [the human being] apprehends nature, not immediately'. What we perceive are 'models and representations' in 'minute portions of matter' in the brain, portions 'beyond the power of the microscope in their minuteness' (*ANE*, 48). These 'brain molecules' do not, however, directly mimic external matter:

> It may be that these brain molecules have the power of four-dimensional movement, and that they can go through four-dimensional movements and form four-dimensional structures. If so, there is a practical way of learning the movements of the very small particles of matter—by observing, not what we can see, but what we can think.
>
> (*ANE*, 49)

This oscillating neurological materiality incorporates the fugitive materiality of the sub-microscopic, locating Hinton's ethereal fourth dimension within the brain, and the brain within it.

A New Era of Thought was completed just as his bigamous marriage was discovered. Hinton had married Maud Florence under the assumed name of John Weldon. There is evidence that certain of his friends and family had known since 1884 of his mistress, with whom he had twins, but the discovery of the affair by his wife led to a trial and conviction for bigamy and Hinton's subsequent departure from England to Japan. He wrote of it to his publisher in intellectual terms:

> I have had to give up everything and go through disgrace such as rarely falls to anyone's lot. But still, although I have lost all outward things I have got on the right side of life. In the book which you have of mine lie the steps of my reasoning.[43]

While in Japan he worked first at a mission before being recruited as the headmaster of the Victoria Public School, a school established by the British expatriate community to commemorate Victoria's jubilee. Records from this period of his life are scarce, and he published no work, but remained interested in both literature and space, briefly hosting a young Lafcadio Hearn and devising cubic climbing structures from bamboo for his sons.[44] While he was abroad Swan Sonnenschein published two more *Scientific Romances*, left by Charles with his editors: 'On the

[43] Reading, Reading University Library, MS Swan Sonnenschein and Co., 4058, Charles Howard Hinton to William Swan Sonnenschein, 22 February 1887.

[44] See Paul Murray, *A Fantastic Journey: The Life and Literature of Lafcadio Hearn* (London: Routledge, 1993), p. 128 and Carleton W. Washburne and Sidney P. Marland, *Winnetka: The History and Significance of an Educational Experiment* (Englewood Cliffs, NJ: Prentice-Hall, 1963), p. 137.

Education of the Imagination', treated in detail below, and 'Many Dimensions', an answer to the question pre-empted by *A New Philosophy*, 'if four dimensions, why not five, or six, or seven?'. Hinton employed a familiar homiletic tone, relating the myth of elephants supporting the universe on a turtle's back, a well-worn illustration of infinite regression, before slipping into his own regressive reveries about the printed page:

> And yet, looking at the same printed papers, being curious, and looking deeper and deeper into them with a microscope, I have seen that in splodgy ink stroke and dull fibrous texture, each part was definite, exact, absolutely so far and no farther, punctiliously correct; and deeper and deeper lying a wealth of form, a rich variety and amplitude of shapes, that in a moment leapt higher than my wildest dreams could conceive.[45]

In 1895, two years after his arrival in the USA and assumption of a teaching post at Princeton, Hinton published *Stella* and *An Unfinished Communication*. The preface announced these pieces as artistic productions continuous with his broader project:

> In the following pages an attempt has been made to dwell upon the wider bearing of conceptions which, whatever their origin, have found more definite expression in the speculations of modern mathematicians than at any other time [...] Just as the study of the minute or the very large requires microscopes, telescopes, and other apparatus, so for the study of the Higher World we need to form within our minds the instrument of observation, the intuition of higher space, the perception of higher matter. Armed thus, we press on into that path wherein all that is higher is more real, hoping to elucidate the dark sayings of bright faith.[46]

These novellas, then, were presented as types of 'apparatus' for the development of higher-dimensional thought. *Stella* was an invisible woman narrative that literalized honesty and openness as corporeal transparency, progressing the vision of higher spatial altruism first essayed in *A New Era*. Stella's guardian has experimented upon her to make the 'coefficient of refraction of the body [...] equal to one'. 'But why should he?,' asks Stella's paramour: 'Don't you see, Hugh, being is being for others. Michael used to say that true life begins with giving up' (*S*, 35). The theories of Michael Graham, the guardian, voiced by Stella and left in a journal, described Hinton's modulating eternal return:

> 'If you feel eternity you will know that you are never separated from any one with whom you have ever been. You come to a different part of yourself each day, and think the part that is separated in time is gone. But in eternity it is always there.
>
> [...]
>
> If you felt it you would know that you are always living in your whole life, that it is always changing, though with your eyes you can only see the part you are in now.' (*S*, 30)

---

[45] C.H. Hinton, 'Many Dimensions', in *Scientific Romances: Second Series* (London: Swan Sonnenschein, 1896), pp. 28–44 (p. 33).

[46] C.H. Hinton, *Stella and an Unfinished Communication: Studies in the Unseen* (London: Swan Sonnenschein, 1895), p. i. All further references to this edition are given in the body of the text using the abbreviation *S*.

The narrative of 'An Unfinished Communication' explicated this model. A young man, wondering through a poor part of New York, sees a sign advertising lessons with an 'Unlearner'. Intrigued, he tracks the tutor down to a seaside village. At an initial encounter he claims that he is shackled by his past. The Unlearner responds: 'But have you ever lived? For life is where man takes up the work of nature and forms a net-work of close personal knowledge, linking each to each, preparing that body in which the soul of man lives' (*S*, 120). A series of stories are told by different characters, and the narrator is directed on to a further village, where he stays with the locals for several days, learning the allegories of their lives. As he makes the journey home the narrator is overwhelmed by the tide and, while drowning, has visions in which he passes between several different first-person consciousnesses:

> And a new consciousness comes over me. I see that, like everything else in Nature, our lives are altering, developing, our whole lives in every event and circumstance. I see my life suddenly transformed from the pitiful thing it is. I see that it is changing—the whole of it. (*S*, 174)

In America, Hinton's work became more focused on popularization and professionalization. In 1897 he achieved some fame by inventing and patenting a cannon for firing baseballs at practising batsmen: even this was conceived in terms of the twists that could be imparted upon the ball. He gave a paper on his higher spatial thought to the Philosophical Society of Washington, moved to jobs at the University of Minneapolis and *The Nautical Almanac*, before settling at the Patent Office in 1904. Essays in *Harper's* magazine discussed his baseball cannon and developed the ideas he had spoken about before the Washington Society, first tentatively sketched in 'A Picture of Our Universe', that electrical current could be represented as 'a vortex sheet whose edge meets the aether along the wire of the circuit'.[47] The mathematical workings behind this idea were demonstrated in an article for the *Proceedings of the Royal Irish Academy*.

In that year *The Fourth Dimension* was published by Swan Sonnenschein, a summary of his work so far that did indeed reveal a complete, flexible, and complex programme of 'thought-mechanics'. It provided colour plates for his system of cubes; a two-part history of higher-dimensional thought, from classical philosophy to the 'meta-geometry' of Bolyai, Lobatchewsky, and Gauss; a 'proof' of his vortex-sheet theory of aethereal electrical phenomena; uses of his system of poiographs for testing logical assertions; an application of this system to Kant's 'Theory of Experience'; a detailed description of the tesseract, the simplest four-dimensional figure, obtained using said systems; detailed instructions for the use of the cubes; and a chapter combining a paper to the Washington Philosophical Society with his paper on Cayley. His earlier work was synthesized and drawn together, the visionary tone calmed and the facts presented as accessibly as possible.

One last work was published in the year of his death. Appropriately enough, *An Episode of Flatland* ended his writing career in tribute to the text that had given it such early impetus. The two-dimensional narrative of *An Episode* contains passages

---

[47] Rudolf v. B. Rucker, 'Introduction', in *Speculations on the Fourth Dimension: Selected Writings of Charles H. Hinton* (New York: Dover Publications, 1980), p. xiv.

that finally approach Bruce Clarke's 'science fiction *in utero*' description of Hinton's work.[48] The idea, first read in *A New Era of Thought*, that thinking higher space would result in structural changes in the brain, comes about in Unaea when the inhabitants begin to realize the existence of the third dimension:

> 'It is undoubtedly the fact,' said the director, 'that this new conception of existence has a marked influence on the power and scope of volition. For one thing when the children get to know that real existence has a dimension they cannot see with their bodily eyes, and has a richness of movement they cannot make with their limbs, they realise that they are beings of this higher kind, directing these extended bodies of a lower plane [...] And thus I find that the very bodies of the children are undergoing modification.'[49]

An Episode inserted recognizable Hintonian practical and theoretical work into the well-honed lower-dimensional setting to predict utopian advance:

> But amongst those who learned by means of models, making visible and tangible the aspects and views of the higher reality, were some who sprang, with a kind of inner awakening, to the knowledge of the third dimension [...] Thus the intimate knowledge of the third dimension was the key which unlocked the mystery of the minute.[50]

While Hinton's mathematical work was of limited consequence in scholarly terms, and his fictional output of limited popularity or effectiveness, his project was always conceived as something para-academic and not strictly literary: an entire system of 'thought-mechanics'. The mechanical system that underwrote it was both highly influential and effective, a hardware and software coupling for the self-helping consciousness expansionist.

'SEEING AS A HIGHER CHILD'

The first instalment in Hinton's programme, 'the means of educating', was a system of cubes described in great detail in the second part of *A New Era*. Bruce Clarke admits: 'The readable portions of his work are essentially prolegomena for and inducements to the further, impenetrable system of hyperspace-instruction.' I want to try to penetrate this system, and to investigate what Clarke describes as 'an arduous playfulness'.[51]

In the first chapter of *A New Era of Thought* Hinton describes the processes by which he came to the philosophy he outlines in this text:

> And so in despair of being able to obtain any other kind of mental possession in the way of knowledge, I commenced to learn arrangements, and I took as the objects to

---

[48] Bruce Clarke, 'A Scientific Romance: Thermodynamics and the Fourth Dimension in Charles Howard Hinton's "The Persian King"', *Weber Studies*, 14 (1997), http://www.altx.com/ebr/w%28ebr%29/essays/clarke.html [accessed 24 February 2010] (para. 1 of 28).

[49] C.H. Hinton, *An Episode of Flatland: or How a Plane Folk Discovered the Third Dimension* (London: Swan Sonnenschein, 1907), p. 149.

[50] Hinton, *An Episode of Flatland*, p. 170.   [51] Clarke, *Energy Forms*, p. 185.

be arranged certain artificial objects of a simple shape. I built up a block of cubes and giving each a name, I learnt a mass of them.   (*ANE*, 12)

He had already rehearsed this process in 'Casting out the Self', the fifth of his *Romances*, where he detailed how the mass he first learnt covered a cubic foot, and he could describe objects in space by referring to the names of the cubes they occupied in his mass. For public consumption this mass was resized to comprise either twenty-seven or eighty-one one-inch cubes. In *Casting* Hinton described the necessity of unlearning up and down and left and right in relation to arrange-ments, as these 'self-elements', 'arising from the particular conditions under which I was placed' (*ANE*, 209), did not give *absolute* knowledge of arrangement. He worked on relearning the cube turned on each of its sides and upside down, only realizing later in his studies the significance of the system of arrangements for understanding higher space: as 'a kind of solid paper' through which to imagine the sides and edges of four-dimensional shapes. Hinton invoked the speculated higher beings of *What is the Fourth Dimension?* to imagine their children: 'and just as children on the earth gain their familiarity with space by means of bricks and blocks and toys, so these higher children must have their own simple objects wherewith they grow into familiarity with their complex world' (*ANE*, 224). His cubes were just such a set of toys, and he went on to describe the process of casting out the self as 'seeing as a higher child' (*ANE*, 227).

On the Education of the Imagination, issued as a pamphlet in 1888 with a brief endnote by its editor, Herman John Falk, also deals with the cubes. Its endnote states that it was written 'some years ago' and 'contains the germ of the work, which is more fully illustrated in his more recent writings, and thus in some respects forms a good introduction to them'.[52] A pedagogical essay, addressed to a fellow educator and referring throughout to a putative pupil, it established a broad theoretical basis.

Hinton wrote that the piece was inspired by a series of extracts from Johannes Kepler's *Cosmographicum Mysterium*. In this text, subtitled 'on the marvellous proportion of the celestial spheres, and on the true and particular causes of the number, size and periodic motions of the heavens, established by means of the five regular geometric solids', Kepler argued that the known planets followed courses through the sky that corresponded in ratio to a nested agglomeration of the Platonic solids: a sphere containing a cube containing a sphere containing a tetra-hedron, and so on. He considered the cube 'the first solid in its class' for nine geo-metric reasons, for example:

1. It alone is generated by its base [...] 2. It alone can be resolved into homogeneous cubes with no prism [...] 3. It alone faces in all directions, and extends in three direc-tions at right angles.

Kepler's final reason for the lofty position of the cube in the heavenly pantheon elevated the human subject and an embodied geometry: 'For a man himself is

[52] C.H. Hinton, 'On the Education of the Imagination', in *Scientific Romances: Second Series* (London: Swan Sonnenschein, 1896), pp. 3–22 (pp. 21–2).

like a cube, in which there are so to speak six regions: upper, lower, fore, hind, right, left.'[53]

Hinton was particularly interested in Kepler's remarks on ratio and the harmony of physical form, the premises for his attempts to arrange the solar system in a harmonic, Platonic fashion. Hinton's interest in Kepler is continuous with his debt to the *Timaeus*, the source text for the Neoplatonic understanding of the natural universe, and its equation of each of the Platonic solids with each of the elements: air, water, fire, earth. Working with the cubes, then, was working with the Platonic basis for earth, a metaphorical grounding when Hinton found himself at an intellectual impasse.

Hinton focused on the cube as an exemplary object for exploring arrangement and form. In such use of the imagination with 'the utmost precision of form' he explicitly connected the generative, eidetic, spatial nature of both mathematical and literary form in the imagination, writing: 'Each line of Dante, for instance, seems to call up a visible image and shape.' As he probed the relationship between the sensations and the mind later fleshed out in *A New Era*, Hinton also cited Goethe: 'Goethe tells us in his *Farbenlehre*, that, when he was studying plants, on shutting his eyes images of flowers would present themselves to him, perfectly distinct in every particular, and would arrange themselves in rosettes or other regular figures.'[54]

He hoped to achieve similar results with his practical course of education: 'The first step, then, in the cultivation of the imagination, is to give a child 27 cubes, and make him name each of them according to its place, as he puts them up.'[55] The author warned against constricting rules, and encouraged exercises and games based on newly acquired spatial skill:

> If, for instance, he is told to put a chair in (1), another in (2), and himself in (11), he is highly amused at having to seat himself in the second chair; and if then he is told to put his hat in (20) he will, after a little consideration, put it on his head.

Hinton remarked that he had also developed a form of cubical chess, although he confessed that none of his pupils were able to play it. He referred to the experimental nature of the work he had undertaken with his pupils, and suggested that he had further research in mind:

> Owing to the co-operation of several of my pupils, who devoted a good deal of their spare time to testing different suggestions, I have been able to work out the application of this method in several directions; and, when certain experiments on colour and

---

[53] Johannes Kepler, *Mysterium Cosmographicum*, trans. A.M. Duncan (Norwalk, CT: Abaris Books, 1999 [1981]), p. 109. Had Hinton read further in Kepler he would have encountered yet more suggestive ideas concerning the Platonic solids. In *Harmonices Mundi* Kepler discussed their sexes and couplings. The cube was, inevitably, male: 'The cube is the outermost and most spacious, because it is the firstborn and, in the very form of its generation, embodies the principle of all the others.' Johannes Kepler, *The Harmony of the World*, trans. E.J. Aiton, A.M. Duncan, and J.V. Field (Philadelphia: The American Philosophical Society, 1997), p. 396.

[54] Hinton, 'On the Education of the Imagination', p. 8.

[55] Hinton, 'On the Education of the Imagination', pp. 12–13.

sound are finished, I hope to give a detailed account of the various ways in which the method may be found serviceable.[56]

What 'On the Education' makes clear is the genesis of Hinton's system of cubes in his teaching. It is devised with, and for, children, and playful elements are stressed. In their preface to *A New Era* Falk and Boole suggest using 'Kindergarten cubes' to follow the exercises. They see active engagement with three-dimensional objects as crucial, noting the limits of the two-dimensional page in dealing with higher space, and advocate a form of what we might now think of as kinaesthetic learning:

> Indeed, we consider that printing, as a method of spreading space knowledge, is but a 'pis aller', and we would go back to that ancient and more fruitful method of the Greek geometers, and, while describing figures on the sand, or piling up pebbles in series, would communicate to others that spirit of learning and generalization begotten in our consciousness by continuous contact with facts, and only by continuous contact with facts vitally maintained.   (*ANE*, vi)

The use of the term 'kindergarten' makes clear another source for Hinton's cubic system. The pedagogical theorization of the imagination in terms of Platonic solids is indebted to Friedrich Fröbel. The German educational theorist and crystallographer devised sets of children's toys, called Gifts, in which cubes were one of the primary constituent parts used to encourage the exploration of form. When Falk and Boole refer in the preface to *A New Era of Thought* to kindergarten cubes, we should recuperate the etymological context: Fröbel's term *Kindergarten* had been in use in English for barely thirty years and was still very much associated with its author.

The Society of Arts Educational Exhibition Collection of Prospectuses of 1854, a collection of catalogues for educational aids, from microscopes to chemicals, had advertised the Gifts for sale in England.[57] Joseph Payne's *Fröbel and the Kindergarten System of Elementary Education*, 'a lecture delivered at the college of Preceptors on the 25th of February, 1874' and published later that year, described the objects:

> The fourth, fifth and sixth gifts consist of the cube variously divided into solid parallelopipeds, or brick-shaped forms, and into smaller cubes and prisms. Observation is called on with increasing strictness, relativity appreciated, and the opportunity afforded for endless manifestations of constructiveness. And all the while impressions are forming in the mind which, in due time, will bear geometrical fruits, and fruits, too, of aesthetic culture. The dawning sense of the beautiful, as well as of the true, is beginning to gain consistency and power.[58]

Fröbel's Gifts were marketed and sold in Britain by, among others, Hinton's publisher Swan Sonnenschein, who also published assorted translations and

---

[56] Hinton, 'On the Education of the Imagination', p. 17.

[57] *A Collection of Prospectuses of the Educational Exhibition of 1854*, 3 vols (London: Royal Society of Arts, 1854).

[58] Joseph Payne, *Fröbel and the Kindergarten System of Elementary Education* (London: Henry S. King and Co., 1874), p. 18.

commentaries on Fröbel's educational work.[59] When Hinton began using cubes in the classroom for thinking, this was a Fröbelian move; when he went into manufacturing his own cubes, there was precedent for the publisher for selling such things.

A New Era detailed the construction of a full set of cubes, including slabs and catalogue cubes, and provided a number of exercises to work with them. These had been assembled by Boole and conspicuously lack the playfulness suggested by Hinton throughout 'On the Education'. In the hands of the adept student, the ludic roots of the system were occluded. Correspondence between Falk and Swan Sonnenschein over the models, which the editors had advised could be purchased through the publisher, shows the difficulties encountered in the realization of Hinton's vision.

On 21 September 1888, some months after publication, William Swan Sonnenschein received an inquiry about the cubes. Sonnenschein wrote: 'It would perhaps be as well, should this gentleman give an order for a set, to have two sets made, as it looks rather bad to have to admit that inquiries for them are unusual.'[60] Another inquiry was received in January of 1889, but it wasn't until February that Falk provided the first sets to the publisher, who returned them, writing: 'The workmanship of the cubes is so rough it would affect sales very badly.'[61] It took Falk until November to source improved sets, with the price set at 17/6 for trade plus 20 per cent for public sales. The models sold very slowly but continued to pique interest. In 1903, Swan Sonnenschein wrote to Hinton, now resident in the USA and once again managing his own affairs: 'Can you send me one set of your models which a lady resident in Nice is very anxious to purchase?'[62] In 1904 a Mr Dyson returned his set.[63]

The cubes are pedagogical objects, kindergarten toys, playthings but in the Hintonian imagination were pre-coded with Platonic atomism. They are very much building blocks; symbolic building blocks of matter, retooled as the building blocks of thought on Neoplatonist basis; an assumption that mathematical ratio and harmony are the fundamentals of the universe. The fact that Hinton's system places a crucial emphasis on memorization is highly significant and also recalls Renaissance Neoplatonism. The art of memory had long recognized the connection of memory and space. In her gloss of the classical text on memory technique, *Ad Herennium*, Frances Yates records:

> The artificial memory is established from places and images [...] A *locus* is a place easily grasped by the memory, such as a house, an intercolumnar space, a corner, an arch, or the like. Images are forms, marks or simulacra [*formae, notae, simulacra*] of what we wish to remember. For instance, if we wish to recall the genus of a horse, or a lion, or an eagle, we must place their images on definite *loci*.[64]

[59] See Baroness Marenholtz-Bülow, *Child and Child-Nature: Contributions to the Understanding of Fröbel's Educational Theories*, trans. Alice M. Christie (London: Swan Sonnenschein, 1879).

[60] MS Swan Sonnenschein and Co. 3282, 21 September 1888.

[61] MS Swan Sonnenschein and Co. 3282, 12 February 1889.

[62] MS Swan Sonnenschein and Co. 3282, 23 January 1903.

[63] MS Swan Sonnenschein and Co. 3282, 11 January 1904.

[64] Frances A. Yates, *The Art of Memory* (London: Peregrine Books, 1969), p. 22.

The most typical form of arrangement employed in the classical art of memory was therefore architectural: a sequence of memory rooms.

In the renaissance, Giordano Bruno travelled in Europe revealing the occult secrets of the Hermetic art of memory, collected in his *De umbris idearum*, prefaced by a passage in which Hermes Trismegistus himself handed a book to Bruno. Bruno's 'shadows of ideas' were in the Platonic tradition: 'shadows of reality which are nearer to reality than physical shadows in the real world'. Yates describes Bruno's theory:

> By imprinting on memory the images of superior agents, we shall know the things below from above; the lower things will arrange themselves in memory once we have arranged there the images of the higher things, which contain the reality of the lower things in a higher form.[65]

In Yates's account, for Bruno memory was a practical means to achieving transcendent knowledge: 'When the contents of memory are unified there will begin to appear within the psyche (so this Hermetic memory artist believes) the vision of the One beyond the multiplicity of appearance.'[66] In his final work on memory, Bruno developed an architectural system of memory, based on a magical geometry. Twenty-seven atria and nine fields were each divided into nine places; thirty cubicles formed a parallel spatial organization.

The Englishman Robert Fludd propounded a similar system of fields, atria, and cubicles, distinguishing between an *ars rotunda* and an *ars quadrata*. The round art located magical, immaterial images: the square, corporeal things and objects. Fludd advocated using real places in which to arrange memory *loci*, and favoured theatres upon which to construct his mnemonic stage; in the drawings of these we see a quadratic arrangement.

The insistent relationship between space and memory had been known since classical times: memory is spatial and space mnemonic. Hinton's arrangement of cubes to be memorized operated within the tradition of these Hermetic systems. It is difficult not to be struck by the parallel hopes of Bruno and Hinton for the results of their mnemonic projects: access to a higher knowledge.

How should these cubes be thought? They were material objects, childish, playful things, repurposed. Abacus or set square? From a contemporary perspective it is tempting to think of them as hardware, and the exercises their author detailed as software, for thinking space. They were a palpable failure as commodities, yet they were simultaneous with, or preceded, the models of four-dimensional projections and cross-sections that now reside in display cases as examples of mathematical models from the halcyon days of their production.[67]

The system of cubes is certainly illustrative of Hinton's *modus operandi* and his relative strengths as a thinker: he took a tactile, hands-on approach to abstraction; he materialized space to enable its thought. His descriptions in chapter three of *What is the Fourth Dimension?*, of thread passing through a thin sheet of wax,

---

[65] Yates, *The Art of Memory*, p. 213.   [66] Yates, *The Art of Memory*, p. 250.
[67] See Gerd Fischer, *Mathematical Models* (Braunschweig, Wiesbaden: Friedr. Vieweg & Sohn, 1986).

seem likely to be derived from sewing cards, such as his mother-in-law Mary Everest Boole had repurposed for geometric education with curve-stitching.[68] Hinton took everyday objects out of their everyday use to educate and in this way was a popularizer of no small skill. His insistence upon the effectiveness of his system was resolute:

> And after a number of years of experiment which were entirely nugatory, I can now lay it down as a verifiable fact, that by taking the proper steps we can feel four-dimensional existence, that the human being somehow, and in some way, is not simply a three-dimensional being. (*ANE*, 46)

TESSERACTS

Hinton's cubes leave a cultural spoor of great interest that allows us to understand how they acted, what they did, and the impact they had. His first dedicated pupils—apart, perhaps, from A Square's nephew—were his sisters-in-law. H.S.M. Coxeter, who was introduced to Mary Ellen's younger sister, Alicia, by her nephew G.I. Taylor, and who also worked with the amateur mathematician towards the end of her life, relates the story:

> He brought a lot of little wooden cubes and piled them up into shapes in his attempt to elucidate the four-dimensional hypercube, or tesseract. He set the three youngest girls the task of memorizing the arbitrary list of Latin words (Decus, Pulvis, etc.) by which he had named the little cubes. Lucy, being a child with a strong sense of duty, worked hard. Ethel found the whole project a meaningless bore and dropped out as soon as she was allowed to do so. But for Alice, age seventeen or eighteen, it was an inspiration, the mainspring of all her research.[69]

Alicia's age, in this account, dates these exercises to the late 1870s. The census of 1882 recorded her as staying with the family of her sister and brother-in-law in Uppingham. In 1887 she co-edited *A New Era* with Herman John Falk before moving to Liverpool in 1889 to work as his secretary. She met and married the actuary Walter Stott in 1890 and became a housewife and mother, but continued to work on the visualization of higher geometry, coining the term 'polytope' to describe a convex solid in four dimensions. She constructed three-dimensional models of their cross-sections and in the late 1890s, through the agency of her husband, photographs of these models were sent to Dutch geometer Pietr Schoute, who recognized similarities with his own work. He came to visit her in England and a decade-long collaboration saw papers published by Boole Stott in 1900, 1908, and 1910.

---

[68] See Shelley Innes, 'Mary Boole and Curve Stitching: A Look into Heaven', *Endeavour*, 28 (2004), 36–8.

[69] H.S.M. Coxeter, 'Alicia Boole Stott', in *Women of Mathematics: A Bibliographic Sourcebook*, ed. Louise S. Grinstein and Paul J. Campbell (Westport, CT and London: Greenwood Press, 1987), pp. 220–4 (p. 221).

A set of her models and many of her diagrams reside at the University of Groningen, which awarded her an honorary doctorate in 1914 in honour of her work with Schoute. She returned to geometry late in life, collaborating with Coxeter, then at Cambridge University, from 1930: a further set of her models are held there. The diagrams accompanying Alicia's work make evident the genealogy from her brother-in-law's system, and the colouring of her models recalls the modified system of *The Fourth Dimension*. Her methods, developed outside formal education, produced original research. They reveal a continuation of Hinton's work, expanding his focus on the cube to the other Platonic solids.

Alicia's research was conducted in isolation, but other researchers in Europe and America were also producing models in the 1880s. W.I. Stringham, working at Johns Hopkins under J.J. Sylvester and alongside Simon Newcomb, rediscovered the six convex polytopes originally discovered by Ludwig Schläfli in 1858. In 1880 Stringham published a paper describing his research featuring images of what appear to be paper models of projections of these solids.[70] Viktor Schlegel built demonstration models of the projections of four-dimensional solids in 1883 which were exhibited at the 1884 meeting of the Society of German Naturalists. These were sold commercially through mathematical catalogues. Sets, in various states of repair, exist at the Smithsonian and the University of Göttingen.[71] In 1900, Basil Wedmore, then a demonstrator at the Finsbury Technical College working under Silvanus P. Thompson, demonstrated models at a Friday evening discourse on 'Transparency and Opacity' at the Royal Institution. 'Mr. Wedmore's cute idea is this,' reported the *Leeds Mercury*:

> He says we can represent, say, a cube on a piece of paper by foreshortened squares, that is a three dimensional figure by two dimensional ones. Correspondingly we ought to be able to represent by foreshortened solid figures inside a three dimensional figure the appearance of one of fourth value.[72]

The visual genealogy of two-dimensional illustrations of fourth-dimensional figures has been studied in depth by Linda Dalrymple Henderson. The plates with which she accompanies her text make her point eloquently: it is easy to see the visual echoes of Jouffret's diagrams of the tesseract in work by Picasso.[73] What of three-dimensional representations of the fourth dimension? Without the need to do so much work, crossing a uni-dimensional representation gap rather than a bi-dimensional, models were a more readily accessible tool: they had not only a visual, but also a tactile efficacy. They were, however, far scarcer than print illustrations and their display was apparently restricted to universities or scientific societies. Where specialist mathematical texts might circulate outside specialist mathematical groupings, models did not.

---

[70] W.I. Stringham, 'Regular Figures in n-Dimensional Space', *American Journal of Mathematics*, 3 (1880), 1–14. These images appear on the cover of this book.
[71] See Fischer, *Mathematical Models*.
[72] 'Scientific and Literary Societies', *Leeds Mercury*, 1 April 1899, 9.
[73] Henderson, *The Fourth Dimension and Non-Euclidean Geometry*, pp. 160–1.

Hinton's mediation of four-dimensional space through material objects, developed in the classroom, enabled and legitimized the work of writers who insisted on the empirical reality of higher space, and provided a practical course for consciousness expansion: a practical course that produced intellectual results. His focus on the eidetic imagination fed into the concerns with visualization of esoteric belief systems. The hybridization of his practical course of four-dimensional instruction with his father's altruistic ethical philosophy nourished these belief systems so profoundly that it is this hybrid whose influence can be discerned most keenly in New Age texts throughout the twentieth century. As Henderson has demonstrated, many of Hinton's followers came to his work through Theosophy. The engagements of key Theosophists with Hinton's work form the core of Chapter 5 but a handful merit mention in the context of this consideration of Hinton's cubes.

First published in the *Occult Review* in 1914, Algernon Blackwood's short story 'A Victim of Higher Space' makes mention of Hinton's system. Blackwood was a committed occultist—contributor to the Theosophical magazine *Lucifer*, investigator alongside Frank Podmore of haunted houses, member of the Golden Dawn (and probably the Esoteric Section of the Theosophical Society)—and drew heavily on his occult reading and experiences for his fiction, turning several cases detailed in Podmore's *Phantasms of the Living* into short stories.

Statically located in the flat of psychic detective John Silence and relating an interview with the improbably named Racine Mudge (shades of Browning's medium Sludge, perhaps, and smudges of corporeal insubstantiality?), it enmeshes a range of Theosophical interests into an account of the experience of higher space explicitly sourced through Hintonian practice—indeed the use of the Hintonian term 'higher space' itself is the first indicator of this source.

Mudge has sought out John Silence in the hope that he might be able to assist him in combating his slipping 'nolens volens' into the world of higher dimensions—not just the fourth dimension but the proliferating higher dimensions of this 'spiritual' and 'mythical state'. Indeed, Mudge is not at first visible in Silence's waiting room. Looking through a spy-hole into the room, Silence at first sees only a thickening line:

> Then suddenly, at the top of the line, and about on a level with the face of the clock, he saw a round luminous disc gazing steadily at him. It was a human eye, looking straight into his own, pressed there against the spy-hole [...] Then, like someone moving out of deep shadow into light, he saw the figure of a man come sliding sideways into view, a whitish face following the eye, and the perpendicular line he had first observed broadening out and developing into the complete figure of a human being.[74]

Telling Silence how he came to be in his current condition, to have contracted his 'disease', Mudge leads the doctor through a well-versed account of Hintonian higher space, referencing the progenitors of non-Euclideanism—'the audacious speculations of Bolyai, the amazing theories of Gauss [...] the breathless intuitions

---

[74] Algernon Blackwood, 'A Victim of Higher Space', in *The Complete John Silence Stories*, ed. S.T. Joshi (New York: Dover Publications, 1998), pp. 230–46 (p. 233).

of Beltrami and Lobatchewsky'—before arriving at the 'dreamer' whose work allowed him to access higher space and some objects with which we are by now very familiar:

> I procured the implements and the coloured blocks for practical experiment, and I followed the instructions carefully till I had arrived at a working conception of four-dimensional space. The tesseract, the figure whose boundaries are cubes, I knew by heart. That is to say, I knew it and saw it mentally, for my eye, of course, could never take in a new measurement, or my hands and feet handle it.[75]

Through use of the 'implements and the coloured blocks for practical experiment', Mudge has fulfilled Hinton's prediction and accessed the universal humanity the author theorized:

> I reached sometimes a point of view whence all the great puzzle of the world became plain to me, and I understood what they call in the Yoga books 'The Great Heresy of Separateness'; why all great teachers have urged the necessity of man loving his neighbour as himself; how men are all really one; and why the utter loss of self is necessary to salvation and the discovery of the true life of the soul.[76]

The knowledge of unity, the attainment of the Theosophical and Hintonian dream of higher space, however, turn sour for Mudge as his knowledge of the higher spatial condition shift gears: 'accidentally, as the result of my years of experiment, I one day slipped bodily into the next world, the world of four dimensions, yet without knowing precisely how I got there, or how I could get back again'.

The idea that once willed access to higher space had been achieved, unwilled or automatic regress would follow was not confined to fiction. In his column on mathematical puzzles in the *Scientific American* for July 1962, Martin Gardner gave a broad overview of four-dimensional geometry, establishing the dimensional analogy and working through the cross-sections and projections of the tesseract. When this article was reprinted as a chapter in his 1965 book *Mathematical Carnival*, Gardner included a sensational letter from Hiram Barton, a 'consulting engineer of Etchingham, Surrey':

> A shudder ran down my spine when I read your reference to Hinton's cubes. I nearly got hooked on them myself in the nineteen-twenties. Please believe me when I say that they are completely mind-destroying. The only person I ever met who had worked with them seriously was Francis Sedlak, a Czech neo-Hegelian Philosopher (he wrote a book called *The Creation of Heaven and Earth*) who lived in an Oneida-like community near Stroud, in Gloucestershire.

> As you must know, the technique consists essentially in the sequential visualizing of the adjoint internal faces of the poly-colored unit cubes making up the larger cube. It is not difficult to acquire considerable facility in this, but the process is one of autohypnosis and, after a while, the sequences begin to parade themselves through one's mind of their own accord. This is pleasurable, in a way, and it was not until I went to see Sedlak in 1929 that I realized the dangers of setting up an autonomous process in one's own

---

[75] Blackwood, 'A Victim of Higher Space', p. 238.
[76] Blackwood, 'A Victim of Higher Space', p. 239.

brain. For the record, the way out is to establish consciously a countersystem differing from the first in that the core cube shows different colored faces, but withdrawal is slow and I wouldn't recommend anyone to play around with the cubes at all.[77]

Strip away the sensational rhetoric—Barton's use of the words 'hooked' and 'withdrawal' and the idea that this form of thought could be 'mind-destroying' clearly reference the risks of drug use and perhaps reflect the date of the composition of the letter—and some very curious concepts remain. Barton describes the process of using the cubes as one of 'autohypnosis'. Facility in the process creates an 'autonomous process' in the brain. In Barton's description it is not so much the subject who is doing the thinking as the object of thought. Despite the implicit analogies to psychedelic or narcotic drugs, the cubes are not ingested: rather, they are thought, contemplated, visualized, and memorized. They enter the mind only through the senses and yet in this account they seem to become in some way structural, seem to become part of the organ of thought itself.

The Francis Sedlak to whom Barton refers was, as well as philosopher and communal liver, a Theosophist, contributing frequent articles to the *Theosophical Review* from 1906 to 1908 and to *The Theosophist* in 1911–12. He contributed an article to Orage's *The New Age* disputing Einstein's Theory of Relativity on the grounds that Einstein was insensible to the dictates of 'Pure Reason'. His partner in a 'free union', Nellie Shaw, wrote about their life together in the Whiteway Colony, including an account of Sedlak's interest in the cubes that beds into the utopian Theosophical version of higher spatial thinking and balances out the sensational tone of Barton's letter. I hope that readers will forgive me for quoting from this long-forgotten text at some length: it is unique as an objective, narrative account of the practice of using Hinton's cubes:

Some readers may be acquainted with a book by C. Howard Hinton, entitled *The Fourth Dimension*, which contains a coloured diagram representing twenty-seven cubes of various colours. This idea was seized upon by Francis, who adapted it to his own ideas.

A box of children's playing blocks was obtained and each one painted a different and nameable shade. So far as I am able to understand, the idea was to build up from the whole twenty-seven cubes one cube, each separate colour being in a particular relation to the next one, and then to gaze fixedly at it until the whole was mentally visualised. This accomplished, the cube was unbuilt and then rebuilt with a different combination of colour, and visualised mentally as before.

This amazing performance required hours of time at first, but gradually the speed quickened, until eventually it became focused upon the mind, and Francis was able to review the blocks in all their twenty-seven positions so swiftly, that it became almost like seeing the cube from all sides at once.

It will be realised that the changes of position were almost innumerable. At first a very hard laborious task, it became an absorbing occupation, to which was given every spare moment. Many persons, not understanding, looked on it as a most unproductive way of spending time. Others admired the wonderful patience, but could see no useful result.

[77] Martin Gardner, *Mathematical Carnival* (Washington, DC: The Mathematical Association of America, 1989 [1965]), pp. 52–3.

Just as the would-be athlete twists and turns on the parallel bars, using time and energy to develop his muscles and gain strength which can be used later in any direction which he may desire, so Francis assumed that this power gained by practice in visualization, seeing mentally the block of cubes on all sides simultaneously, could also be used in any sphere and on any subject; in fact, it was ability to see through anything, and must eventually lead to clairvoyance.

This study of the cubes was followed intermittently, since it was not a mental exercise calling for philosophic reasoning or mental effort whatever. So, after devoting many months to the cubes and having an urge in another direction, Francis would drop them again for several years.

The extraordinary thing was that afterwards he could resume the practice without difficulty. He did not lose the power; indeed, he seemed to have a positive affection for these bits of wood, which he would tenderly dust and preserve.

Towards the end of his long and trying illness, when terrible coughing prevented him from sleeping at night, the long silent hours seemed interminable. On my enquiring one morning as to what sort of a night he had had, he said almost joyfully, 'Oh, being awake does not trouble me now. I do the cubes, and the time flies.' So I thanked God and blessed the cubes, for which had been found a utilitarian use at a most desperate psychological juncture. Power won cannot be lost, and will some day be utilised.[78]

In this more detailed account we read of the repetitive and arduous nature of the practice of the cubes, its reliance upon memory, and how facility improved with practice. Sedlak is the most dedicated student of the cubes since Alicia, and his practice, while it does not ultimately lead to clairvoyant ability, gives him great comfort and pleasure as a meditative device, a process of unselfing.

The process of 'casting out the self', a form of subjective decentring, is morphological in the Goethean sense. Beyond the quotation taken directly from *Farbenlehre*, other sections of 'On the Education' are derived from Goethe's *Theory of Colour*: Hinton cites the experiment of looking through a prism to examine rays of light, before coming to conclusions about the purpose of analogy and the generation of original thought that echo Goethe. Hinton writes:

Undoubtedly, every fresh structure must grow out of some previously existent one; and every idea must spring from others already in existence. This is felt by the consciousness as analogy. Thus, we see that imagination, which consists in calling up images and in superposing them, as it were, is a necessary factor in the process of thought; for, without this superposition or juxtaposition, it would be impossible to form analogies.[79]

Goethe's account of the imagination places similar weight on repetitive procedures and the catalytic power of analogy:

Imagination is first re-creative, repeating only the objects. Furthermore, it is productive by animating, developing, extending, transforming the objects. In addition, we can postulate a perceptive imagination which apprehends identities and similarities

[78] Nellie Shaw, *A Czech Philosopher on the Cotswolds; Being an Account of the Life and Work of Francis Sedlak* (London: C.W. Daniel Co., 1940), pp. 107–9.
[79] Hinton, 'On the Education of the Imagination', pp. 6–7.

[...] Here it becomes evident how desirable analogy is which carries the mind to many related points, so that its activity can unite again the homogenous and the homologous.[80]

Henry Bortoft's detailed account of Goethean 'seeing' describes elements that are yet closer to the practice of using the cubes as we have read described, particularly as related to the process of removing the 'self-elements' of up and down and left and right. Goethean seeing is also a process of imaginary inversion:

> Observing the phenomenon in Goethe's way requires us to look, as if the direction of seeing were reversed, going from ourselves towards the phenomenon instead of vice-versa. This is done by putting attention into seeing, so that we really do see what we are doing instead of just having a visual impression [...] He would then repeat the observations he had made, but this time doing so entirely in his imagination without using the apparatus. He called this discipline *Exakte sinnliche Phantasie*, which can be translated 'exact sensorial imagination'. In this case it would mean trying to visualize making the observations with the prism, and seeing the qualities of the different colours in the right order at a boundary as if we were producing them. This would then be transformed in imagination into an image of the colours with the boundary in the opposite orientation, and then transformed back again. This process can be repeated several times.[81]

When we read Goethe's statement that 'every new object, clearly seen, opens up a new organ of perception in us' we can better understand Hinton's account of neurological function in *A New Era of Thought*, a knottedness between thought and matter that makes for material change in the organ of thought.

We find profound correlation between thought and matter, such that the act of thought can bring the matter of the thinking thing into being, an extraordinary sensitivity of co-creation. In light of our consideration of Zöllner's knots as quasi-objects, the cubes seem to be quasi-objects too: they bring into play a collective of higher-dimensional thinkers, clustered around and contemplating the cubes, passing them on and recommending them to others. The act of engaging with the cubes occasions a mode of thought inaccessible to those without them: they have enabled the development of a body of theory—Hinton's higher spatial philosophy—that could not exist without them, and in this theory thinking the cubes alters the material structures of the brain. They make blurry the distinction between subject and object, thing and thought.

Michel Serres has repeatedly considered the mythical origins of geometry in the story of Thales of Miletus who, standing in the shadow of the pyramids, realizes that he can calculate their height by means of a ratio, knowing his own height and that of his own shadow. Reading this classical fable repeatedly, attending particularly

---

[80] Gerhard M. Vasco, *Diderot and Goethe: A Study in Science and Humanism* (Geneva: Librairie Slatkine, 1978), p. 88 (trans. of Goethe, *Werke* (Weimar: Hrsg. Im Auftrage der Grossherzogin Sophie von Sachsen, 1877–1919), sec. 4, XXXIV, pp. 136–7).

[81] Henry Bortoft, *Goethe's Scientific Consciousness* (London: Institute for Cultural Research, 1986), p. 14.

to the formal organization of its central image, Serres wonders where this moment of inspiration comes from.

In the essay 'Gnomon: The Beginnings of Geometry in Greece', Serres focuses his attention on the pyramid, the thing that casts a shadow: 'We do not really know why the shaft or pin is called a gnomon, but we do know that this word designates that which understands, decides, judges, interprets or distinguishes, the rule which makes knowledge possible.'[82] The gnomon itself is a machine, argues Serres, producing automatic knowledge. It defines no position for an operator, inscribing its knowledge directly onto the sand. What does this mean for the human subject? Serres explains:

> The world represents itself, is reflected in the face of the sundial and we take part in this event no more and no less than the post, for standing upright, we also cast shadows, or, as seated scribes, stylus in hand, we too leave lines. Modernity begins when this real world space is taken as a scene and this scene, controlled by the director, turns inside out—like the finger of a glove or a simple optical diagram—and plunges into the utopia of a knowing, inner, intimate subject.[83]

The human subject becomes equivalent to the gnomon and its sublimation of the gnomon marks the emergence of the modern subject.

Serres goes on to trace the development of the gnomon. He describes the knowledge that it produces as 'algorithmic': series of numbers that can be plotted onto tables. This mode of knowledge production, he argues, has two components:

> One which could be called mechanical and the other which could be called mnemonic. These can be described as the accumulation or recapitulation of the results of mechanical procedures or conditions of their repetition; the automaton and tables or dictionaries; hardware and software.[84]

This gives us a useful working description of what the cubes are: both gnomonic and mnemonic. The cubes cast their shadows in the mind. They produce knowledge without us, within us, and through us. They are thinking things, things thought upon and things that think. And when we make fluid this relationship between subject and thinking thing, in Hintonian and Goethean tradition, we better understand Hinton's conjecture that the brain *must* exist in four dimensions because we can think in four dimensions. We are encountering an example of 'a delicate empiricism which makes itself utterly identical with the object, thereby becoming true theory'.

Serres goes on to distinguish between the gnomonic thinking of the pre-Socratics and the geometry of the post-Socratic thinkers. He considers the *Meno* and notes the distinction between the slave who can recall learned multiplication tables, a form of algorithmic thinking, and Socrates's demonstrative thought. He routes through the apagogic disproof as the first demonstration: 'To invent geometry and demonstration consists of filling the gaps of the gnomon, those of

---

[82] Michel Serres, 'Gnomon: The Beginnings of Geometry in Greece', in *A History of Scientific Thought*, ed. Michel Serres (Oxford: Blackwell, 1995), pp. 73–123 (p. 79).
[83] Serres, 'Gnomon', p. 80.   [84] Serres, 'Gnomon', p. 84.

knowledge, of artificial intelligence, of algorithmic thought.'[85] The cube users, collected around and with the cubes, a community of quasi-subjects and quasi-objects, begin to fill out the gaps from the gnomon Hinton discerned and in so doing intuit not just an alternate geometry, but an alternate spatial and social imaginary. Hinton's discovery might be the recognition of a knowledge hidden in three-dimensional shadows.

BEYOND THE DIMENSIONAL ROMANCE

Hinton's contribution to the idea of higher space is a catalysing hybridization. Working on a space that oscillated between empirical and ideal, he mediated this idea through the material. Bringing ideas from ethical philosophy into speculative mathematical and physical treatises, he produced work that provided a treasure trove of ideas for artists, fiction writers, satirists, and esoteric theorists. His work was 'popular' in its insertion into print contexts that aimed to democratize scientific knowledge, the potency of its hybridity for appropriation by and redeployment within popular social/religious movements, and its provision of a practical course of application.

The work of Hinton indicated a utopian possibility for higher space, a possibility while not endorsed by Abbott's ambivalent satire, nevertheless gestured towards by dint of its socially and intellectually progressive mores. Bruce Clarke sees Hinton's 'desire not just to imagine but to inhabit the fourth dimension' as a quest to escape from the entropic doom of thermodynamics, but reads the scopic regime of Hinton's hyperspace as paranoid and suggests the manipulation of cubes as a defence against exterior control.[86] I tend towards the former observation: Hinton's higher space can usefully be reconfigured as a fugitive space; a space escaping but also of escape. And perhaps it is here that its roots in the kindergarten are best understood: Hinton's higher space is accessed through disciplined playfulness and inflected with liberation from adult concerns and morality. Its utopia is as much Neverland as Spaceland. Yet space, no matter how abstract, is inherently socio-political, as *Flatland* makes clear. Hinton's escape into a space of thought operates as a bulwark, and cartographers and colonists follow. At the fin de siècle these concerns and an analogy between higher space and colonial space become clearer. A reading of the fictions of higher space demonstrating such features follows in Chapter 6.

The dual sense of both play and learning makes Hintonian theory an attractive source for an emergent generic form that has frequently been characterized as childish and somehow insufficiently serious. For the science fictions of H.G. Wells, who relentlessly portrayed his speculations as grounded in scientific materialism, the fourth dimension as theorized in the work of Hinton provided a switch through which to mediate the transcendental: a narrative sleight of hand with which to set loose the imagination into freer realms where individuals might slip between

[85] Serres, 'Gnomon', p. 108.   [86] Clarke, *Energy Forms*, p. 180.

worlds and return altered. Writers of 'weird' fictions following Wells recognized
and developed the non-human potential skulking in spaces extended beyond
thought and apparently accessible only through a form of willed automatism such
as Sedlak's. As Kantian space was threatened, so was the central position of the
human in the universe. Furthermore, as the subject was made permeable, so were
possibilities suggested for psychological disturbances that registered this shifting of
the ground beneath coherent, correlated subjectivity.

The years surrounding the publication of *A New Era of Thought* witnessed
responses to or quotation of the idea of the fourth dimension in an array of cultural
contexts and demonstrate the currency and popularity of the idea. On 14 January
1887 E.A. Hamilton Gordon gave a paper at the Science Schools Debating Society
in South Kensington on the fourth dimension, citing an article by R.A. Proctor
that first appeared in the *Gentleman's Magazine* in 1880. When his paper was
reprinted in the *Science Schools Journal* Hamilton Gordon also acknowledged
Hinton but claimed not to have read the *Romances* prior to his speculations.[87]
The following year a story by a student who had heard Hamilton Gordon's paper
was serialized in the same journal. 'The Chronic Argonauts', by H.G. Wells,
although only the first of several versions of what would become *The Time
Machine*, already contained the idea of a geometry of four dimensions, with time
occupying the *y*-axis.[88]

The poem 'A Pure Hypothesis' was published in May Kendall's 1887 collection
Dreams to Sell alongside comic verses that had appeared in *Punch*. Kendall extrapolated upwards where Abbott had extrapolated downwards: her poem imagines a
lover in four-dimensioned space for whom the idea of a lower space is confounding
and 'unutterably wrong':

> He told us: 'Science can conceive
> A race whose feeble comprehension
> Can't be persuaded to believe
> That there exists our Fourth Dimension,
> Whom Time and Space for ever baulk;
> But of these things being incomplete,
> Whether upon their heads they walk
> Or stand upon their feet—
>
> We cannot tell, we do not know,
> Imagination stops confounded;
> We can but say "It *may* be so,"
> To every theory propounded.'
> Too glad were we in this our scheme
> Of things, his notions to embrace,—
> But—I have dreamed an awful dream
> Of *Three-dimensioned* Space![89]

[87] E.A. Hamilton Gordon, 'The Fourth Dimension', *Science Schools Journal*, 5 (1887), 145–51.
[88] See Harry M. Geduld, *The Definitive Time Machine* (Indianapolis: Indiana University Press, 1987).
[89] May Kendall, *Dreams to Sell* (London: Longmans, Green and Co., 1887), p. 11.

In February 1887 Oscar Wilde's 'Canterville Ghost', serialized in *Court and Society Review*, playfully juxtaposed contemporary spiritualist obsessions with the older gothic tropes of the ghost story. Early in the story, the titular ghost is set upon by the children of the family who have moved into Canterville Chase: 'Hastily adopting the Fourth dimension of Space as a means of escape, he vanished through the wainscoting, and the house became quite quiet.'[90] The family rapidly acclimatize to the ghost and write a letter to the SPR.

In 1888, H.P. Blavatsky's *The Secret Doctrine* was published to a rapturous reception by the many members of the Theosophical Society, and the yet more numerous readers of esoteric journals. Over two pages Blavatsky addressed 'the fashion of speculating on the attributes of the two, three, and four or more "dimensional Space;"', paying particular attention to the theories of Zöllner. Inimical towards the very idea of space, she shifted emphasis onto matter, and argued that what was in fact under discussion was 'a sixth characteristic of matter', concluding with the prophecy 'that in the progress of time—as the faculties of humanity are multiplied—so will the characteristics of matter be multiplied also'.[91] Blavatsky's account was brief, but its material emphasis echoed and amplified Hintonian higher space and the prophetic character of her writing chimed with Hinton's tone in *A New Era of Thought*. The responses to higher space of her followers in the Theosophical Society and more or less loosely associated occultists and esoteric thinkers will be assessed in Chapter 5.

[90] Oscar Wilde, 'The Canterville Ghost', in *The Canterville Ghost, The Happy Prince and Other Stories* (London: Penguin, 2010), p. 197 (repr. from *Court and Society Review*, 23 February 1887).
[91] H.P. Blavatsky, *The Secret Doctrine*, 2 vols (London: The Theosophical Publishing Company, 1888), I, p. 252.

5

Through

The Theosophical Society, Authority, and Mediation

In *Isis Unveiled*, the foundational text of the Theosophical Society, Helena Petrova Blavatsky described space as 'boundless' with 'neither distant nor proximate places'. Not yet obscuring her sources, she quoted Dr Thomas Young's invitation to 'speculate with freedom on the possibility of independent worlds; some existing in different parts, others pervading each other, unseen and unknown, in the same space, and others again to which space may not be a necessary mode of existence'.[1]

Blavatsky's doctrine as presented in *Isis* was amalgamated from various esoteric and philosophical traditions—Rosicrucianism, Spiritualism, Neoplatonism—and Eastern sacred texts, primarily Hindu and Buddhist texts of Indian origin. Incorporated into this were scientific sources, from Young to Tait and Darwin, philosophers all in the Theosophical parlance. While the British academy had been involved in disputes over Kant, older philosophical views of space had flourished in the context of increased interest in mystical and esoteric ideas. Alex Owens writes: 'The term *mystical revival* was used across the ideological board to identify a range of spiritual alternatives to religious orthodoxy that sprang up in the 1880s and 1890s and gained momentum and prominence as the old century gave way to the new.'[2] This revival brought increasing numbers into contact with Pythagorean ideas, primarily through Renaissance Neoplatonism and its subsequent commentators.

The motion to found the Theosophical Society, for example, was carried among those present at a lecture given in the New York lodgings of Madame Blavatsky by George Henry Felt in 1878. His subject had been the 'geometrical figures of the Egyptian Cabballa', his discovery of which was 'regarded as among the most surprising feats of the human intellect'.[3] Other groups held great stock in the ideas of geometry. In 1875, posthumously published from the manuscripts of the prolific antiquarian and author on Freemasonry, Rev. George Oliver DD, *The Pythagorean Triangle* described the ritual significance to Freemasonry of the point, the line, the triangle, the square, the cube, the circle, and the numbers related to each:

> In the following pages, the doctrines and references which necessarily result from a minute consideration of the Science of Numbers, as enunciated in the Pythagorean

[1] H.P. Blavatsky, *Isis Unveiled*, 2 vols, 2nd edn (Point Loma: Theosophical Publishing Co., 1910), I, p. 185.

[2] Alex Owens, *The Place of Enchantment: British Occultism and the Culture of the Modern* (Chicago: University of Chicago Press, 2004), p. 20.

[3] Henry Steel Olcott, *Old Diary Leaves*, 6 vols (New York and London: G. Puttnam and Sons, 1895), I, p. 119.

Triangle, will be subjected to a scientific analysation; for it is a remarkable fact, that although the institution of Freemasonry is based upon it, we have no authorised lecture to illustrate its fundamental principles, or display its mysterious properties.[4]

For both Freemasonry and Felt, the nature philosophy of Plato's *Timaeus* was the ultimate source, as seen through the lens of Renaissance Neoplatonism. The Egyptian Pythagorean Timaeus's geometrically conditioned account of the origin of the universe equated the elements of earth, water, air, and fire with the Platonic Solids, the cube, the tetrahedron, the octahedron, and the isocahedron. The dodecahedron was universal: 'There still remained a fifth construction, which the god used for embroidering the constellations on the whole heaven.'[5] The earth was a sphere inside the sphere of the universe on which the stars were arranged. All was according to the perfect ratios of Pythagorean geometry, all was created by the demiurge, artisan, craftsman, or architect of the universe.

Reaching back to the *Timaeus* is instructive on several counts. The version of the text experienced by the late nineteenth-century reader depended greatly upon the translation. Benjamin Jowett's translation, published in 1871, considered the *Timaeus* among Plato's work as 'the most obscure and repulsive to the modern reader, and has nevertheless had the greatest influence over the ancient and mediaeval world'.[6] Indeed, until the thirteenth century, Cicero's translation of the *Timaeus* was the only Platonic text available in Latin.[7] In German intellectual life, the influence of the dialogue lasted well into the nineteenth century and it had important adherents, according to Paul Friedländer:

> The *Timaeus* was taken very seriously in the circles around the philosopher Schelling. In 1804, it was translated into German by the Mainz court physician Windischmann as 'A Genuine Document of True Physics.' Somewhat later (in 1826), Liechtenstaedt, a professor of medicine in Breslau, wrote a book on 'Plato's doctrines in natural science and medicine.' Goethe read both of these works and, in his *Geschichte der Farbenlehre*, he refers to Plato in words that we might well adopt again and again in what follows: 'It is still more welcome to see in Plato how we encounter every earlier type of doctrine in purified and elevated form.'[8]

Jowett was concerned to reconstruct the epistemological condition of the 'ancient physical philosopher', the confusions and shortcomings of his understanding of physics and cosmology, his ignorance of experiment or induction: this figure was 'dreaming of geometrical figures lost in a flux of sense'.[9]

    [4] Rev. George Oliver, *The Pythagorean Triangle, or the Science of Numbers* (London: John Hogg and Co., 1875), p. xviii.

    [5] Plato, 'Timaeus', in *Timaeus and Critias*, trans. Desmond Lee (Harmondsworth: Penguin, 1987), pp. 27–124 (p. 78).

    [6] Benjamin Jowett, 'Introduction' to 'Timaeus', in *The Dialogues of Plato*, ed. and trans. Benjamin Jowett, 5 vols (Oxford: The Clarendon Press, 1871), II, pp. 467–510 (p. 467).

    [7] See Stephen Gersh, 'The Medieval Legacy from Ancient Platonism', in *The Platonic Tradition in the Middle Ages*, ed. Stephen Gersh and Maarten J.F.M. Hoenen (Berlin and New York: Walter de Gruyter, 2002), pp. 3–30.

    [8] Paul Friedländer, *Plato: The Dialogues, Second and Third Periods*, trans. Hans Meyerhoff (London: Routledge and Kegan Paul, 1969), pp. 355–6.

    [9] *Dialogues of Plato*, II, p. 467.

Not only did the nineteenth-century reader need to reconstruct the ancient physical philosopher, but he also had to contend with the shadow of Neoplatonism:

> The influence which the *Timaeus* has exercised upon posterity is due partly to a misunderstanding. In the supposed depths of this dialogue the Neo-Platonists found hidden meanings and connections with the Jewish and Christian Scriptures, and out of them they elicited doctrines quite at variance with the spirit of Plato. Believing that he was inspired by the Holy Ghost, or had received his wisdom from Moses, they seemed to find in his writings the Christian Trinity, the Word, the Church, the creation of the world in a Jewish sense, as they really found the personality of God or of mind, and the immortality of the soul.[10]

Neoplatonism filtered ancient Egyptian and pagan Greek philosophy through a Christian theology. Jowett noted that so too did it attempt a further accommodation, between Plato and his pupil Aristotle. He stressed that the mysticism of Plato was 'the growth of an age in which philosophy was not wholly separated from poetry and mythology'.[11] He warned of the danger with modern translators of 'the tendency to regard the *Timaeus* as the centre of [Plato's] system'.[12]

One such translator was Thomas Taylor, 'the Platonist', who at the beginning of the nineteenth century published numerous translations of the tenth-century Alexandrian Neoplatonists. His *magnum opus*, a translation of Proclus's commentary on the *Timaeus*, informed his translation of Plato's source text that was prone to all the distortions highlighted by Jowett. Introducing a later reprint of Taylor's translation, R. Catesby Taliaferro repeated Jowett's observation: 'The Plato of Taylor is seen through the eyes […] of men who were attempting a synthesis of his pupil and opponent Aristotle with religious emanation theories from the East.'[13] While Taylor was not well regarded as a translator in Britain, however, as Ralph Waldo Emerson remarked to Wordsworth, 'in every American library his translations were found'.[14] Certainly, when American Theosophists such as William Olcott read the *Timaeus*, it was to Taylor's translation that they turned.

The syncretic doctrine of Theosophy laid claim to this body of Neoplatonic thought. Jowett's hope that 'we know that mysticism is not criticism' was largely irrelevant in the context of a movement whose interests did not lie in critical thought: the *Timaeus* contained Plato's version of the Atlantis myth, after all, a story more potent in a Theosophical context than warnings from any number of masters of Oxford colleges.[15] George Oliver's history of Pythagorean theory, meanwhile, made clear Freemasonry's conviviality with this body of thought, a conviviality barely coded beneath its inscription of the creator as an architect. An uncritical and extra-academic tradition of mystical geometry blossomed in the late 1870s and 1880s in which the Pythagorean ideas of the *Timaeus*, defining the universe in

---

[10] *Dialogues of Plato*, II, p. 468.   [11] *Dialogues of Plato*, II, p. 469.
[12] *Dialogues of Plato*, II, p. 469.
[13] R. Catesby Taliaferro, 'Foreword', in *The Timaeus and the Critias, or Atlanticus: The Thomas Taylor Translation* (New York: Pantheon Books, 1952), pp. 9–36 (p. 9).
[14] Quoted in Kathleen Raine and George Mills Harper (eds), *Thomas Taylor the Platonist: Selected Writings* (New York: Bollingen Foundation, 1969), p. 123.
[15] *Dialogues of Plato*, II, p. 468.

geometric terms, were the subject of discussion in parlours, at meetings, and at seances, and particularly in Blavatsky's Theosophical Society.

PRECIPITATION AND APPORT

Blavatsky's peripatetic reading reflected her life—born in Russia, Blavatasky had travelled in Russia, Mongolia, India, Ceylon, Turkey, Greece, Egypt, Syria, the Americas, Italy, France, Germany, Belgium, and England—and the Theosophical Society was from its origin an international organization, its transatlantic reach assured by the presence of three British signatories to the Society's foundation in New York. Shortly after completing *Isis*, Blavatsky and her partner Colonel Henry Steel Olcott left New York for India, via London, establishing bases as they went. The Theosophical mission soon expanded throughout the colonies and into Europe, through well-established occult networks, and into South-East Asia through active proselytizing.

If the global travels of the Society's founder suggested a physical reflection of one aspect of Theosophy's theory of space, this was underscored by 'phenomena', manifestations of the purportedly spiritual abilities of mediums, which frequently transgressed standard spatial constraints. The 'precipitation' or 'apport' of physical objects, their apparently spontaneous appearance in, or disappearance from, a location, was a feature of mediumistic practice that Madame Blavatsky continued to produce in a Theosophical context.

Communication produced by such 'precipitations' in late 1880 formed a core canon of Theosophical thought. While Blavatsky and Olcott were staying with A.P. Sinnett, the editor of *The Pioneer* newspaper, in Simla, India, letters purporting to come from the 'Mahatmas' Khoot Hoomi and Morya, spiritual adepts based in Tibet, an 'imperial terra incognita' in Roger Luckhurst's phrase, would precipitate in the Sinnetts' house.[16] The contents of these letters, which would flutter through gaps in the ceiling, appear on desks or, at HQ in Adyar, in a specially constructed shrine, formed the basis for Sinnett's *The Occult World* (1881) and *Esoteric Buddhism* (1883), the popularity of which contributed considerably to the growth of the Society in England.

'Precipitation' indicates the close imbrications of Theosophical space with matter. Matter could pass through this rarefied space, like a cloud or mist, to recongeal elsewhere, becoming rain of a typically communicative order. Precipitations were not even confined to the same landmass: Theosophical space allowed for the translocation of matter across continents. A glove given to Blavatsky by Charles Carleton Massey when she passed through London made the return journey from India unaccompanied. Massey's letter to Blavatsky recounting the experience from his end was leaked to the *Bombay Gazette*:

> A letter from the barrister, dated London 18th February, says, that on getting to his chambers in the Temple the day before—the 17th—he found a telegram from a certain lady of good education and the highest respectability, who is what is termed a powerful

[16] Roger Luckhurst, *The Invention of Telepathy* (Oxford: Oxford University Press, 2002), p. 144. See A.P. Sinnett, *The Occult World*, 3rd edn (London: Trübner and Co., 1883), pp. 54–9.

'medium', but above suspicion of trickery, to be at her house at 6 p.m., as her familiar 'spirit' has a message for him from Madame Blavatsky. He was punctual to the appointment, and was received by the lady and her husband, who presently ushered him into a darkened room. What happened we will let the barrister himself describe. 'Truth to say,' he remarks, 'I did not expect much, but prompt to the appointment the "spirit" came, loading the air with sweet perfume, and commencing the interview (which did not last five minutes) by flinging something light and soft in my face—a good shot in the dark. From this proceeded the perfume aforesaid. Directly I handled it I knew what was up, without being told. The glove! the glove! from *you*, from Bombay, when the papers had already informed us your ship arrived on Sunday, two days ago. What can I say—what think?'[17]

The published account of 'this new trans-atmospheric mail service' leaves little doubt that Theosophical space was a medium of communication that outstripped the telegraph networks and ships relied upon by the uninitiated. With a human medium at either end of the spatial medium, communications technology was bettered and things, objects, or possessions could be made to dissipate and precipitate over vast distances.

In her second *magnum opus*, Blavatsky refused the association of Theosophical space with the fourth dimension. 'In passing,' she wrote,

> it is worth while to point out the real significance of the sound but incomplete intuition that has prompted—among Spiritualists and Theosophists, and several great men of Science, for the matter of that—the use of the modern expression, 'the fourth dimension of Space.'[18]

Speculating on the subject was a 'fashion', 'assuming that Space itself is measurable in any direction' was a 'superficial absurdity', and in her final analysis the idea of a fourth dimension of space was dismissed because 'these terms, and the term "dimension" itself, all belong to one plane of thought, to one stage of evolution, to one characteristic of matter'. She echoed Hinton in arguing that the phrase 'the fourth dimension' should in fact be considered a contraction of the phrase 'the fourth dimension of matter in space', but that matter itself should be considered in relation to the human senses:

> Matter has extension, colour, motion (molecular motion), taste, and smell, corresponding to the existing senses of man, and by the time that it fully develops the next characteristic—let us call it for the moment Permeability—this will correspond to the next sense of man—let us call it 'Normal Clairvoyance'.

In the next stage of evolution 'the faculties of humanity [would be] multiplied'.[19] Permeability, and the ability to sense it—clairvoyance—was an evolutionary inevitability.

Blavatsky's doctrinal dismissal of the fourth dimension was an obstacle to development of the idea in a Theosophical context but there were Theosophists who

[17] C.C. Massey, 'Letter', *Bombay Gazette*, 13 March 1879, 3.
[18] H.P. Blavatsky, *The Secret Doctrine*, 2 vols (London: The Theosophical Publishing Company, 1888), I, p. 251.
[19] Blavatsky, *The Secret Doctrine*, I, p. 252.

made the connection between a kind of space that enabled co-location and the permeability Blavatsky predicted. After her death in May 1891 such views found more frequent voice and often claimed to witness the fulfilment of her predictions of normal clairvoyance. In the fin de siècle the Theosophical Society and its adherents became the most prominent and broadest vector through which permeating higher spatial thought was propagated. For Linda Dalrymple Henderson, Theosophical interest was a 'second major impetus' in the popularization of the fourth dimension after the 'hyperspace philosophy' exemplified by Hinton.[20]

In the early twentieth century Hinton and his publisher noticed the popularity of his work in such circles. Sonnenschein replied to Hinton at the offices of the *Nautical Almanac*, writing: 'No doubt the help of the Theosophical people would be very useful. Right or wrong, they are at any rate open to consider new propositions, which is the reverse of the case with the ordinary mathematician.'[21] Hinton evidently courted this Theosophical interest, becoming friendly with the San Francisco-based writer and Theosophist Gelett Burgess. In 1901, Sonnenschein replied to Hinton's friend Herman John Falk about an increase in demand for sets of his cubes: 'We believe the new impetus given to the study is due chiefly to the Theosophical Societies, who have taken the subject up lately.'[22]

The chief conduit for this higher spatial revival within Theosophy was Charles W. Leadbeater. This chapter examines several influential engagements with the idea of the fourth dimension, all accented towards implications for matter, but an analysis of Leadbeater's repeated approaches to the idea, the most concentrated and focused versioning of higher spatial theory in the context of occult societies and groups, binds it.[23] This is not to suggest that Leadbeater's work was universally accepted within the Theosophical Society: there remained dissident voices, and these too will be heard in order to fill out the full spectrum of responses. In this, the Theosophical Society's engagement with the idea of higher space reflected that of society as a whole; characterized by disagreement, dispute, and occasionally outright confusion.

Indeed, higher space stages a paradox or tension at the core of Theosophical thought. Almost from its outset, the Society was bureaucratically organized with a rigid, top-down structure. The spiritual leadership of Helena Petrova Blavatsky was absolute, and fellow founder Colonel Henry Steel Olcott performed a secretarial function, assisting and supporting the great mystic. The Society's adherents sought to ascend through levels of enlightenment as described to them by Blavatsky's masterworks *Isis Unveiled* (1877) and *The Secret Doctrine* (1888). Only those who studied and achieved could ascend. The Society was frequently riven by disputes over doctrine, fraud, or political wrangling and after Blavatsky's death became yet more fragmented. Unpicking the groupings within the groupings, the proliferating titles or the splinter groups produced by schisms—the British Theosophical

---

[20] Linda Dalrymple Henderson, *The Fourth Dimension and Non-Euclidean Geometry in Modern Art* (Princeton, Princeton University Press, 1983), p. 31.

[21] Archives of Swan Sonnenschein & Co., MS 3282, 13 December 1901.

[22] Archives of Swan Sonnenschein & Co., MS 3282, 29 April 1903.

[23] Bragdon and Uspensky's books *Four Dimensional Vistas* (1923) and *Tertium Organum* (1923) were widely read in the early twentieth century. See Henderson, *The Fourth Dimension and Non-Euclidean Geometry*, for more on these.

Society, the London Lodge, the Inner Group, the Esoteric Section, the Eleusinian Society—can become a distraction.

It is more useful to plot the graph of the Society's sense of hierarchy. At one point in the late 1890s, in Joy Dixon's telling account, the London Society was 'safely ensconced in Albemarle Street, in the heart of London's clubland [where] these mostly male Theosophists conducted learned discussions of the Baghavad Gita, the Egyptian roots of Masonic ritual, the occult significance of four-dimensional hyper-solids'.[24] Members and society figures invited by writing to attend meetings by A.P. Sinnett were required to wear full evening dress.

Such social elitism seems at odds with a spiritual philosophy at whose core was the One, a Buddhist unity of consciousness. Democratic currents and elitist counter-currents whirled through Theosophical thought and solidified in the arena of higher space. The push and pull of pure idealism and finer-grained materialism, the hylic pluralism of one later Theosophical theorist, echoed this swirl.[25] Was Theosophical higher space ultimately a social democratic utopia or a gentleman's club, accessible only to the sufficiently well-connected member, a spiritual, and perhaps also a material, aristocrat?

Elitism was reflected in the modes of certain texts. I refer throughout to Leadbeater's interest in higher space as an appropriation. This is not only because he borrowed Hinton's theoretical approach in its entirety, but because Leadbeater's textual production in the late nineteenth century seems to me a form of intellectual land-grab, in which he described new spaces in order to gain authority over them, and, by extension, over his Theosophical brothers and sisters. This process has explicit resonances with the external realities of the imperial mapping of space. In an analysis of the prevalence of spatial metaphors in the language of nineteenth-century science, Alice Jenkins has noted that surveyors often marched alongside soldiers.[26] This observation applies to Leadbeater's cartographic expeditions into astral spaces, in which he functioned as the avant garde, reporting back to artists who would map what he described. Perhaps yet more suggestive in light of the Theosophical Society's ambiguous relationship with its oriental masters was the Geographical Survey's employment of indigenous Buddhist monks to map Tibet, a space to which it was denied entry. The spies of empire went garbed in the robes of oriental religion, opening up the imaginary archive of the space beyond reach.

Another obstacle to assessing the work of the Theosophical Society is the need to negotiate highly ambiguous texts, and in their complications these contribute to the paradoxes of hierarchy. Where the Society for Psychical Research endeavoured to apply the techniques of scientific naturalism to the research of phenomena we now term paranormal, and produced reports that were appropriately forensic, the

[24] Joy Dixon, *Divine Feminine: Theosophy and Feminism in England* (Baltimore: Johns Hopkins University Press, 2001), p. 42.
[25] See J.J. Poortman, *Vehicles of Consciousness*, 4 vols (Utrecht: The Theosophical Society in the Netherlands, 1978).
[26] Alice Jenkins, 'Spatial Imagery in Nineteenth Century Representations of Science: Faraday and Tyndall', in *Making Space for Science: Territorial Themes in the Shaping of Knowledge*, ed. Crosbie Smith and Jon Agar (New York: Palgrave Macmillan, 1998), pp. 181–91.

doctrine of the Theosophical Society in the Blavatsky years was predicated upon the repeated production of phenomena. It did not investigate, or question, but developed an entire theology on the basis that such events were facts of life. In short, you either had it or you didn't. Madame Blavatsky claimed that her major works were transmitted automatically from the Masters; Leadbeater made his name describing the conditions of life on spiritual planes of existence. Rather than trying to find for one side or the other, to judge Blavatsky a fraud or declare Leadbeater a charlatan, I am interested to consider what we can learn from this ambiguity, what it internalizes and reproduces.

THE GLOBAL SPACE OF THE THEOSOPHICAL SOCIETY

A most striking feature of the Theosophical Society was the rapidity with which it grew from a salon-based occult society to being an international socio-political enterprise. The Theosophical Society's own statistics show that there were lodges in eight countries by 1900, eighteen by 1910 and forty-one by 1925. The first year for which they claim accurate membership figures, 1907, there were 14,863 active members of 567 lodges in thirteen countries, including Cuba, Hungary, and Finland.[27]

The Theosophical Society's relationship with proselytizing was defined by its belief in a Universal Brotherhood of religions. By taking Buddhist vows immediately on their arrival in Ceylon, Blavatsky and Olcott marked themselves as very different missionaries to their Christian counterparts. Like the Christian missions, however, the Theosophical Society also founded schools, but these were Buddhist organizations in Ceylon and Hindu colleges in India. Olcott became immediately involved with Ceylonese independence, campaigning against Christian missionaries and contributing to the design of a Buddhist flag for Ceylon, while both founders were active in India.[28]

Paul Johnson, unusually among historians of the Theosophical Society, has stressed the social networks underlying the Society's mythology. He argues that 'HPB's adept sponsors were a succession of human mentors rather than a cosmic hierarchy of supermen'.[29] He identifies as adepts a roster of transnational occultists encountered during Blavatsky's travels in the 1850s and 1860s. On arrival in India, Indian religious and political figures were honoured as gurus and became the basis, in Johnson's reading, for the masters Koot Hoomi and Morya. Johnson is careful to articulate that he does not think that these figures actually wrote the Mahatma letters, but regards Blavatsky's belief in the Mahatmas as a belief in the political and religious views of real people, obfuscated by the haze of the mythical Tibetan

[27] See C. Jinarajadasa, *The Golden Book of the Theosophical Society: A Brief History of the Society's Growth from 1875–1925* (Adyar, Madras: Theosophical Publishing House, 1925), pp. 263–4.

[28] See Stephen Prothero, *The White Buddhist: The Asian Odyssey of Henry Steel Olcott* (Bloomington and Indianapolis: Indiana University Press, 1996).

[29] K. Paul Johnson, *The Masters Revealed: Madame Blavatsky and the Myth of the Great White Lodge* (Albany: State University of New York Press, 1994), p. 244.

origin of the Masters in order to protect and mystify the true identities of individuals opposed to British rule.

Swami Dayananda Sarasvati, founder of the group the Arja Samaj, an advocate of social and political reform on the basis of his Vedic philosophy, was the founders' first contact in India, before a bitter falling out left them estranged and opposed. A Sikh reform organization, the Singh Sabha, became the new ally, and contacts were maintained with a number of Sikh and Hindu maharajas. As in Ceylon, the Theosophical Society was united to indigenous groups because they too opposed the Christian mission in the colonies. Johnson tracks contact with a number of reform organizations and maharajas 'devoted to revival of Indian culture and the eventual attainment of national self-determination'.[30]

The British authorities in India certainly recognized the early Theosophical Society as a concrete socio-political entity. Olcott and Blavatsky were put under constant surveillance by a police spy during a long journey in the north of India in 1879. Johnson cites a chain of letters passed between imperial diplomats and bureaucrats held in archive at the India Office in London, describing suspicions emanating from the Turkish consul to Washington that Blavatsky was a Russian spy. This research reveals a communicative counterpart to the Mahatma letters, a shadow network of imperial correspondence criss-crossing secret diplomatic space by traditional establishment routes no less occult than those of the Theosophical Society.

While Blavatsky had offered her services to the Russian government as a spy as early as 1870, the real threat she posed to British rule in India was not directed from the government of her homeland but inspired by the Theosophical Society's integration into, and acceptance of, Indian society. Where imperial rule was distant from indigenous Indians, the Theosophical Society was intimately involved. Leading Anglo-Indian Theosophist and former colonial civil servant A.O. Hume established the Indian National Congress in 1885 after a successful but curtailed career in Indian government in which he notably campaigned for the rural poor. Hume had been a combative and engaged member of the Theosophical Society since the early 1880s, frequently challenging Blavatsky and deferring to his own 'Mahatmas', who may or may not have been real political figures.

After a period of infighting and consolidation in the late 1880s and early 1890s, the Theosophical Society began to look outwards to international affairs once again. Following intense diplomatic lobbying the Society was invited to host a Theosophical Congress at the Parliament of Religions taking place alongside the World Fair in Chicago in 1893. Delegations were sent to the International Moral Education Congress in 1908, the International Anti-Vivisection and Animal Protection Conference in 1909, and the Universal Races Congress in 1911.[31]

The Theosophical Society's involvement in Indian politics intensified with Annie Besant's energetic and determined championing of Indian Home Rule from 1913 to 1917, culminating in her election to the Chair of the Indian National Congress in 1917. Three-quarters Irish, Besant had established interests in both

---

[30] Johnson, *The Masters Revealed*, p. 5.   [31] See Dixon, *Divine Feminine*, pp. 140–1.

Home Rule and Indian politics, but on her accession to the presidency of the Theosophical Society she became a practical and determined advocate of Indian independence within the aegis of a federated British Empire. She founded the Central Hindu College in Benares, launched the weekly newspaper *Commonweal*, published articles that were collected as *Nation Building*, gave lectures collected as *Wake Up India!*, even bought the *Madras Standard* and renamed it *New India*. Joanne Stafford Mortimer summarizes:

> The tactics that Annie Besant used were similar to those she had learned years before in England during her various campaigns for reform. They included wide newspaper and periodical publicity, mass meetings, whistle-stop speaking tours, posters, pamphlets, petitions, and local organizations.[32]

Her election to the presidency followed a period of house arrest by the Madras government, who saw her activities as dangerously inflammatory. Within the year the British government had changed its tack and was considering Home Rule in earnest.

While the Theosophical Society operated in independence politics in South-East Asia, its network expanded throughout the English-speaking world. The split of the Society in the USA did nothing to stop a proliferation of lodges; Leadbeater went to Australia; and before his death, Olcott toured the world. Crucial to the establishment of this international structure was publication and the organization of knowledge. Theosophical presses were established in Adyar, London, and New York before more distant outposts contributed to the proliferation of Theosophical printed matter. Another measure of the international reach of the Theosophical Society can be seen in the boom in Theosophical publishing in the 1890s. Journals were launched in Melbourne, Sydney, Paris, Leipzig, Adyar, Bombay, Dublin, London, Toronto, New York, Boston, and San Francisco. Some were short-lived, but for every venture that died, another was launched in a yet further-flung outpost of Theosophical thought—by the turn of the century *New England Notes* was no longer published from Boston, but the *New Zealand Theosophical Magazine* printed its first number in Auckland.

This boom can in part be ascribed to the founder's death. While alive, Blavatsky exerted a tight editorial control over Theosophical textual production, writing articles directly into proofs of *Lucifer*, and this was inevitably relaxed when her charismatic and controlling presence no longer defined doctrine. Doctrinal, social, and political differences often resulted in the establishment of new publications supporting divergent agendas within the movement. Reading these journals we discover a variety of positions on the subject of higher space.

In 1888 Charles Johnston, an American scholar of Sanskrit married to one of Blavatsky's nieces, preceded the founder into print on the matter. Publishing an article in *The Theosophist* on science's 'one veritable romance', Johnston gave a standard version of the dimensional analogy and an extended discussion of

---

[32] Joanne Stafford Mortimer, 'Annie Besant and India 1913–1917', *Journal of Contemporary History*, 18 (1983), 61–78 (p. 71).

Zöllner's experiments, adding: 'That the old philosophers of India were familiar both with the fact and theory of the fourth dimension, some of their metaphysical conceptions leave us small room to doubt.' He concluded that the fourth dimension was

> an idea harmonising perfectly with the Indian idea of innumerable *lokas* filling the universe. Space, being merely a form of *Maya*, it is evident that its varying dimensions are only phases of perception, and not realities, and that every added conception is a fresh state in our divine unfolding, a new phase of the absorption of the finite into the INFINITE, of the expansion of the unit to the ALL.[33]

This argument founded a stream of Theosophical response that might be considered an orientalist synthesis, relating higher-dimensioned spaces to Hindu conceptions of space returning, in their expansion, to an essential unity of consciousness. Here, the aspect of higher space that was so important was its dissolution of standard space.

While *The Theosophist*, based in Adyar, was a house journal for Blavatsky's confidants, the American Society, under William Q. Judge, went its own way even before its official split in 1895. A series of articles in the Californian journal *The Path* reproduced a talk given to Society members in 1889, providing a synopsis of ideas from Hinton's *Scientific Romances*. The author, F.S. Collins, was circumspect in his connection of Theosophical thought and the theory of the fourth dimension, writing 'that most persons would regard both as being vaguely mysterious, and many persons would consider both as arrant nonsense'. The connection was, nevertheless, 'not preposterous, but quite natural'.[34] When Henry T. Edge responded, basing his critique in idealism and highlighting the analogical nature of 'forms', this discussion echoed similar debates in the British scientific community of the 1870s:

> Mr. Hinton's reasoning is very specious and his deductions from his premises are very correct; but it must be borne in mind that forms are merely symbolical and not real, and should be relegated to the same category as algebraical expressions.[35]

Edge's Hintonian reading was more current than that of Collins and he noted admiringly *A New Era of Thought*'s 'arduous system of mental discipline', suggesting, however, that Hinton was 'developing his astral senses, and that, instead of being able to travel mentally in four directions, he will find that there is no necessity to travel in any direction at all, extension having been entirely abolished'.

It is interesting to note that Hinton's work had found readers on the West Coast of America as early as 1889, and that in Theosophical outposts distant from Blavatsky discussions were unconcerned with contradicting her doctrinal pronouncements. Indeed, closer to the hub of Theosophical activity in Europe and India, discussion of the fourth dimension was subdued while Blavatsky remained

---

[33] Charles Johnston, 'Psychism and the Fourth Dimension', *The Theosophist*, 9 (1888), 423.

[34] Frank S. Collins, 'The Fourth Dimension, a Paper read by Frank S. Collins, Part 1', *The Path*, 4 (1889), 17–19 (p. 17).

[35] H.T. Edge, 'Popular Misconceptions about the Fourth Dimension', *The Path*, 4 (1889), 252.

alive. A single article in *Lucifer* in 1890 deals with the subject. In this, Frederick J. Dick creatively repurposed the dimensional analogy to imagine conscious leaves on a tree, the consciousnesses of which would become three-dimensional if they travelled back up the branches on which they grew, and which would become one with all other leaf consciousnesses doing the same. This nature metaphor was but a precursor to the journey required of human consciousnesses:

> As with the leaves, so ordinary human skulls only see around them what is like them-
> selves: in this case three dimensional. And ordinary modes of walking or day-time
> skull-consciousness are of two kinds, forward and backward, and from side-to-side.
> These two notions must be stilled, and the conscious life pass through the point of
> attachment backwards by supreme concentration, before height and depth can be
> added to the consciousness, and the far more REAL empire of four dimensions be
> explored.[36]

For another Theosophical correspondent the same 'far more REAL empire' was approached from the side of 'illusion'. A regular correspondent to the *Boy's Own Paper*, the Rev. J.B. Bartlett, considered the 'glimpse of "the fourth dimension"' provided by experiments with a Möbius strip (without calling it such). The Rev. Bartlett remarked that the twisted strip of paper had 'been thought to afford some sort of illustration of what is called the fourth dimensions [*sic*]'. Glossing this 'condition of existence' and grasping something of its prepositional diffi-culty, he suggested that it was 'a fourth direction (which we may describe as inwardness)'.[37]

Lucifer alerted its readers to the article and W. Kingsland responded with some observations that followed their own Möbius-like logic:

> the *twist* in the paper which causes an *illusive* idea as to what will happen when it is
> cut, is very suggestive of the *mental twist* which is productive of so many illusive ideas
> in respect of the so-called supernatural [...] One may indeed trace some sort of analogy
> here, as to what may occur when we have a clear conception of four-dimensional
> space; for undoubtedly we shall then find that much of the idea of *separateness*,
> produced by the aforesaid mental twist, has proved itself to be—an illusion!

This trickle-down from a popular youth publication into an occult journal gives a glimpse of interplay between disparate print vectors. This recurrence of knotty thought in the context of higher space also invites conjecture. Kingsland's sense of the 'mental twist' grasps key features of the tension in the Theosophical context between exclusivity and inclusivity, unity and separateness. Should we read his 'mental twist' in the inversions and confusions of social hierarchies of Theosophy and their re-establishment at the imperial centre, the illusion of separateness prov-ing more resilient than hoped?

[36] Frederick J. Dick, 'The Meaning of Separated Life—A Mathematical Story of 2, 3 & 4 Dimensions', *Lucifer*, 6 (1890), 243–5 (p. 243).
[37] Bartlett, 'A Glimpse of the "Fourth Dimension"'.

SPIRITUAL DEMOCRATS OR
SPIRITUAL ARISTOCRATS?

Oscillations and tensions between hierarchy and openness in Theosophical doctrine were profound and continuous. The Society initially met on a democratic model but rapidly reverted to a hierarchical system based on Freemasonry. George Felt, a signatory at the foundation, described the problem in a letter to *The Spiritualist*:

> The Theosophical Society was started under the mistaken impression that a fraternity of that kind could be run on the modern mutual admiration plan for the benefit of the newspapers, but very soon everything was in confusion. There were no degrees of membership nor grades, but all were equal. Most members apparently came to teach, rather than to learn, and their views were thoroughly ventilated on the street corners. The propriety of making different degrees was at once apparent to the real Theosophists, and the absolute necessity of forming the Society into a secret body.[38]

With levels and degrees established, Felt's 'real Theosophists' could be certain that no neophytes would be present when they displayed the results of advanced occultist practice. In 1876 a circular described for members the 'origins, plans and aims' of the Theosophical Society, illustrating this essential paradox. 'Section V' of the circular detailed sections and degrees and the requirements for advancement—'all candidates for active fellowship are required to enter as probationers'. 'Section VI' described the various objects of the society, 'finally, and chiefly, to aid in the institution of a Brotherhood of Humanity, wherein all good and pure men, of every race, shall recognize each other as the equal effects (upon this planet) of one Uncreate, Universal, Infinite and Everlasting Cause'.[39]

As doctrine developed, this torsion became more complex. In *The Secret Doctrine* Blavatsky described an occult vision of evolution based on the idea of 'root races', an idea that had circulated in occult groupings throughout the nineteenth century. Blavatsky predicted seven root races with numerous sub-roots, locating much of nineteenth-century humanity within the fifth, or Aryan, root race, and claiming that the sixth sub-root would emerge in California in the early years of the following century. The hierarchy of the secret society was thus reflected in the hierarchy of human racial evolution.

In the Indian context, this hierarchy mapped onto the caste system, and while the Theosophical Society preached universal brotherhood, it was with recognition that, for example, the Brahmin was a spiritually advanced class of the Aryan race. With Leadbeater's theorization of granular materiality in the hierarchically mapped system of planes, in which certain thoughts or spiritualities were coarser and others finer, some thoughts materially destructive, others generative, all permeable to each other, there were physical repercussions of this mode of thought. Joy Dixon writes:

> These claims were inserted into an evolutionary schema that consolidated gendered and racialized associations between refinement and purity, on the one hand, and coarseness and (sexual) impurity on the other. Just as the 'families of Aryan India' were

---

[38] Olcott, *Old Diary Leaves*, I, pp. 127–8, n. 1. [39] Jinarajadasa, *The Golden Book*, p. 26.

believed to possess finer physical bodies than Europeans, other races possessed physical and astral bodies made of coarser material [...] The working classes were also believed to inhabit 'coarser' physical bodies.[40]

Despite an emphasis on connectedness and permeability, then, all were not spiritually—or materially—equal.

Annie Besant too, while re-emphasizing the universal connectedness of all people in the Theosophical schema, at the same time stressed the spiritual aristocracy of some: Atlantis and Lemuria were ruled by highly evolved King-Initiates. Joy Dixon notes that British Theosophists saw a correspondence between such hierarchies and their own social class:

> The truly ideal state was a spiritual aristocracy that recognized that all men were not born equal, that there were older as well as younger brothers in the human family. Many Theosophists felt no embarrassment about identifying themselves and the middle class more generally with the 'elder brethren'.[41]

The confusion of this situation was most profound on the ground in India, where the Theosophical Society adopted an integrationist approach. The traditional colonial hierarchy was disregarded socially but reproduced in textual productions in which the white Theosophists voiced their spiritual teachers. Gauri Viswanathan's analysis highlights the complexities of the social relations between colonizer and colonized bound up in the experience of Anglo-Indian Theosophists:

> Since it required intermediaries who, as spiritual adepts, led them like guides to the astral knowledge they craved, proprietorship of the occult was split unevenly between spiritual teacher and Anglo-Indian disciple. Thus, the very word master acquired an ironic twist. In the colonial situation it was inevitably conjoined to hierarchical relationships. Yet in the practice of the occult the relations of domination and subordination were necessarily inverted, and masters were those who guided initiates into unseen phenomena, which remained the uncolonized space resisting the bureaucratic compulsions of colonial management.[42]

Yet this inversion is unstable, particularly in Theosophical texts which bring questions of authority to the foreground. Through her reading of the Mahatma letters transmitted to Sinnett, Viswanathan asks: 'Whose voice is to predominate; the voice of the spiritual master conveying the deepest secrets of the unknown, or the voice of the interpreter and receiver through whom the last impregnable barrier of the unseen and unknown is breached?' Viswanathan identifies 'fissures' in Theosophical texts 'most transparent in the disjunctive visions of history, memory, time, and knowledge'. To this list we might add the category of space. It is particularly noticeable that such fissures were largely smoothed over in the later Theosophical works of Leadbeater, written from a base back in London, the voice of the oriental master subsumed once more by the colonizing student.

[40] Dixon, *Divine Feminine*, p. 132.   [41] Dixon, *Divine Feminine*, p. 144.
[42] Gauri Viswanathan, 'The Ordinary Business of Occultism', *Critical Inquiry*, 27 (2000), 1–20 (p. 3).

The theoretical positing of a progressive union of consciousness for all living things aligned the Theosophical Society with socially and politically progressive bodies of thought at the start of the 1890s and the resolutely middle-class organizations, based upon the tried and tested model of the club, that promoted them. The proximity and fluidity of movement between members of groups such as the Fabian Society and the Society for Psychical Research meant an overlap and cross-fertilization of ideas among educated, progressive thinkers. In 1890 Edward Carpenter's collection of essays *Civilisation: Its Cause and Cure* received a rapturous review in *Lucifer*. Two essays, 'Modern Science: A Criticism' and 'The Science of the Future: A Forecast', were highlighted by the anonymous reviewer, most likely Blavatsky herself: 'Pure Theosophy; the purest Occultism, say you! Yes, of the purest; each page carries home to the Theosophical reader the conviction that here is a comrade, here a fellow-worker, and of the most excellent.'[43]

The review concluded by recommending the book for every Theosophical bookshelf, and noting its particular utility in proselytizing the message of Theosophy to those who were disinclined to engage with the movement by reputation. Carpenter's monistic world view and his criticism of the methodology and exclusionary language of professional scientific naturalism made him friend to Theosophy whether he realized it or not. Later that year, perhaps by way of thanks, Carpenter contributed a poem to *Lucifer*. 'Underneath and After All' assumed the voice of God to present a progressive, mystical theology. At the heart of the poem was a vision of space as eternal, immutable, a priori, and theologized, and it began an exchange between Carpenter and Theosophy that orbited the nexus of space and soul: 'As space spreads everywhere and all things move and change within it, but it moves not nor changes / So I am the space within the soul, of which the space without is but the similitude and mental image.'[44] Space, static and immutable, contains 'all things', things which in contrast to unchanging space are prone to change. And yet, in Carpenter's verse, external space is a copy, a mental image of 'the space within the soul' that is God, the more perfect form of space.

Carpenter was later ambivalent about his dalliance with Theosophy: in his autobiography *My Days and Dreams* he included Blavatsky's Society in his account 'of a number of new movements or enterprises tending towards the establishment of mystical ideas and a new social order' and claimed such movements 'marked the coming of a great reaction from the smug commercialism and materialism of the mid-Victorian epoch, and a preparation for the new universe of the twentieth century'.[45] He could not, however, return the compliment paid to his work in *Lucifer* to that magazine's editor's *magnum opus*: he wrote that 'no words can describe the general rot and confusion of Blavatsky's *Secret Doctrine*'.[46]

Sheila Rowbotham suggests that Carpenter was playing down what was a more significant influence on his thought than he cared to admit.[47] Certainly, when he

---

[43] Anon., 'Civilization: Its Cause and Cure and Other Essays', *Lucifer*, 6 (1890), 159–60 (p. 160).
[44] Edward Carpenter, 'Underneath and After All', *Lucifer*, 6 (1890), 248.
[45] Edward Carpenter, *My Days and Dreams* (London: G. Allen & Unwin, 1916), p. 240.
[46] Carpenter, *My Days and Dreams*, p. 244.
[47] See Sheila Rowbotham, *Edward Carpenter: A Life of Liberty and Love* (London: Verso, 2009).

travelled to Ceylon and India on a voyage of self-discovery in 1892, he visited the headquarters of the Theosophical Society, staying at Adyar, and reporting on the activities there. Carpenter's account of his travels, *From Adams Peak to Elephanta*, published on his return, described also his experiments in meditation with a *gnáni* in Ceylon, and offered a more extended exploration of the Eastern mystical idea of space that establishes a trans-generational Hintonian connection:

> There is another idea, which modern science has been familiarising us with, and which is bringing us towards the same conception—that, namely, of the fourth dimension. The supposition that the actual world has four space-dimensions instead of three makes many things conceivable which otherwise would be incredible. It makes it conceivable that apparently separate objects, e.g. distinct people, are really physically united; that things apparently sundered by enormous distances of space are really quite close together; that a person or other object might pass in and out of a closed room without disturbance of walls, doors, or windows, etc.; and if this fourth dimension were to become a factor of our consciousness it is obvious that we should have means of know-ledge which to the ordinary sense would appear simply miraculous. There is much apparently to suggest that the consciousness attained to by the Indian gnánis in their degree, and by hypnotic subjects in theirs is of this fourth-dimensional order.
>
> As a solid is related to its own surfaces, so, it would appear, is the cosmic conscious-ness related to the ordinary consciousness. The phases of the personal consciousness are but different facets of the other consciousness; and experiences which seem remote from each other in the individual are perhaps all equally near in the universal. Space itself, as we know it, may be practically annihilated in the consciousness of a larger space of which it is but the superficies; and a person living in London may not unlikely find that he has a backdoor opening quite simply and unceremoniously out in Bombay.
>
> 'The true quality of the soul,' said the Guru one day, 'is that of space, by which it is at rest, everywhere. But this space (Akasa) within the soul is far above the ordinary material space. The whole of the latter, including all the suns and stars, appears to you then as it were but an atom of the former' and here he held up his fingers as though crumbling a speck of dust between them.[48]

In the last paragraph of this account we find the echo of *Underneath and After All*'s static, spatial soul, but also the 'unbounded' spaces of Blavatskyan doctrine. Unpicking this passage from the beginning, though, we read influences from projective geom-etry, from stage conjuring, from higher spatial theory and from psychology.

In its first paragraph Carpenter's account draws on and extends Charles Howard Hinton's theorization of the fourth dimension: 'distinct people are really physically united', he writes, reifying the universal consciousness to which Hinton declared himself inclined. This idea is further developed, extending the familiar dimen-sional analogy to the consciousness: as a solid is related to its own surfaces 'so, it would appear, is the cosmic consciousness related to the ordinary consciousness'.

Linda Dalrymple Henderson situates Carpenter's idea of 'cosmic consciousness' in relation to Frederick Myers's theorization of the 'subliminal consciousness', to

---

[48] Edward Carpenter, *From Adams Peak to Elephanta* (London: Swan Sonnenschein, 1892), pp. 161–2.

which Carpenter referred in later work.[49] We should consider also the theories of Carl du Prel, whose *The Philosophy of Mysticism* was translated by Massey and published in England in 1889. Du Prel's work, like that of Myers, operated on disciplinary boundaries; the Munich Psychological Society, to which du Prel belonged, was modelled on the Society for Psychical Research. Psychology and what would become parapsychology were as yet the same, and discussion of consciousness could not take place without discussion of the spirit and the soul.

In his preface to du Prel's *The Philosophy of Mysticism*, Massey summarized du Prel's work, making reference to Fechner's idea of the 'threshold of sensibility':

> Behind the phenomena of consciousness, both objective and subjective—thus, behind the consciousness itself—must certainly be placed the ultimate reality or being of which consciousness offers only a reflection or representation. This inscrutable being is therefore termed 'the Unconscious'. But now the question arises whether this 'Unconscious' lies immediately behind our physically conditioned consciousness, or may be pushed back indefinitely, so that there is room for a root of conscious individuality, only relatively unconscious for the organism of sense. Du Prel finds an answer to this question in the recognition and significance of what is now known as the psycho-physical 'threshold of sensibility,' and in its occasional mobility or displacement.[50]

Leaving aside for a moment the seeming inescapability of spatial metaphors in language—what lies 'behind' consciousness?—the connection of the subliminal consciousness or unconscious with the threshold of sensibility has higher spatial repercussions. Because higher space can, as in Carpenter's account, provide 'means of knowledge which to the ordinary sense would appear simply miraculous', it exceeds the same threshold of sensibility. For du Prel, the mobility of the threshold of sensibility was a product of evolution: the senses were expanding; the transcendental subject was emerging and becoming more conscious, as evidenced by the phenomena of hypnotism and somnambulism.

In Carpenter we read Charles Howard Hinton's higher spatial consciousness, creating an extended sensibility, and ideas about sub- and supra-liminal consciousness connected to hypnotism from sources also current in progressive thought in this period.[51] The extended sensory reach of higher space, as suggested by Carpenter, was quite familiar to occult readers by the early 1890s, and as well rehearsed for the Theosophical reader, at least, was the notion 'that things apparently sundered by enormous distances of space are really quite close together'. Carpenter's stressing of the malleability of physical space is explicitly orientalized—'a person living in London may not unlikely find that he has a backdoor opening

---

[49] See Linda Dalrymple Henderson, 'Mysticism as the "Tie That Binds": The Case of Edward Carpenter and Modernism', *Art Journal*, 46 (1987), 29–37.

[50] C.C. Massey, 'Translator's Preface', in Carl du Prel, *The Philosophy of Mysticism*, trans. C.C. Massey, 2 vols (London: George Redway, 1889), I, pp. ix–xxii.

[51] In the light of the permeability of and social proximity between groups such as the Fellowship of New Life, the SPR, and the London Lodge of the Theosophical Society, the cross-fertilization of ideas grounded in psychological studies with those of utopian political hue or derived from oriental mysticism is unsurprising.

quite simply and unceremoniously out in Bombay'—shrinking the physical space of empire and mirroring Blavatskyan phenomena, the production of letters from the Masters that purported to have travelled around the globe on the astral plane. Carpenter's reportage also enacted the curious ventriloquism of 'received' Theosophical texts, giving voice to his *gnāni* in inversion of the traditional hierarchies of the colonial experience: in Ceylon, the Guru was the master.

At the same time Theosophical journals began to re-engage with the idea dismissed by Blavatsky. The notion of consciousness was now to the fore, as Theosophical discourse sought to free the ego from its earthly trappings in the quest for enlightenment. So a correspondent writing with a question on the nature of intuition to Theosophical Q&A magazine *The Vahan*, an *Ariel* for late nineteenth-century occult media workers, was directed towards Charles Howard Hinton's *Scientific Romances*: 'As for the means of cultivation of the intuitional faculties, I speak with great diffidence, but it is probably safe to say that these means lie in the direction of the casting out of the lower self, more especially on the intellectual plane.'[52] Hinton's theory and cubic regime gave the practical occultist a hands-on methodology for achieving higher spirituality; the steady worker within the Theosophical bureaucracy might attain heady heights by applying to the right sources and asking the right questions of the more adept.

In more popular forums there was no attempt to restrain permeating higher space within hierarchical structures. In the twelve months immediately following the publication of *From Adam's Peak* an explosion of fourth-dimensional discourse sensationally extended the idea of an augmented sensorium available through higher space. The Blavatskyan prophecy of permeability was declared imminent with greater urgency. The campaigning new journalist and editor William T. Stead was only tangentially connected to the Theosophical Society but was something of a fellow traveller: having commissioned his much-admired colleague Annie Besant to review Blavatsky's *The Secret Doctrine* for the *Pall Mall Gazette* in 1889, he had witnessed her subsequent conversion from social materialism and rise through the ranks of the Theosophical Society. Stead, despite many shared interests, was an expansive and prolific popularizer and publisher whose urge to share and democratize information was at odds with Theosophy's elitism. The size of Stead's readership for his journal *Review of Reviews*—estimated at 300,000 in 1890—made his engagement with higher space, however brief, of considerable import for its influence alone.

THE FOURTH DIMENSION AS MEDIUM

Roger Luckhurst's account of William Stead's advocacy of telepathy stresses the 'shifting formations of high and low' in fin de siècle culture, and particularly

---

[52] Anon., 'What is the Nature of Intuition, and Which are the Best Means of Developing this Faculty?', *Vahan*, 2 (1892), 5.

print publishing.[53] The influential radical journalist whose turn to spiritualism dominated the second half of his life allows Luckhurst to negotiate 'a large network of cultural connections across divergent cultural terrains'.[54] Luckhurst analyses the popularization of knowledge, questioning passive, simplified knowledge transfer from expert to amateur, and probes Stead's 'affective journalism' in the context of his urge to democratize knowledge, his obsession with electrical technologies, and the way these came together in his internationalism and vision of a technologically connected empire.

Stead's engagement with the fourth dimension could be considered a sub-chapter to Luckhurst's account. It lasted for only three consecutive issues of the *Review of Reviews* in 1893 but constitutes a fascinating case study of the oscillatory cultural operation of higher space, bouncing between high and low culture, and its vagaries even within the field of psychical research.[55] Stead introduced the idea in a sensational mode typical of his journalistic practice, and accented its potential with the same obsessions and interests Luckhurst investigates with relation to thought transference: technology, affective reach, empire. Even the curiosity of the brevity of this engagement is continuous with the slipperiness of higher space.

In March 1893, Stead recommended a cluster of books of interest to psychic researchers. Frederick Myers's essay on the 'Subliminal Consciousness' in the *Proceedings of the Society for Psychical Research* was commended for explaining 'so clearly and exhaustively the method by which the psychologist is learning to evolve a new science of the hitherto invisible and unknown world'. A brief paragraph suggested as complementary Arthur Willink's *The World of the Unseen*: 'Mr. Willink holds that the unseen world is of four dimensions, and into this space of four dimensions or Higher Space, as he calls it, the dead pass, and from which they can communicate with us.' A final trio of books included *Do the Dead Return? A Record of Experience of Spiritualism by a Clergyman of the Church of England*; Mr Carlyle Petersilea's *Discovered Country*, 'which is said to have been written automatically, describing life on the other side'; and *Dreams of the Dead* by Edward Stanton: 'It is very curious and more Theosophical than Christian. The writer holds that we are on the advent of the sixth race. A new physical sense is developing in the nerve constitution of man. The time is at hand when a new civilization will be founded by a select amalgam.'[56]

Stead had been interested in psychical research for over a decade; he attended his first seance in 1881, and in 1884 he had hosted the thought-reader Stuart Cumberland at the offices of the *Pall Mall Gazette*. By 1893 he had been editing and publishing the *Review of Reviews* for three years and his absorption in spiritualism was increasingly evident to his readers, who encountered frequent articles and editorials on thought-reading, ghosts, and the afterlife. He had published *Real Ghost Stories* in November 1892, in which he aimed, in Roger Luckhurst's phrase,

---

[53] Luckhurst, *The Invention of Telepathy*, p. 118.
[54] Luckhurst, *The Invention of Telepathy*, p. 121.
[55] I have been unable to find mention of the fourth dimension anywhere else in Stead's work, despite his lifelong interest in the supernatural.
[56] W.T. Stead, 'Some Books of the Month', *Review of Reviews*, 7 (1893), 325.

to 'democratise psychical research by appropriating the sober SPR "Census of Hallucinations" Project'.[57] He had become increasingly interested in automatic writing following the death of American journalist Julia Ames, and had suggested that the readers of the *Review of Reviews* investigate the phenomenon in 1892, by which stage he was already collecting the automatic scripts which would later be published as *Letters from Julia* (1897).

The synthesizing function of *Review of Reviews* was exemplified by an early account of these experiments in automatism appearing in the April issue. Stead prefaced these with two pages combining elements from a number of the books he had recommended the previous month. His 'Throughth: Or, On the Eve of the Fourth Dimension: A Record of Experiments in Telepathic Automatic Handwriting' appropriated the visionary style and evolutionary claims of *Dreams of the Dead* and the theoretical ideas of *The Unseen World* to ground the practice of automatic writing and announce the advent of a new age in communications.

The fourth dimension, wrote Stead, could be imagined by those 'with a vivid imagination', but 'has never been seen by mortal man':

> We however get glimpses of it in clairvoyance, in the phenomena of hypnotism, and in all the experiments which are known as telepathy, crystal-gazing, thought-reading, and all things in which we see, hear or communicate through things, which according to the known laws of third dimensional space, would render communication impossible.[58]

Stead was in no doubt about the correct prepositional description for higher space, indeed this prepositional action was its very essence: 'Hence, Throughth.' The interpenetrative qualities of higher space, short-circuiting the very materiality of matter, were central to Stead's throughth because they enabled unhindered communication; the extended sensorium of the fourth-dimensional vista was supplemented by a fourth-dimensional auditorium. Above all, throughth was a medium, and its primary characteristic recalled a section quoted from Myers the previous month: 'The possible law of which I speak is that of the Interpenetration of Worlds.'[59]

Stead's claim that 'it is becoming more and more evident to those who observe and note the signs of the times that we are in very deed and truth on the eve of the fourth dimension' exceeded the standard sensationalism he had defended as journalistic practice. Not only did he assume a position of observational authority from which to pronounce, but the detailing of his millenarian revelation was yet more wild-eyed and visionary than it was melodramatic:

> In the new world which opens up before life becomes infinitely more divine and miraculous than it has ever been conceived by the wildest flights of imagination of the poet. Many attributes which have hitherto been regarded as the exclusive possession of the Deity will be shared with His creatures. The past mingles with the present, and the future unfolds its secrets. Death loses its sting, and parting its sadness. The limitations

[57] Luckhurst, *The Invention of Telepathy*, p. 121.
[58] W.T. Stead, 'Throughth : Or, On the Eve of the Fourth Dimension: A Record of Experiments in Telepathic Automatic Handwriting', *Review of Reviews*, 7 (1893), 426–32 (p. 426).
[59] Stead, 'Some Books of the Month', 325.

of time and space—three-dimensional space, that is—furl up and disappear. Spirit is manifested through matter, and we enter into a new heaven and a new earth.[60]

Such an existence was promised to all by Stead, whose democratic, affective urge extended to his broad readership. Willink's *The World of the Unseen*, published in Macmillan's theological series and indebted, as its title indicates, to *The Unseen Universe*, indicated the reach of the idea of higher space: 'There can be few persons who have not at least heard of the Higher Space, or, to use the more familiar expression, the Fourth Dimension of Space'; following Stead's intervention there can have been fewer still.[61]

Yet Stead's democratic embrace of the evolutionary paradigm ran counter to that which was becoming prevalent in occult groupings; the evolution of spirit frequently carried with it an implication of hierarchy; there were spiritual aristocrats who, like Stanton's sixth race, would be a 'select amalgam'. In this, Stead's version of fourth-dimensional capability as a product of evolutionary advance was aligned with Hintonian universality.

The revelationary zeal with which Stead embraced this vision of a fourth-dimensional future is in part a product of his approach to sensationalist writing, but also surely owes much to his sources. Indeed, to an observer so attuned, the early months of 1893 readily suggested the inevitability of this vision. One of the more excitable texts of this period, *I Awoke! Conditions of Life on the Other Side Communicated by Automatic Writing* (1893), sold at one shilling net, and reprinted and extended two years later, described a higher-dimensional afterlife. It referred throughout to 'the Master', a Christ, of 'a form which is in four dimensions, and which cannot be seen by ordinary earthly vision'.[62] An appendix described the conditions of the various dimensions in which the dead lived, and is notable for its embrace of a full range of higher dimensionalities, a fleshing out of *Flatland*'s suggestions of fifth, sixth, and beyond.

The Appendix, 'received' in 1891, claimed that 'there is a fourth dimension [...] which represents what you might call the inter-penetrative sphere'. It continued: 'This fourth dimension, only guessed at by you, is our first, the other three fall from us as crude and imperfect.' The inhabitants of this dimension were capable of impossible feats of transportation:

> This power, when perfected, would give man absolute power of progression in every direction and in every part of the universe. He could pass through the heart of mountains, or could rise into the atmosphere to any height by altering, as it were, his own density, and the density of his path; nothing would prove a hindrance.[63]

---

[60] Stead, 'Throughth', 427. For Stead's use of melodrama see Judith R. Walkowitz, *City of Dreadful Delight: Narratives of Sexual Danger in Late-Victorian London* (London: Virago, 1992).

[61] Arthur Willink, *The World of the Unseen: An Essay on the Relation of Higher Space to Things Eternal* (London: Macmillan, 1893), p. 13.

[62] Anon., *I Awoke! Conditions of Life on the Other Side Communicated by Automatic Writing* (London: Simpkin and Marshall, 1893), p. 25.

[63] Anon., *I Awoke!*, p. ii.

Perhaps unnecessarily, the fifth dimension extended these capabilities to cosmic space: 'Let us call the fourth dimension inter-progression, then the fifth might be called trans-progression. From sphere to sphere, from star to star, and from star to sun shall the children of men wander at free will.' The less than complete under-standing of astronomy demonstrated by the dictating intelligence did not deter further revelations. 'As men rise from dimension to dimension their powers are changed and increased in many ways.'[64] The sixth dimension was the first 'time-dimension' in which linear time was infinitely malleable, could be slowed down or sped up. In the seventh 'time may be said to have no existence': the past was as accessible as the present; only the future remained hidden.[65] Those who had access to dimensions beyond the seventh would enter into a vague form of omnipotence: 'After the time-dimensions come those that belong more directly to the human will, its powers and its limitations.'[66]

I Awoke!, like any text that purports to be transmitted through automatic writing, occupies a curious cultural position. To proponents of the practice the text's very existence offers evidence of the phenomenon of automatism and legitimates content offering mediated access to the mysterious unseen. A more distanced analysis would observe that such texts reflect the conditions of their composition; regard-less of their origin, the 'medium' through which they are channelled is inevitably embedded in an occult network; they tend synoptically to appropriate (or confirm) current occultist or scientistic thought. The observations of Gauri Viswanathan about the inversions of the colonial social order instantiated by such ventriloquism require some distortion in the instance of a text such as *I Awoke!*, whose primary and secondary authors remain anonymous. There is certainly a question of the location of authority when the one voice the text reproduces purports to be late: the authorities of the present time and the material are disturbed; the past and the immaterial asserted.

Stead's democratized fourth dimension did not last long. In the May issue of *Review of Reviews* he summarized Professor Hermann Schubert's paper published in *The Monist*, 'The Fourth Dimension: Mathematical and Spiritualistic'. Professor Schubert was 'very hostile to spiritualism' and stressed the need for 'slow, unceasing research' rather than 'the thoughtless employment of fanciful ideas'. Stead retreated from his previous enthusiasm with an unconvincing objection to an 'unscientific' line of argument.[67]

Encapsulated in Stead's engagement with higher space is the idea's status as a rhetorical ping-pong ball: no sooner did a supernaturalist account of the dimen-sionality of 'the other side' appear to offer a millenarian vision of the future, than a hard-headed philosopher cut down speculations. Little wonder that the general public frequently expressed confusion at the idea. Yet as is evidenced by Stead's reading list, millennial visions were enjoying some currency in 1893, and although many drew directly from *Revelations*, they managed to maintain optimism about

---

[64] Anon., *I Awoke!*, p. iii.   [65] Anon., *I Awoke!*, p. v.   [66] Anon., *I Awoke!*, p. iv.
[67] W.T. Stead, 'On the Eve of the Fourth Dimension, Mathematical and Spiritualistic', *Review of Reviews*, 7 (1893), 542. See also Hermann Schubert, 'The Fourth Dimension: Mathematical and Spiritualistic', *Monist*, 3 (1893), 435–49.

the changes in store in the new century. A parallel tradition was simultaneously emerging in fiction—particularly among the texts we now read as the first science fiction and fantasy stories and novels—that represented higher space in an entirely different fashion, picking up perhaps on the paranoid possibilities of extended space diagnosed by Bruce Clarke.

PLANES AND STRATA

Theosophists responded to the popular and millennial idea of a space of permeation with a retrenchment of the orientalist synthesis that emphasized a schematized planar system incorporating both the ideas of higher space and the permeability of matter. Herbert Coryn's 1893 piece in *Lucifer*, premised upon an ideal genesis for space, described the passage of the monad, Theosophy's spiritual essence, derived more from Neoplatonism than from Leibniz, through planes of existence correspondent to alchemical elements. Within this schema, he wrote:

> This next step is the return of the plane of water, the astral plane, with the now well acquired physical intellectuality. The new sense is clairvoyance; the new property of matter is its complete permeability to itself, answering to the Fourth Dimension; the new power is that of effecting this, of doing actually what appears to be done when a solid is reflected in a mirror.[68]

Coryn's argument gathered together the diverse elements that contributed to Blavatskyan doctrine and applied them to higher space, tweaking Blavatsky's conclusion in *The Secret Doctrine*, allowing her doctrinal superiority but lining up permeability directly with the fourth dimension:

> Visual and practical transparence [*sic*] is the 'fourth Dimension'; not the taking on by matter of any stature in some inconceivable direction, but the taking on by human consciousness of a new sense and power. The term 'Fourth Dimension' is therefore, as H.P.B. points out, incorrect.[69]

The use of the geometric term plane to describe a level, or phase, of the existence of consciousness, was popularized within Theosophy. Its first recorded English-language use was in the 1836 novel *Philothea: A Grecian Romance* by Lydia M. Child, and Child herself seems likely to have encountered the usage in the work of the late Neoplatonist Proclus, whose *The Elements of Theology* contained the phrase 'εν το ψυχικω πλατει', 'in the plane of existence': in his translation of *The Elements*, E.R. Dodds commented that Proclus's term *to platos* was 'the literal equivalent of the "planes" of modern Theosophy'.[70]

In Child's tale, dedicated to 'a few kindred spirits, prone to people space "with life and mystical predominance"', Philothea and her grandfather, Anaxagoras,

---

[68] Herbert Coryn, 'The Fourth Dimension', *Lucifer*, 12 (1893), 326–32 (p. 329).

[69] Coryn, 'The Fourth Dimension', 331.

[70] E.R. Dodds, 'The Astral Body in Neoplatonism', in Proclus, *The Elements of Theology*, ed. and trans. E.R. Dodds (Oxford: Clarendon Press, 1933), p. 303.

are joined in discussion by Plato himself, who discusses the construction of the universe in planar terms:

> There are doubtless men in other parts of the universe better than we are, because they stand on a higher plane of existence, and approach nearer to the idea of man. The celestial lion is intellectual, but the sublunary irrational; for the former is nearer the idea of a lion. The lower planes of existence receive the influences of the higher, according to the purity and stillness of the will. If this be restless and turbid, the waters from a pure fountain become corrupted, and the corruption flows down to lower planes of existence, until it at last manifests itself in corporeal forms. The sympathy thus produced between things earthly and celestial is the origin of imagination; by which men have power to trace the images of supernal forms, invisible to mortal eyes. Every man can be elevated to a higher plane by quiescence of the will; and thus may become a prophet.[71]

Child's novel inhabits the Neoplatonic tradition, stressing perfect Platonic forms and ideals, and embedding these in a modern, progressive view of the mystical East. The use of a geometric metaphor to describe an elevated existence emerges from a philosophical world view in which geometric forms are 'supernal'.

Blavatsky, who quoted Child in *Isis Unveiled*, had by the 1880s taken the monopoly on this metaphorical usage, which the *OED* records occurring in both the *Transactions* of the London Lodge and Blavatsky's *Key to Theosophy*. In the Blavatskyan version, the idea of *lōkas* was borrowed from Hinduism and the geometric term 'plane' mapped onto it. The emphasis on the supernal power of geometric forms in Neoplatonic philosophy was a catalyst for the adoption of geometric terms, just as such terms suggested affinities with contemporary geometric notions such as the idea of the fourth dimension. The suggestion of earlier Theosophical commentators that the *lōkas* corresponded to higher dimensions of space was developed and most completely worked through in the writing of C.W. Leadbeater.[72]

Invited by A.P. Sinnett to speak at a black-tie meeting of the members of the London Lodge on 21 November 1894, Leadbeater gave a paper on the subject of *The Astral Plane*, which was transcribed for the *Transactions* of the Society and published the following year as a book. *The Astral Plane* described the conditions of this 'realm', 'its inhabitants and phenomena'.[73] It adopted a scientific tone for its observations, achieved through clairvoyance, which Leadbeater claimed he had first experienced in 1885, while he had been staying at Adyar. He recounted in *How Theosophy Came to Me* a visit by one of the Masters who advised him on meditation techniques. Leadbeater 'took the hint' and began practising as advised:

> Certain channels had to be opened and certain partitions broken down; I was told that forty days was a fair estimate of the average time required if the effort was really

---

[71] Lydia Maria Child, *Philothea: A Romance*, 2nd edn (Boston: Otis, Broaders and Company, 1839), pp. 149–50.

[72] See Johnston, 'Psychism and the Fourth Dimension'.

[73] C.W. Leadbeater, *The Astral Plane: Its Inhabitants and Phenomena* (London: The Theosophical Society Publishing Company, 1895), pp. 3, 4.

> energetic and persevering. I worked at it for forty-two days, and seemed to myself to be on the brink of the final victory when the Master Himself intervened and performed the final act of breaking through which completed the process, and enabled me thereafter to use astral sight while still retaining full consciousness in the physical body—which is equivalent to saying that the astral consciousness and memory became continuous whether the physical body was awake or asleep.[74]

In the intervening years Leadbeater, a quondam Anglican vicar, had applied his efforts to working his way up through the Theosophical Society with some success.

His first approach to the fourth dimension was made in his lecture on *The Astral Plane*. He recorded difficulties in describing the appearance of the Astral Plane, 'or *Kâmalôka* as it is called in Sanskrit', due to an altered vision: 'Sight on that plane is a faculty very different from and much more extended than physical vision. An object is seen, as it were, from all sides at once, the inside of a solid being as plainly open to the view as the outside.'[75] His apologia continued:

> When we add to this that every particle in the interior of a solid body is as fully and clearly visible as those on the outside, it will be comprehended that under such conditions even the most familiar objects may at first be totally unrecognizable. It is this characteristic of astral vision which has led to its sometimes being spoken of as sight in the fourth dimension—a very suggestive and expressive phrase. But in addition to these possible sources of error, matters are further complicated by the fact that astral sight cognizes forms of matter which, while still purely physical, are nevertheless invisible under ordinary conditions.[76]

Leadbeater's mapping of the fourth dimension specifically onto the astral realm was combined with his description of a finer-grained materialism that became apparent in this medium. As one ascended the planes vision became yet more granular and microscopic, enabling an adept to see 'the still finer order of matter that is sometimes described as ethereal'.

However this lecture might have been received on the evening of its delivery, by the time Leadbeater had transcribed it for publication he had received notice from the Master Khoot Hoomi that it was a highly significant work in the history of human knowledge. The manuscript was spirited away to be archived in Tibet. The introduction of Leadbeater's Sri Lankan acolyte Jinarajadasa to later editions reported this sequence of events:

> For this reason the little book, *The Astral Plane*, was definitely a landmark, and the Master as Keeper of the Records desired to place its manuscript in the great Museum. This Museum contains a careful selection of various objects of historical importance to the Masters and Their pupils in connection with their higher studies, and it is

[74] C.W. Leadbeater, *How Theosophy Came to Me* (Adyar: Theosophical Publishing House, 1930), pp. 157, 158. Leadbeater also thanked the Master Djwal Kul, who assisted in his further training and with whom he had a 'close association [...] in my last life, when I studied under Him in the Pythagorean school which He established in Athens, and even had the honour of managing it after his death' (p. 159), thereby laying claim to mystical geometric pedigree of the most elevated kind.

[75] Leadbeater, *The Astral Plane*, p. 4.

[76] Leadbeater, *The Astral Plane*, pp. 8–9.

especially a record of the progress of humanity in various fields of activity. It contains, for instance, globes modelled to show the configuration of the Earth at various epochs of time; it was from these globes that Bishop Leadbeater drew the maps which were published in another transaction of the London Lodge, that on *Atlantis* by W. Scott-Elliot.[77]

Having explored, mapped, and reported back on the Astral Plane, Leadbeater turned his attention to the plane beyond, the Devachanic. Despite yet more confusing conditions—'it would seem as though in Devachan space and time were non-existent, for events which here take place in succession and at widely-separated places, appear there to be occurring simultaneously and at the same point'—here, too, higher spatial conditions presided over altered materiality. In the Devachanic Plane processes of dematerialization and permeation were inverted as the immaterial coalesced and became visible. An experiment to view 'thought forms', theorized by Annie Besant in *Lucifer*, was undertaken:

> One investigator remaining on the lowest subdivision to send out the thought-forms, while others rose to the next higher level, so as to be able to observe what took place from above, and thus avoid many possibilities of confusion. Under these circumstances the experiment was tried of sending an affectionate and helpful thought to an absent friend. The result was very remarkable; a sort of vibrating shell, formed in the matter of the plane, issued in all directions around the operator, corresponding exactly to the circle which spreads out in still water from the spot where a stone has been thrown into it, except that this was a sphere of vibration extending itself in three (or perhaps four) dimensions instead of merely over a flat surface.[78]

Leadbeater and Besant jointly authored *Thought Forms* nearly a decade later, in which they acknowledged the significance for their research of the discovery of X-rays in 1895: 'Röntgen's rays have rearranged some of the older ideas of matter, while radium has revolutionized them, and is leading science beyond the border-land of ether into the astral world. The boundaries between animate and inanimate matter are broken down.'[79]

In a remark that resonates with the twenty-first-century reader through whose home and body wirelessly transmitted information ebbs and flows, Steven Connor has noted: 'The predominating modern experience, I want to say, is that of being permeated.'[80] For Connor, this experience germinated in the crucible of fin de siècle occultism as the discovery of X-rays lent legitimacy to the ancient notion that matter could be made permeable to vision.

Connor remarks that 'occultists, spiritualists and supernaturalists seized upon the Röntgen rays as the ocular proof of the powers they had been claiming for

[77] C. Jinarajadasa, 'Introduction', in C.W. Leadbeater, *The Astral Plane* (Adyar: Theosophical Publishing House, 1970), pp. vi–xx (p. xiv).

[78] C.W. Leadbeater, *The Devachanic Plane* (London: The Theosophical Publishing Society, 1896), pp. 4, 19.

[79] Annie Besant and C.W. Leadbeater, *Thought Forms* (London and New York: The Theosophical Publishing House, 1905), p. 11.

[80] Steven Connor, 'Pregnable of Eye: X-Rays, Vision and Magic', http://www.stevenconnor.com/ xray/ [accessed 22 October 2012] (para. 3 of 94).

decades'.[81] These powers of clairvoyance had been associated with the fourth dimension since the Zöllner event and suddenly, in 1895, here were images that fixed penetrated matter onto paper and plate. Linda Dalrymple Henderson has described how significant such images were for Modernist artists: 'By pointing to the limited extent of the visible spectrum, x rays established unquestionably the relativity of perception and turned the attention of artists away from the visual world towards an invisible, immaterial reality, at times associated with the fourth dimension.'[82]

As matter gave way to vision, so the immaterial congealed. Occultists sought to capture images of souls and thoughts. Leadbeater and Besant had been inspired in their experiments by the work of Hippolyte Baraduc, a French doctor who was 'well on the way towards photographing astro-mental images, to obtaining pictures of what from the materialistic standpoint would be the results of vibrations in the grey matter of the brain'.[83] Baraduc's theorization of his work emphasized fluidity between the thing thought and the material object of perception. 'He quite rightly says that the creation of an object is the passing out of an image from the mind and its subsequent materialization, and he seeks the chemical effect caused on silver slats by this thought-created picture.'[84]

Leadbeater and Besant recast this process in Theosophical terms: their investigation took place on the astral plane. The permeability of boundaries demonstrated by scientific discoveries gave their experiments in thought transmission urgency. Besant had begun recording 'thought forms' in 1895 and the pair had undertaken experiments to transmit thoughts to each other and to describe how these thoughts appeared, dictating their visions to specialist artists who would produce their best impressions: the astral plane, like a court of law, a place in which photography was not possible. The authors were grateful to the artists, but acknowledged obstacles too profound to be overcome that might have found agreement from one A Square:

There are some serious difficulties in our way, for our conception of space is limited to three dimensions, and when we attempt to make a drawing we practically limit ourselves to two. In reality the presentation even of ordinary three-dimensional objects is seriously defective, for scarcely a line or angle in our drawing is accurately shown [...] If to this difficulty we add the other and far more serious one of a limitation of consciousness, and suppose ourselves to be showing the picture to a being who knew only two dimensions, we see how utterly impossible it would be to convey to him any adequate impression of such a landscape as we see. Precisely this difficulty in its most aggravated form stands in our way, when we try to make a drawing of even a very simple thought-form. The vast majority of those who look at the picture are absolutely limited to three dimensions, and furthermore, have not the slightest conception of that inner world to which thought-forms belong, with all its splendid light and colour.[85]

---

[81] Connor, 'Pregnable of Eye' (para. 33).

[82] Linda Dalrymple Henderson, 'X Rays and the Quest for Invisible Reality in the Art of Kupka, Duchamp, and the Cubists', *Art Journal*, 47 (1988), 323–40 (p. 336).

[83] Besant and Leadbeater, *Thought Forms*, p. 12.

[84] Besant and Leadbeater, *Thought Forms*, p. 14.

[85] Besant and Leadbeater, *Thought Forms*, pp. 16–17.

The tone of this passage is that of warning from on high rather than of an apology. Thoughts could be discerned only by those who were able to access the astral plane. While the human 'medium' remained significant for spiritualism, in Theosophical and other groupings, mediumship seemed to have dispersed into spaces of hierarchical access. The all-permeating 'throughth' was, counter-intuitively, an exclusive space.

The tensions and undertows of Theosophical approaches are evident and perhaps best understood as appropriations of the location of authority itself. The approach of Leadbeater and Besant to higher spatial phenomena—to report them as witnessed in a certain locale, as might a naturalist—and the language used—referring to planes as 'realms' and those who visited them as 'explorers'—represented higher space as real and accessible to colonization. Leadbeater's work was an intellectual colonization of these imaginary spaces and his cartographic enterprise was co-extensive with his interest in higher space. His copied maps of Atlantis showed his endeavour to fix extra-factual spaces into legitimate textual forms. *Thought Forms* was a kind of field guide, a classificatory and morphological report on thoughts as vibrating objects, brought back from the *terra incognita* of higher space. It summarized findings in three general principles:

1. Quality of thought determines colour.
2. Nature of thought determines form.
3. Definiteness of thought determines clearness of outline.[86]

To bolster this authority on matters higher spatial, Leadbeater referred again and again to the work of Hinton. Replying to a question about the ego in *The Vahan* in April 1898, he borrowed an image from *A New Era of Thought*: 'Let us suppose the grey matter of the brain to be laid out upon a flat surface, so that the layer is only one particle thick—that is to say, let us suppose ourselves looking down on it from the "fourth dimension," since that is exactly the appearance that it would present if regarded form that point of view.'[87] Again, the interactions of space and matter are inverted, as the organ of thought is flattened, dimensionally reduced.

In *Clairvoyance* his project was more developed, his approach a more confident taxonomical assertion of the differences between planar visions, and a bold correction of Blavatskyan dogma on the topic. He quoted his own most recent approach to the subject in *The Vahan*, which began: 'There is a distinct difference between etheric sight and astral sight, and it is the latter which seems to correspond to the fourth dimension.' The transparency and interpenetrability that characterized his earlier account had been augmented and clarified. In a description that owed to Hintonian practice, Leadbeater materialized higher space through the cube, inevitably made porous to vision:

---

[86] Besant and Leadbeater, *Thought Forms*, p. 31.
[87] C.W.L., 'Does a Highly Developed Ego, That of a Master, for Instance, Put on the Limitations of the Physical Brain When it Descends to Work on the Physical Plane?', *Vahan*, 8 (1898), 7.

If you were looking etherically at a wooden cube with writing on all its sides, it would be as though the cube were glass, so that you could see through it, and you would see the writing on the opposite side all backwards, while that on the right and left sides would not be clear to you at all unless you moved, because you would see it edgewise. But if you looked at it astrally you would see all the sides at once, and all the right way up, as though the whole cube had been flattened out before you, and you would see every particle of the inside as well—not through the others, but all flattened out. You would be looking at it from another direction, at right angles to all the directions that we know.[88]

He recorded also that Hinton's tesseract was 'quite a familiar figure upon the astral plane'.

With Annie Besant accompanying him in his clairvoyant explorations and experiments, Leadbeater was increasingly central to the London branch of the organization, and therefore a celebrity within it, with increased visibility and extended reach for his pronouncements. He was confident enough to baldly contradict Blavatskyan dogma now that it interfered with his spatial scheme:

> I would therefore venture deferentially to suggest that, when Madame Blavatsky wrote as she did, she had in mind etheric vision and not astral, and that the extreme applicability of the phrase to this other and higher faculty, of which she was not at the moment thinking, did not occur to her.

In the first year of the new century Leadbeater lectured on the fourth dimension to the Amsterdam Lodge. He repeated his etheric/astral clarification and recommendations of Hinton, arguing that:

> We, however, as students of Theosophy, know that it [the fourth dimension] is within our reach; that everywhere around us is this unseen, this astral world. So when a man first learns to understand the fact in this abstract way, he can leave that, to proceed to real experience and practice.[89]

By the 1903 publication of *The Other Side of Death*, the idea of the fourth dimension was so central to Leadbeater's theorization of planar consciousness that he devoted an entire chapter, 'An Extension of Consciousness', to the subject. At its heart was a detailed synopsis of Hinton's work, promoted as a way of exceeding the limitations of both space and physical consciousness, including descriptions of the tesseract, the image of threads passing through a sheet of wax, the idea of gases extending into the fourth dimension, and examples of counterpart gloves. Leadbeater described the study of the fourth dimension as 'the best method that I know to obtain a conception of the conditions which prevail upon the astral plane', but noted that:

> Mr Hinton himself does not treat the matter as an approach to the comprehension of the astral plane; indeed, I am not certain that he so much as believes there is an astral

---

[88] C.W. Leadbeater, *Clairvoyance* (London: The Theosophical Publishing Society, 1899), p. 32.

[89] C.W. Leadbeater, *The Fourth Dimension (A lecture given by C.W. Leadbetter [sic] before the Amsterdam Lodge T.S. in April, 1900. Stenographic notes in Dutch by J.J. Hallo, Jr.; translated into English by Mrs. Marie Knothe)* (San Francisco: Mercury Publishing Office, 1900), p. 31.

plane; he is regarding it simply as a higher conception of physical space, as a truth existing in the physical world, and to be recognized by those who will take the trouble of studying it sufficiently deeply.[90]

There were, however, significant correspondences with Theosophical thought on at least one very important count:

> We who are Theosophical students realize that the first step on the path of true progress is to get rid of the self, to cast aside the delusion of separateness, and thus to develope [*sic*] perfect unselfishness and learn to work for the benefit of humanity. It is surely more than a coincidence that the first step necessary for the successful and practical study of the fourth dimension is what Hinton calls 'casting out the self.'[91]

Later Theosophical admiration for the arduousness of Hinton's system is continuous with its exclusionary approach to space. The messianic overtones of Leadbeater's forty days and nights of seclusion and meditation in Adyar underscore the inaccessibility of such experience to the uninitiated. Where Carpenter and Stead were interested in the possibility of universal communion offered by this space, and by sharing this possibility, Leadbeater's insistence on selflessness was accented with an insistence on work: 'the trouble of studying it deeply'. Nevertheless, his extensive and repeated use of higher spatial ideas nested Hintonian thought within NeoTheosophy and reinvigorated the idea of the fourth dimension. He was Hinton's most prolific propagandist at the fin de siècle.

The pedagogical mode of Hinton's work was yet more suited to occult writings, in which it was frequently internalized, though in particularly convoluted ways that inflected their relationship to authority. Occult movements, while often structured as hierarchies that reflected the social context from which they emerged, nevertheless displaced the location of the authoritative voice; and so it was with mediated occult texts. Who was the true authority? The medium or the spirit she channelled? Hintonian writing presented rigorous thought in a marginalized intellectual tradition, grounded in a comfortingly authoritative pedagogical voice, and created a formal geometric hinge between the transcendental and the material that was incorporated into Theosophy. Through this route it would recur, occluded, in the Theosophist W.B. Yeats's gyre and sphere and Ezra Pound's vortex, a dynamic form that perfectly combined the scientific and occult producing 'equations' for the emotions.[92] Despite Wyndham Lewis's insistence on vorticism as a purely English innovation, Theosophical geometry also resurfaced in *Blast* through a more circuitous route, via Edward Wadsworth's partial translation of Wassily Kandinsky's 'Concerning the Spiritual in Modern Art'.[93]

In literary criticism, the engagement with occult imagery and ideas of key Modernists—Yeats, H.D., Eliot, and Pound—has been taken increasingly seriously

---

[90] C.W. Leadbeater, *The Other Side of Death* (London: Theosophical Publishing Society, 1903), p. 109.

[91] Leadbeater, *The Other Side of Death*, p. 123.

[92] See Timothy Materer, *Modernist Alchemy: Poetry and the Occult* (Ithaca, NY: Cornell University Press, 1995), p. 33.

[93] Edward Wadsworth, 'Inner Necessity', *Blast*, 1 (1914), 119–25.

since the scholarship of Leon Surette and Timothy Materer in the 1990s.[94] At the same time, critical work on literature and science has read the same material as emerging from within the realm of the scientific.[95] Just as assorted traditions of esoteric and occult geometries eddied and swirled through European artistic and literary thought, so they recombined with the more formal approaches of science: a posteriori disaggregation of these fields is neither accurate nor appropriate.

In occultist discourse in the 1890s higher space became a medium in which immaterial objects would clot and congeal into material forms, while material forms would dissipate and atomize. In performing such a mediating role emphasis fell on higher space as an agent of permeability, a space to which access allowed ingress into not just solid objects, but thoughts. There is a sense in which the permeability of the 1890s was finer-grained than that of the 1880s: accessing the interior of closed boxes from the fourth dimension in the Zöllnerian and Hintonian accounts was not so much a movement through, as ingress from, an entirely new direction; the granular material permeability of Theosophy coincided with greater public awareness of discoveries about the nature of matter, and radiation in particular.

Steven Connor comments on Besant and Leadbeater's theory of thought forms that 'this is the fantasy of the thing the thingness of which has been entirely purged, its place taken up completely by the thought that doubles and determines it'.[96] What better space for this process to take place than a space in this period frequently thought but nowhere found? Theosophical theory contributed to a profoundly altering conception not just of objecthood in the fin de siècle. The disturbance of social and racial hierarchies in the Society's narratives of its own originary myths contributed also to a reconfigured subjecthood.

Adopted and adapted by the membership of the Theosophical Society, higher space also became a turbo-powered analogue for global space. The global span of the Theosophical Society surely supported this abstract analogue: here was an organization whose network expanded within a subsection of global colonial space. Perhaps, too, the dissipation of matter leant itself to this more attenuated higher space, allowed it to be spread more broadly.

These features of higher space were all embraced in fictional encounters with the idea in the fin de siècle, encounters that evidenced and evoked optimism, anxiety, and innovation in equal measure. In reading such encounters we read of the imaginative reshaping in a cultural context of ideas concerning the physical world. Here, misreading and translations are as generative and catalysing for thought as accurate portrayals, and language, narrative, content, and form respond and act in network.

---

[94] Leon Surette, *The Birth of Modernism: Ezra Pound, T.S. Eliot, W.B. Yeats, and the Occult* (Montreal/Kingston: McGill-Queen's University Press, 1994).

[95] See Ian F.A. Bell, *Critic as Scientist: The Modernist Poetics of Ezra Pound* (London: Methuen, 1981).

[96] Steven Connor, 'Thinking Things', http://www.stevenconnor.com/thinkingthings [accessed 22 October 2012] (para. 30 of 94). This is an expanded version of the essay that was published in *Textual Practice*.

6

Fictions
The Spaces of Literature after *n*-Dimensions

The spaces of fiction are multidimensional before higher space. Most concretely, the printed text is a spatial object. A book is, Michel Butor has observed, a 'volume', a three-dimensional agglomeration of two-dimensional sheets of paper and, once we open its covers, a diptych, a visual plane divided into two equal sections.[1] On each of these sections printed words and images occupy space: the Greek alphabet runs these words in horizontal lines, from left to right; East Asian scripts traditionally run vertically downwards, from right to left; other writing systems demonstrate other directionalities. Before printing, though, the planes of the diptych are plain: typography can be organized with freedom upon this two-dimensional white space, as the experiments of the Futurists sought to demonstrate.

The reader, meanwhile, occupies a space in relation to the text. For the relationship to be fruitful the two spaces must be sufficiently proximate to allow page-turning and reading, but each text remains a volume wherever each reader may be in relation to it. Given the mobility a reader enjoys in relation to the resolutely static printed object, we might characterize this relationship as higher dimensional, with the reader occupying a higher-dimensional space to the text. Mark McGurl has made a similar argument with regards to *Flatland* which he argues 'encourages the fantasy of multidimensionality': 'even as it emphasizes the literal spatial relation of the reader's body to the physical ground of representation, readers of *Flatland* are encouraged in the fantasy that their position puts them in a higher dimension altogether'.[2] If so, we should recall that the dimension on the w-axis occupied by the reader, beyond the x, y, and z of the volume, is characterized by freedom of movement; a relative dimension of space-*time*.

More complicated are the multiple spaces of reading, the spaces we encounter within the text. Butor describes the novel's space as 'complementary' to the space in which its reader exists; it displaces the space of the reader only so long as it holds her interest. 'We will need details; show us a sample of this decor, object, furnishing, which will serve as a token. What kind of room?' Butor describes how objects 'can assume a signal value' in this act of narrative detailing:

Up to this moment, the room which is becoming explicit before our eyes remained an amorphous container, a kind of sack in which the objects were stuffed pell mell; now

---

[1] See Michel Butor, 'The Space of the Novel' and 'The Book as Object', in *Inventory*, ed. Richard Howard (London: Jonathan Cape, 1961), pp. 31–8, 39–56.

[2] Mark McGurl, 'Social Geometries: Taking Place in Henry James', *Representations*, 68 (1999), 59–83 (p. 64).

the narrator extracts them one by one, at random. Soon we will want to know their locations [...] In order to achieve such and such an arrangement, it will be necessary to introduce certain details, or objects, which are not usually mentioned, so as to create in the imagined space certain precise and stable figures.[3]

The role of objects in making present and concrete abstract space is something with which the current argument is familiar but in narrative fiction they assume a role of even greater significance as instigators of space. Things in texts make space in the mind. What sort of objects, or things, mediate the higher-dimensioned spaces of fiction in the way that knots did for Zöllner, or cubes for Hinton? How do they bring about the higher spaces of fiction?

Thing-interest apart, we are moving beyond surface descriptions and into what Butor describes as the 'imagined space' of the reader created by chains of signification and representation. We might distinguish between the represented space of the text—the described space of a room, a field, or a city—and the produced space—our sense of space produced by language. Isobel Armstrong seeks to understand how 'a writer convince[s] us through the abstractions of language that spatial experience created in a text is, by an extraordinary transposition, recognizable, vivid, a *lived experience*'.[4] Considering Kantian space 'foundational' for the nineteenth-century novel, she begins her analysis from a summary of the four propositions that ground Kantian space. Firstly, that space is external, outside the subject. Secondly, that space cannot be thought away; that we can think of an absence of objects in space, but not an absence of space. Thirdly, that space is single, a unity that can be parcelled up but does not pre-exist as sections. And finally, that space is infinite, containing everything. Armstrong argues that 'a novel needs not simply to evoke space, as it would a smell or the taste and sight of a madeleine (famously in Proust) or a jam puff (in George Eliot's *The Mill on the Floss*). It has to make space a constitutive element, to *produce* it.' She argues that this production can be achieved

> by alternately negating and confirming the four principles that make spatial experience possible. Take away the body, reintroduce it: empty space of objects, restore objects; obliterate partitioned space, reinstate division; all of these are procedures that can be reversed or combined with one another. We are prompted into intensified spatial imagining when the novelist signals a change in spatial relations.[5]

Given Armstrong's focus on the nineteenth-century novel and the concern with Kantian spatiality that has been apparent among many higher spatial thinkers of the fin de siècle, Kant will be a useful point of orientation for thinking how the fourth dimension changed the produced spaces of the novel in this period, for after *n*-dimensional space, writers are operating in a universe at least partially post-Kantian.

Armstrong's set of practices for producing space in the novel will need to be recalibrated to account for the post-Kantian space of *n*-dimensions. Higher space

[3] Butor, 'The Space of the Novel', p. 34.
[4] Isobel Armstrong, 'Spaces of the Nineteenth-Century Novel', in *The Cambridge History of Victorian Literature*, ed. Kate Flint (Cambridge: Cambridge University Press, 2012), pp. 575–97 (p. 575).
[5] Armstrong, 'Spaces', p. 578.

is no longer space external, outside me, but is also through me. In higher spatial fiction the 'inter-spatial subject' that interests Armstrong must be reconfigured as a hyper- or higher-spatial subject. The theoretical possibilities of bi- and co-location afforded by higher space can place two subjects in the same space or the same subject in two spaces. The subject might even occupy a part of the four-dimensioned space such that it does not impinge on a particular three-dimensional section of that space at all, might slip out of three-dimensional space altogether but be close to it, like the Sphere before he enters Flatland or A Square hovering above Lineland.

As this last feature indicates, higher space also makes problematic the notion of space as a unity. Space of dimensions $n-1$ is by geometric definition a 'section' of n-dimensional space: in the case of three-dimensional space a two-dimensional plane is a cross-section of said space; three-dimensional space has an analogous relationship to four-dimensional space. We become aware of the logical problems of applying geometrical ideas to space itself. The cross-section is an abstraction. Any object in three-dimensional space that approaches a plane remains three-dimensional. Its thickness may be infinitely small, but it has thickness. If space is to remain a unity, if there is to be only one kind of space, mustn't it be of consistent dimensionality?

Higher space also makes slippery the notion that space cannot be thought away. If we can think space expanded and extended beyond the sensible, are we still so certain that we cannot think standard space away? Here we must dive deeper into Kant's categories of thought. According to Kant, space is a priori because it comes before the empirical, it is not sensed. Higher-dimensioned space, while it is not sensed, does not come to us in the same way. It is a product of the understanding alone, an a posteriori concept without empirical content. We may concede that it is possible to imagine higher space but there remain significant obstacles to representing what has never been sensed.

We might ask, then, how fiction produces higher space, how it makes higher space a constitutive element; what is the nature of the hyper-spatial subject as opposed to the inter-spatial subject Armstrong sees as germane to the novel? How are *its* relations structured? What is clear from the outset is that higher space suggests forms of embodiment (or disembodiment) beyond those of standard space.

Armstrong writes that 'Kant added a further element to his spatial thought, the body'.[6] In his inaugural address Kant contemplated the subjective experience of handedness, noting that right-handed solid objects could not be made to coincide with left-handed. These 'incongruent counterparts' demonstrated for Kant that space was external and absolute; since they could not be made to coincide they revealed that space was independent of the objects within it. Kant had used human hands as his example of incongruent counterparts, an example that emphasized the embodied nature of the subjective experience of spatiality. To reproduce the embodied nature of space the novelist 'needs not only to produce a mimesis of the

---

[6] Armstrong, 'Spaces', p. 577.

a prioris of space—boundaries and horizons, windows and walls—but to create the concrete fully inhabited world of the situating body'.[7]

To read this fictional body in a nineteenth-century novel we should pay close attention to the information provided by all the senses. Each is more or less extensive, and we may arrange them hierarchically according to their reach: vision, hearing, smell, touch, and taste. The senses embody the fictional character. After *n*-dimensions the matter of embodiment is less simple. In the foundational text of nineteenth-century mathematical higher spatial theory, Möbius's 1827 paper on Barycentric Calculus, it was demonstrated that a putative space of four dimensions would enable the orientation into congruence of incongruent solid objects.[8] Higher space in its germinal form therefore suggested alternative embodiment, making possible the inversion of the body, the transportation of left into right. The *a prioris* of space listed by Armstrong—boundaries and horizons, windows and walls—can no longer be taken for granted after the imagination of a kind of space that dissolves such structures. In reading higher spatial fictions we both encounter alternative embodiments and abandon these *a prioris*. The senses will be put to the test.

Of these, we will most frequently begin with vision. Michel Butor writes:

> One of the most effective means [of producing space] is the introduction of an observer, of an eye, which can be motionless and passive, in which case we get passages which will be the equivalent of photographs, or in motion, in which case we get film or painting.[9]

The observing eye of the reader is yet more mobile than that of the character. Armstrong writes: 'But whereas the fictional character is fixed along a trajectory of moves by the author, the reader can be granted a mobile situatedness, sometimes identifying with the fictional character's spatial experience and perspective, sometimes external to it, looking *at* rather than *with* them.'[10] The reader has the same advantage over the character she has over the material text: mobility.

At its plainest level, narrative voice is central to the constituted spatiality of a text. In the broadest terms, a first-person narrative locates the reader in a determinate subjective space while the third person allows for greater mobility of perspective. Such mobility finds its apogee in the intermediate voice that uses aspects of both third- and first-person, a style that is most frequently termed free indirect discourse after the French 'style indirect libre'. In the English language this technique was frequently evident in the nineteenth-century novel—Austen and Eliot give us rich examples—but was theorized most comprehensively by Henry James. Gérard Genette writes that 'for post-Jamesian partisans of the mimetic novel (and for James himself)' it was 'the best narrative form'.[11]

Genette reads James's theory of the novel as a resurgence of the Platonic struggle between 'pure narrative' and mimesis. 'We know how this contrast [...] abruptly

---

[7] Armstrong, 'Spaces', p. 578.

[8] See August Möbius, 'On Higher Space', in *Sourcebook in Mathematics*, ed. D.E. Smith, trans. Henry P. Manning (New York: McGraw Hill Book Company, 1929), pp. 525–6 (first publ. in *Der barycentrische Calcul* (Leipzig: [n. pub.], 1827)).

[9] Butor, 'The Space of the Novel', p. 34.   [10] Armstrong, 'Spaces', p. 578.

[11] Gérard Genette, *Narrative Discourse: An Essay in Method*, trans. Jane E. Lewin (Ithaca, NY: Cornell University Press, 1983), p. 168.

surged forth again in novel theory in the United States and England at the end of the nineteenth century and the beginning of the twentieth, with Henry James and his disciples, in the barely transposed terms of showing vs. telling.'[12] Genette considers the 'focalization' of narration and reads Jamesian technique as a 'fixed internal focalization', maintaining its centre in one character as opposed to moving between many (variable) or many at the same time (multiple).[13]

Dorrit Cohn terms the technique 'narrated monologue' and offers a clear description of how it is achieved: the dropping of the third-person verbs of narration, or thought, as the passage proceeds. Cohn makes the clear point that this technique fuses both thought and vision:

> The narrated monologue—in contrast to the quoted monologue—suppresses all marks of quotation that set it off from the narration, and this self-effacement can be achieved most perfectly in a milieu where the narrative presentation adheres most consistently to a figural perspective, shaping the entire fictional world as an uninterrupted *vision avec*. The narrated monologue itself, however, is not *vision avec*, but what we might call *pensé avec*: here the coincidence of perspectives is compounded by a consonance of voices, with the language of the text momentarily resonating with the language of the figural mind.[14]

As an example of the internal spatiality of narrative perspective, free indirect discourse might owe much to the kinds of mobility offered by a fourth dimension. The narrated monologue has a 'now-you-see-it, now-you-don't quality' due to its commingling of 'narratorial and figural language'.[15] Cohn extends its ambiguity to the presentation of character and context. Where narration imposes a boundary between the internal and external, the narrated monologue 'can reflect sites and happenings even as they show a character reflecting on these sites and happenings'.[16] This is a technique that uses language to produce spatial effects, to manoeuvre the readerly consciousness into communion with the figural consciousness, and allows a perspective that it is at the same time single and double, a liminal, intermediate voice. Cohn describes this as 'imperceptibly integrating mental reactions into the neural-objective report of actions, scenes, and spoken words'.[17] A reading of James, and attention to his interest in the fourth dimension, is a particularly useful way to investigate this.

Twentieth-century literary criticism had been very interested in narrative space before structuralism, under the rubric of 'spatial form'. Joseph Frank's essay 'Spatial Form in Modern Literature', first published in the *Sewanee Review* in 1945 and anthologized numerous times since, gave rise to many followers and several continuations and instigated lively debate, not least with Frank Kermode in the pages of *Critical Inquiry*. Joseph Frank built on a reading of Lessing's organization of aesthetic theory, that

---

[12] Genette, *Narrative Discourse*, p. 163.   [13] Genette, *Narrative Discourse*, p. 189.
[14] Dorrit Cohn, *Transparent Minds: Narrative Modes for Presenting Consciousness in Fiction* (Princeton: Princeton University Press, 1978), p. 111.
[15] Cohn, *Transparent Minds*, p. 109.   [16] Cohn, *Transparent Minds*, p. 132.
[17] Cohn, *Transparent Minds*, p. 115.

time and space were the two extremes defining the limits of literature and the plastic arts in their relation to sensuous perception; and it is possible, following Lessing's example, to trace the evolution of art forms by their oscillations between these two poles.

He argued that modern literature, as exemplified by several canonical works—*The Waste Land, The Cantos, Swann's Way, Ulysses*, and, in a lengthy analysis, *Nightwood*—was 'moving in the direction of spatial form. This means that the reader is intended to apprehend their work spatially, in a moment of time, rather than as a sequence.' In other words, these key Modernist works attempt 'to remove all traces of time-value'.[18]

In later accounts Frank has argued for the commonality between his essay and the structural approaches of Jakobson, Shklovsky, and Genette. Frank sees the Formalist distinction between story and plot as indicating the spatial organization wrought on the text by the novelist and highlights Genette's description of Saussurean linguistics as 'a mode of being of language that one must call spatial'.[19] Frank's spatial form is a unifying mode of analysis that accounts for 'three fundamental aspects of narrative: language, structure, and reader perception'.[20] For this reading it not only gathers together the spaces of literature we have already sketched but indicates another layer in how we might consider the space of higher spatial fictions: their formal organization.

Perhaps the ultimate space of the novel is the idea of space prevalent at the time in which a text is written. The ways in which this cultural space was extended or altered by higher-dimensional thinking have been the subject of the present work and it is through analysis of literary texts that I hope to reach some kind of summary of the condition of cultural space after the *n*-dimensional turn.

Higher-dimensional spaces recur with some frequency in the literature of the fin de siècle. H.G. Wells's early 'scientific romances' used the idea again and again—Bernard Bergonzi has described it as Wells's 'favourite motif' in his writing of the 1890s—in the novels *The Time Machine* (1894–5), *The Wonderful Visit* (1895), *The Invisible Man* (1897), and the short stories 'The Remarkable Case of Davidson's Eyes' (1895) and 'The Plattner Story' (1896).[21] Charles Howard Hinton published 'Stella' (1895), an invisible person novella, and a work that nourished Wells, 'An Unfinished Communication' (1895), and *An Episode of Flatland* (1907). *Lilith* (1895), by George MacDonald, owing more to the tradition of the fairy story and biblical myth than the more modern scientific fantasies of Wells, also featured at

---

[18] Joseph Frank, 'Spatial Form in Modern Literature: An Essay in Three Parts', *Sewanee Review*, 53 (1945), 221–40, 433–56, 643–53 (pp. 225, 650).

[19] Joseph Frank, 'Spatial Form: Thirty Years After', in *Spatial Form in Narrative*, ed. Jeffrey R. Smitten and Ann Daghistany (Ithaca and London: Cornell University Press, 1981), pp. 202–43 (p. 241).

[20] Jeffrey R. Smitten, 'Introduction: Spatial Form and Narrative Theory', in *Spatial Form in Narrative*, ed. Jeffrey R. Smitten and Ann Daghistany (Ithaca and London: Cornell University Press, 1981), pp. 15–36 (p. 15).

[21] Bernard Bergonzi, *The Early H.G. Wells* (Manchester: Manchester University Press, 1961), p. 71.

its core a space, at least nominally, higher dimensional—'the region of the seven dimensions'.[22]

The first of three collaborations between Joseph Conrad and Ford Madox Hueffer (later Ford) *The Inheritors* (1901), centred on the machinations of the Dimensionists, a race of humanoids from the fourth dimension. Henry James made brief allusion to the fourth dimension in *The Spoils of Poynton* (1897), a text of considerable interest also for its thingly concerns, while his story 'A Great Good Place' (1899) models a spatial elsewhere with higher dimensional overtones.[23]

George Griffith, a mass-market generic writer who had enjoyed huge success with the future war novel *The Angel of the Revolution* (1893), used the idea of the fourth dimension in two short stories around the turn of the century before placing it centre stage in the last novel he published before his death, *The Mummy and Miss Nitocris* (1906). A profoundly flawed but terrifically fun pulp read, Griffith's novel is a treasure trove of higher-dimensional tropes. The John Silence story by Algernon Blackwood, 'A Victim of Higher Space', was not published until 1911, but was most likely written around 1907.

The American satirist and journalist Ambrose Bierce offered an account of higher-dimensioned space in 'Science to the Front', first published in 1910 as the epilogue to three tales of mysterious disappearances first published in 1893. Another American magazine humorist, Hinton's literary executor Gelett Burgess, touched on the idea briefly in a comic ghost story, 'The Ghost Extinguisher' (1905), while the novelist Mary Wilkins Freeman employed it for a supernatural tale, 1905's 'The Hall Bedroom'.

Perhaps most notoriously, H.P. Lovecraft was drawn again and again to the ideas and terminology of *n*-dimensional space. Lovecraft's vision of 'unplumbed space' was undoubtedly indebted to William Hope Hodgson, a British author whose work he much admired. Both Hodgson and Lovecraft fall strictly beyond the period terms of this book, but their work demands acknowledgement in this chapter. In this canon, highly admired Modernist novelists find themselves on equal footing with authors of pulp fictions.

MIRRORS: INVERSION AND THE LITERARY VISUAL

In his advice to the investigator of things as participants in networks of action, Bruno Latour suggests:

> Finally, when everything else has failed, the resource of fiction can bring—through the use of counterfactual history, thought experiments, and 'scientification'—the solid objects of today into the fluid states where their connections with humans may make sense.[24]

[22] George MacDonald, *Lilith* (London: Chatto & Windus, 1895; repr. Grand Rapids, MI: Wm. B. Eerdmans Publishing Company, 2000), p. 21.

[23] I owe acknowledgement to Elizabeth Throesch for drawing my attention to this James story in the context of higher-dimensional thinking.

[24] Bruno Latour, *Reassembling the Social: An Introduction to Actor Network Theory* (Oxford: Oxford University Press, 2008), p. 82.

Latour's advice is useful here. I am interested in the fluidity of the things of fiction in specifically spatial terms: how they make space in fiction, and what sorts of thing make what sorts of spaces. The things of fiction are already more fluid than the solid objects they stand in for. Let us note but defer the fluidity of the language in which fictional things are rendered, the levels of signification that provide each reader with different mental constructs from the same words, and focus instead on Butor's objects of 'signal value' in creating space in the mind of the reader.

George MacDonald's *Lilith* provides a wealth of spatializing objects in complex arrays that lead the narrator Vane into 'the region of the seven dimensions'. Vane's first glimpse of Mr Raven, a guide who takes him across the threshold between worlds, occurs as sunlight shines into the library through a gap in the clouds and falls upon a painting of an ancestor. 'The direct sunlight brought out the painting wonderfully,' comments Vane:

> With my eyes full of the light reflected from it, something, I cannot tell what, made me turn and cast a glance to the farther end of the room, when I saw, or seemed to see, a tall figure reaching up a hand to a bookshelf.[25]

Vane later follows the same figure to the main garret in the house, where he finds another mirror: 'A few rather dim sunrays, marking their track through the cloud of motes that had just been stirred up, fell upon a tall mirror with a dusty face, old-fashioned and rather narrow—in appearance an ordinary glass.' Only in appearance is the mirror ordinary, however. Looking into it Vane becomes 'aware that it reflected neither the chamber nor my own person', seeing in place of his reflection a wild landscape. Vane reaches forward, stumbles, and finds himself 'on a houseless heath'.[26]

Yet another mirror features when Vane is returned from this landscape to the library. 'In the library was one small window to the east, through which, at this time of the year, the first rays of the sun shone upon a mirror whence they were reflected on the masked door.' Vane's eyes follow the reflected light which indicates the door masked by book spines; he opens the door and the light, 'like an eager hound', indicates a book within which he finds a manuscript written by his father giving a detailed account of how the mirror portal in the garret is operated.[27] In this, Mr Raven remarks that 'its doorness depends on the light'. Chains are pulled to revolve the roof of the garret to face the sun:

> A moment more and the chamber grew much clearer: a patch of sunlight had fallen upon a mirror on the wall opposite that against which the other leaned, and on the dust I saw the path of the reflected rays to the mirror on the ground. But from the latter none were returned, they seemed to go clean through, there was nowhere in the room a second patch of light.[28]

---

[25] MacDonald, *Lilith*, p. 6. [26] MacDonald, *Lilith*, p. 11.
[27] MacDonald, *Lilith*, p. 37. [28] MacDonald, *Lilith*, p. 41.

Vane returns to the garret and attempts the operation himself.

> Suspecting polarisation as the thing required, I shifted and shifted the mirrors, changing their relation, until at last, in a great degree, so far as I was concerned, by chance, things came right between them, and I saw the mountains blue and steady and clear. I stepped forward and my feet were among the heather.[29]

It is a weaving chain of object and object-relations that leads Vane into the mythical region of the seven dimensions—'or call it a state of things, an economy of conditions, an idea of existence'—but the repeated mirrors are impossible to ignore.[30] At least four different reflective glasses polarize and bounce rays of light before opening onto entirely new and alternate spaces, open heathland contrasting starkly with the confined architectural spaces of garrets and libraries.

The mirror of higher-dimensional fiction may be a door or a window. In *The Mummy and Miss Nitocris* Professor Marmion looks towards a wardrobe mirror: 'To his amazement he did not see himself reflected in it. The mirror seemed to have vanished, and in its place was a window looking into his study.'[31]

The mirror has an immediate relationship to higher space, its inversion of the visual plane giving an image of the incongruent counterparts we cannot invert in three-dimensioned space. Indeed, we might read the looking glass as the higher spatial portal through which we can never pass in lived space. As Carroll's Alice declares:

> Oh, Kitty! how nice it would be if we could only get through into Looking-glass House! I'm sure it's got, oh! such beautiful things in it! Let's pretend there's a way of getting through into it, somehow, Kitty. Let's pretend the glass has got all soft like gauze, so that we can get through.[32]

Barely has she wished it than Alice is 'through the glass'. For Vane the pretence is not explicit: the space of identification between MacDonald's character and reader is not quite as intimate as that between Alice and Kitty (or Kitty and the reader). Vane must puzzle and deduce his way through the network of mirrors and beams of light, with the reader following close behind him.

Vane's other objects—books, paintings, and manuscripts—are also portals. The things Vane encounters are all spatializing in Butor's sense—we amass a mental stock of images of portraits, libraries with secret doors, and garrets containing mirrors. We picture ourselves in a grand house and recreate the rooms within it. The mirrors are spatializing in another sense, presenting on their plane a space that is our own, but not our own. 'In itself, the reflection is an equivocal theme,' writes Genette:

> the reflection is a *double*, that is to say at the same time an *other* and a *same*. This ambivalence provokes in baroque thought an inversion of significations which makes

---

[29] MacDonald, *Lilith*, p. 43.   [30] MacDonald, *Lilith*, p. 12.
[31] George Griffith, *The Mummy and Miss Nitocris* (Milton Keynes: Tutis Digital Publishing, 2008; repr. of London: T. Werner Laurie, 1906), p. 23.
[32] Lewis Carroll, *Through the Looking Glass and What Alice Found There* (London: Academy Editions, 1977), p. 12.

identity fantastic (I am an other) and otherness reassuring (There is another world, but it is similar to this one).[33]

The space through the mirror is simultaneously alienating and comforting; it reflects and multiplies. The mirror itself, like the knot, is a form through which matter and thought are made fluid. In a text giving detailed instructions on the theory, use, and construction of magnetic mirrors for occult ritual and scrying, Paschal Beverly Randolph described the source for the form of the reflective glass, or crystal ball:

> It was found, also, that the shape of the brain, at the foundation-point, was of the same general form or shape as the earth on which we dwell; that is to say, an oblate spheroid, whence, by experiment, it was deduced that such section of a figure, oblately spheroidal, was also the very best form of a magic mirror. Such a figure having two mathematically true and absolutely certain *foci*, so that a stream of magnetism being thrown upon one focus slid along the surface and returned to the centre of the other focus, from the centre of the fore-brain, thus completing a magnetic circuit and rendering the portion of brain in the line of contact exceedingly active, by reason of its increased magnetic play and motion of the brain-particles there situate.[34]

The mirror becomes materially one with the self, a magical prosthetic channelling the matter of the brain. Pancho, the hero of Franz Hartmann's *The Talking Image of Urur*, finds himself in a hall dominated by such an occult mirror:

> This light was reflected in all the mirrors; but more especially from one great concave mirror in the middle of the front wall, which caught the rays of the light and threw them into the little mirrors, where they sparkled like so many diamonds.
>
> Full of surprise Pancho approached that mirror and saw his own image reflected, although magnified into superhuman dimensions. While his attention was directed intensely toward that mirror, and while he was wondering about this strange phenomenon, his consciousness became suddenly centred in that image, and then it seemed to him as if he himself were that image, looking out of the mirror, and he beheld his figure reflected from all the little mirrors along the walls.[35]

The occult mirror channels and multiples the self as the textual mirror multiplies its internal spaces, the spaces that displace our reading space; both leave us feeling discombobulated. We should note that all three structures through which Vane is led to the region of the seven dimensions—mirrors, paintings, and texts—are capable of creating the effect of *mise en abîme*, of locating the subject in spaces of thought that appear to have the potential to recede ad infinitum. In the text of the

[33] Gérard Genette, 'Complexe de Narcisse', in *Figures I* (Paris: Editions de Seuil, 1966), pp. 21–8 (p. 21).

[34] B. Randolph, *Seership! The Magnetic Mirror* (Toledo, OH: K.C. Randolph, 1884), p. 55.

[35] Franz Hartmann, *The Talking Image of Urur* (New York: John W. Lovell and Company, 1890), p. 290.

manuscript discovered in the library by Vane, Mr Raven explains the science, or meta-science, of the portal to Vane's father:

> He then talked of the relations of mind to matter, and of sense to qualities, in a way I could only a little understand, whence he went on to yet stranger things which I could not at all comprehend. He spoke much about dimensions, telling me that there were many more than three, some of them concerned with powers which were indeed in us, but of which as yet we knew absolutely nothing.[36]

This tone of incomprehension is struck again and again in higher-dimensional narratives. Greville MacDonald has glossed his father's terminology, indicating the German mystic Jakob Boehme as George's source for extra dimensions:

> The fourth dimension, comprising four others of elemental, illimitable reach—like the four points of the compass, or the Four Elements in spiritual aspect—is the one in which we have to find sustenance and get emancipation from the plumb-line that holds us to earth and its brute-laws. Yet it pervades and uses that inseparable three for work and expression, and through these for art and salvation. Our bodies are not more vile than our souls, for we are they: only we have to get home with them—not by going in at any door, but by finding the door out, on every side of us, that frees us of our self.[37]

PORTALS: BI-LOCATION AND THE OBJECT

The fourth dimension itself becomes a portal in George MacDonald's version of higher space, opening onto a further four dimensions that free the soul from the material earth and its 'brute-laws', such as gravity, that arbiter of three-dimensional space, its 'plumb-line'.

If *Lilith*'s garret containing an array of mirrors capable of some recondite polarization of light reads like an ad hoc laboratory, H.G. Wells's 'The Remarkable Case of Davidson's Eyes' finds the protagonist surrounded by the fragments of a smashed electrometer following an explicit laboratory accident. At the end of the tale, Davidson's supervisor, Professor Wade, suggests that the apparatus has been instrumental in the event: 'His idea seems to be that Davidson, stooping between the poles of the big electro-magnet, had some extraordinary twist given to his retinal elements through the sudden change in the field of force due to the lightning.'[38] In Hinton's novella *Stella* Hugh Churton discovers

> a series of rooms fitted up as laboratories. The apparatus was in good order. There were many instruments with the use of which I was not acquainted. Whatever Michael Graham had required for the prosecution of his researches—apparently in the border

---

[36] MacDonald, *Lilith*, p. 41.
[37] Greville MacDonald, 'Introduction', in *Lilith* (London: George Allen and Unwin, 1924), pp. ix–xx (p. xv).
[38] H.G. Wells, 'The Remarkable Tale of Davidson's Eyes', in *Selected Short Stories* (London: Penguin, 1968), pp. 174–83 (pp. 182–3).

land between chemistry and physics—he had obtained regardless of expense. One of the finest spectroscopes I had ever seen was there.   (*S*, 16–17)

Optical apparatus can be just as easily replaced by equations, algebra, the double writing of mathematics standing in for the material objects. Wells's invisible man Griffin records: 'I found a general principle of pigments and refraction, a formula, a geometrical expression involving four dimensions.'[39] Stephen Lorne, the protagonist of George Griffith's novel *The Justice of Revenge*, is recovering from 'a terrible shock of such intensity that it would have driven most men insane':

> Sitting in the little room that was half study and half laboratory, the words and formulae on the sheet before him began to move about and form themselves into strange shapes and combinations, which gave him visions of amazing possibilities, and even glimpses into that fourth dimension of space which has been the dream of philosophers for ages.[40]

Mary Wilkins Freeman's story 'The Hall Bedroom' saves its higher-dimensional twist for its final paragraph, in which a bricked-in room is discovered containing nothing but 'a sheet of paper covered with figures, as if somebody had been doing sums. They made a lot of talk about those figures, and they tried to make out that the fifth dimension, whatever that is, was proved, but they said afterward they didn't prove anything.'[41]

The continuity between apparatus and equations is clear in thematic terms: both come from the realm of science. It is when we think them as thinking things that they give us more useful information. Both structures are Latourian black boxes, containing condensed within them the hard-won results of earlier thought: the polishing of lenses or prisms to achieve the correct refractive index, for example; a single algebraic notation representing the solving of prior geometrical problems. Both have already done some of our thinking for us, are things that make the divide between thinking subject and inert object more fluid than we first read it; both are things that telescope when we unpack them, containers and mediators for extended spaces and times. In Algernon Blackwood's story 'A Victim of Higher Space', the thinking thing with which the victim has thought higher space is already familiar to us: 'I procured the implements and the coloured blocks for practical experiment, and I followed the instructions carefully till I had arrived at a working conception of four-dimensioned space.'[42]

Portals are not the only permeable things in higher spatial texts. Professor Marmion watches as his daughter's 'hand passed without any resistance through the jar' containing his drinks. The narration locates us within the Professor's thoughts: 'There in material form on the corner of his table was a point blank, tangible contradiction of the universally held axiom that two bodies cannot occupy

---

[39] H.G. Wells, *The Invisible Man: A Grotesque Romance* (London: J.M. Dent, 1995), p. 81.
[40] George Griffith, 'The Justice of Revenge', *Newcastle Weekly Courant*, 28 April 1900, 2.
[41] Mary E. Wilkins Freeman, 'The Hall Bedroom', in *Short Story Classics (American)*, ed. William Patten (New York: P.F. Collier and Son, 1905), IV, pp. 1275–99 (p. 1298). All further references to this edition will be given in the body of the text using the abbreviation THB.
[42] Algernon Blackwood, 'A Victim of Higher Space', in *The Complete John Silence Stories*, ed. S.T. Joshi (New York: Dover Publications, 1998), pp. 230–46 (p. 238).

the same space.' The Professor reaches for the decanter but grasps a flagon placed there by his daughter's double. To test the notion forming in his mind, that these occurrences might result from his higher-dimensional speculations, he succeeds in removing a scarab ring *through* his finger.

For Bill Brown the word 'things' 'index[es] a certain limit or liminality, [...] hover[s] over the threshold between the nameable and unnameable, the figurable and unfigurable, the identifiable and unidentifiable'.[43] In *The Spoils of Poynton* this liminality reaches over the text in Mrs Gereth's

> strange, almost maniacal disposition to thrust in everywhere the question of 'things'; to read all behaviour in the light of some fancied relation to them [...] 'Things' were of course the sum of the world; only, for Mrs Gereth, the sum of the world was rare French furniture and oriental china.[44]

The spoils are emoting things and act as if endowed with consciousness. They define the behaviour of the characters in the novel through their relations to these characters. Mrs Gereth's 'mania' is even more pronounced in her own words:

> There isn't one of them I don't know and love—yes, as one remembers and cherishes the happiest moments of one's life. Blindfold, in the dark, with the brush of a finger, I could tell one from another. They're living things to me; they know me, they return the touch of my hand.[45]

Fleda, the 'central intelligence' of the novel, is Mrs Gereth's only true confidante, and the only other character who understands the strength and source of this thingly relationship. Appraising the collection of objects she exclaims:

> 'If there were more there would be too many to convey the impression in which half the beauty resides—the impression somehow of something dreamed and missed, something reduced, relinquished, resigned: the poetry, as it were, of something sensibly gone.' Fleda ingeniously and triumphantly worked it out. 'Ah, there's something here that will never be in the inventory!'

Mrs Gereth asks her to 'give it a name'. She replies: 'I can give it a dozen. It's a kind of fourth dimension. It's a presence, a perfume, a touch. It's a soul, a story, a life. There's ever so much more here than you and I.'[46]

BODIES: CORPOREAL SUBJECTIVITY

Fluidity between the minds and the things of higher spatial fictions is reproduced in bodies. At the level of representation the fourth dimension enables the body to be inverted, absented, or doubled in three-space. Incongruent counterparts can be made to coincide, space is no longer a unity, and two objects or consciousnesses can occupy the same space. As for things, so for bodies. The 'Foreword' to *The*

---

[43] Bill Brown, 'Thing Theory', *Critical Inquiry*, 28 (2001), 1–22 (p. 5).
[44] Henry James, *The Spoils of Poynton* (Oxford: Oxford University Press, 1982), p. 16.
[45] James, *The Spoils of Poynton*, p. 20.   [46] James, *The Spoils of Poynton*, p. 172.

Mummy and Miss Nitocris, excerpted, it is claimed, 'from the "Geometrical Possibilities," of Abd'el Kasir, of Cordoba, circa. 1050 A.D.', summarized the conditions of 'another world, or state of existence' beyond the 'tri-dimensional'. A dweller in such 'would be freed from those conditions of Time and Space':

> For example, he would be able to make himself visible or invisible to us at will by entering into or withdrawing himself from this State, and returning into that of Four Dimensions, whither our eyes could not follow him—even though he might be close to us in our sense of nearness. Moreover, he could be in two or more places at once, and cause two bodies to occupy the same space—which to us is inconceivable. Stranger still, he might be both alive and dead at the same time—since Past, Present, and Future would be all one to him; the world without beginning or end.[47]

Hinton explored the inversions of the fourth dimension in his first collection of *Scientific Romances* in the pieces 'A Plane World' and 'A Picture of Our Universe'. 'A Plane World' borrowed *Flatland*'s conceit of personified geometric figures to conceive of a male and female pair of triangles of opposite orientation. 'It is said that two beings, the most ideally perfect Vir and Mulier, were once living in a state of most perfect happiness when, owing to certain abstruse studies of the Mulier, she was turned irremediably into a man' (*SR*, 146). Mulier would not tell the secret of her inversion but pledged to either die or return to her prior orientation. Hinton continues the metaphoric inversion of gender in his explanation: 'From our point of view it is easy to see what had happened. If the figure Mulier be taken up and turned over it will be easy to see that, though still a woman, her configuration has become that of a man' (*SR*, 147).

The idea of gendered congruence was not isolated to the work of Hinton. The 1904 novel *Eduards Traum* by the humorist Wilhelm Busch, an heir to the tradition of the geometric satire, riffed extensively on incongruent geometric couples and the annihilation of their differences in higher space:

> I just saw two spherical triangles, one the exact reflected image of the other. They returned in tears from the congruence office where they had been refused. A pair of infinitely delicate gloves, one left one and one right one, were groomsman and brides-maid, and they comforted the unfortunate couple, saying that they were in the same predicament and if there was no other hope they could after all elope into the fourth dimension, where nothing was impossible. 'Alas!' sighed the bride, 'who knows what the fourth dimension is like?' One might have pitied the poor people but we must not be too quick with our sympathy, for the inhabitants of this unsubstantial country are hollow, sun and moon shine through them, and any one who stands behind them can easily count the buttons on their vests in front.[48]

For higher spatial thinkers with progressive views about gender and sexuality, the potentials of higher spatial union generated optimism or humour and the dissolution of fixed oppositions. In physical mode, though, Hinton speculated

---

[47] Griffith, 'The Justice', 2.
[48] Wilhelm Busch, *Edward's Dream*, ed. and trans. Paul Carus (Chicago: Open Court Publishing Company, 1909), pp. 36–7.

alternate outcomes to the meeting of doubles. 'A Picture of Our Universe' invited its reader to picture coils of opposite orientation, 'twists' and 'image twists'. Relating these to the negative and positive charges of static electricity, it imagined that for every twist the image twist, its 'simulacrum', was somewhere produced. Having rehearsed various conditions for the conduction and discharge of electric current and extrapolated to larger, more complex bodies, Hinton transposed his speculations onto the human body in terms that reproduced the annihilation of the idea of incongruent counterparts:

> We must conceive that in our world there were to be for each man somewhere a counter-man, a presentment of himself, a real counterfeit, outwardly fashioned like himself, but with his right hand opposite his original's right hand. Exactly like the image of the man in a mirror. And then when the man and his counterfeit met, a sudden whirl, a blaze, a little steam, and the two human beings, having mutually unwound each other, leave nothing but a residuum of formless particles.   (*SR*, 172)

Gottfried Plattner, the subject of Wells's short story 'The Plattner Story', does not meet but becomes his own 'counterfeit'. Our narrator—the voice of Wells at barely one remove—informs us: 'Never was there a more undeniable fact than the inversion of Gottfried Plattner's anatomical structure.'[49] Plattner, a teacher at a private school, has disappeared for nine days after igniting a mysterious green powder brought to the school by a pupil. On his return he has become a mirror image of himself:

> Gottfried's heart beats on the right side of his body [...] all other unsymmetrical parts of his body are similarly misplaced. The right lobe of his liver is on the left side, the left on his right; while his lungs, too, are similarly contraposed.[50]

Wells offers the following technical explanation for the fate that has befallen Plattner:

> There is no way of taking a man and moving him about in space, as ordinary people understand space, that will result in our changing his sides [...] Mathematical theorists tell us that the only way in which the right and left sides of a body can be changed is by taking that body clean out of space as we know it—taking it out of ordinary existence, that is—and turning it somewhere outside space [...] To put the thing in technical language, the curious inversion of Plattner's right and left sides is proof that he has moved out of our space into what is called the Fourth Dimension, and that he has returned again to our world.[51]

Plattner is fortunate to return as more than 'a residuum of formless particles'. He has undergone a higher-dimensional accident but has been able to come back, inverted: a curious body, certainly, a corporeal oddity, but not impossible—organ inversion occurs in nature. He has gone suddenly, like Mulier—'she disappeared absolutely; although she was surrounded by her friends, she absolutely vanished'

---

[49] H.G. Wells, 'The Plattner Story', in *Selected Short Stories* (London: Penguin, 1968), pp. 193–211 (p. 193).
[50] Wells, 'The Plattner Story', p. 194.   [51] Wells, 'The Plattner Story', pp. 195–6.

(*SR*, 147)—but just as suddenly returned. Wells's invisible man Griffin, on the other hand, struggles to make the return and the lodger in 'The Hall Bedroom' is forever lost to the higher space of the dark, like two lodgers before him.

Ambrose Bierce's 'Mysterious Disappearance' stories present unornamented descriptions of characters slipping forever out of space. In 'The Difficulty of Crossing a Field' a planter named Williamson vanishes from the view of his wife, child, a neighbour, and his son. The neighbour reports: 'My son's exclamation caused me to look toward the spot where I had seen the deceased an instant before, but he was not there, nor was he anywhere visible.'[52] In 'An Unfinished Race' James Burne Warson bets that he can run from Leamington to Coventry and back. He is followed in a cart by three men.

> Suddenly—in the very middle of the roadway, not a dozen yards from them, and with their eyes full upon him—the man seemed to stumble, pitched headlong forward, uttered a terrible cry and vanished! He did not fall to the earth—he vanished before touching it. No trace of him was ever discovered.[53]

'Charles Ashmore's Trail' tells of a young man who goes out in a light snowfall to the spring on a family farm to retrieve a bucket of water. When he doesn't return the family go out to look for him:

> After going a little more than half-way—perhaps seventy-five yards—the father, who was in advance, halted, and elevating his lantern stood peering intently into the dark- ness ahead [...] The trail of the young man had abruptly ended, and all beyond was smooth, unbroken snow. The last footprints were as conspicuous as any in the line; the very nail-marks were distinctly visible.[54]

The young man's voice remains like his footprints and is heard 'at irregular intervals of a few days [...] The intervals of silence grew longer and longer, the voice fainter and farther, and by midsummer it was heard no more.'[55]

When these three stories were republished in Bierce's *Collected Works* in 1910 they were appended by a piece entitled 'Science to the Front'. This epilogue described the work of Dr Hern of Leipsic, who theorized 'vacua' in space 'through which animate and inanimate objects may fall into the invisible world and be seen and heard no more'. Hern's geographical location was no accident: his theories were popular

> 'particularly,' says one writer, 'among the followers of Hegel, and mathematicians who hold to the actual existence of a so-called non-Euclidean space: that is to say, of space which has more dimensions than length, breadth, and thickness—space in which it would be possible to tie a knot in an endless cord and to turn a rubber ball inside out without "a solution of its continuity," or, in other words, without breaking or cracking it.'[56]

[52] Ambrose Bierce, *The Collected Works*, 12 vols (New York: The Neale Publishing Company, 1909–12), III (1910), p. 417.
[53] Bierce, *The Collected Works*, III, p. 420.   [54] Bierce, *The Collected Works*, III, p. 422.
[55] Bierce, *The Collected Works*, III, p. 424.   [56] Bierce, *The Collected Works*, III, p. 425.

The continuum of corporeal possibilities afforded by higher space run the gamut from total absence to full (or doubled) presence. 'A Victim of Higher Space' embodies all the stages along the spectrum, an expanding cross-section in one figure:

> First he saw a thin perpendicular line tracing itself from just above the height of the clock and continuing downwards till it reached the woolly fire-mat. This line grew wider, broadened, grew solid. It was no shadow; it was something substantial. It defined itself more and more. Then suddenly, at the top of the line, and about on a level with the face of the clock, he saw a round luminous disc gazing steadily at him. It was a human eye, looking straight into his own, pressed there against the spy-hole [...] Then, like someone moving out of deep shadow into light, he saw the figure of a man come sliding sideways into view, a whitish face following the eye, and the perpendicular line he had first observed broadening out and developing into the complete figure of a human being.[57]

Both the Victim and Griffin are desperate to return to full corporeal form. Hinton's Stella, however, prefers incorporeality. She describes to Hugh Churton how the scientist Michael Graham had succeeded in making her invisible:

> 'But he found out how to alter the coefficient of refraction of the body. He made my coefficient equal to one.'
> 'But why should he?'
> 'Don't you see, Hugh, being is being for others. Michael used to say that true life begins with giving up.' (*S*, 35)

The absenting of the corporeal self in the Hintonian vision of higher space is an act of pure altruism. For the egotists Griffin and the Victim, the loss of the corporeal self, the spatial body of the three-dimensioned space, is terrifying: selfishness is self-fullness; altruism is an act of spatial voiding, of occupying no space, even that of one's own body. Mind, however, remains for Griffin, Stella, and the Victim, while the lodger in 'The Hall Bedroom' leaves behind a journal; only Bierce's characters disappear without trace.

Corporeal inversion and dissolution caused by exposure to higher-dimensioned spaces provoke polarized representations, but responses to the implications of permeability for the body tend towards unanimity. The possibilities of co-presence are uncomfortable at best. When the strictures of Kantian space are violated in 'The Plattner Story', Wells uses typographic emphasis to make his point:

> Then came a thing that made him shout aloud, and awoke his stunned faculties to instant activity. *Two of the boys, gesticulating, walked one after the other clean through him!* Neither manifested the slightest consciousness of his presence. It is difficult to imagine the sensation he felt.[58]

---

[57] Blackwood, 'A Victim of Higher Space', p. 233.
[58] Wells, 'The Plattner Story', p. 202.

POSSESSION: VISION 'AVEC'

The co-location of consciousnesses maps onto the well-rehearsed supernatural trope of possession. In George du Maurier's *The Martian*, the hero Barty Josselin is possessed by a Martian intelligence who writes through him, an interplanetary fictional equivalent of the 'communicating intelligences' of many an automatically produced occult text. Martia gives Barty written instructions: 'First of all, I will write out for you a list of books which you must study whenever you feel I'm inside you—and this more for me than for yourself.' Martia produces best-selling works in two languages, the most celebrated of which is '*La quatrième Dimension* in French'.[59] Possession makes Barty an authority on the very higher space that legitimizes the concept of co-location.

The gendered possession of Barty repeats the physical intimacy hoped for by our incongruent couples. The doubling of embodiment of possession is unavoidably intimate. George Griffith's novel *The Mummy and Miss Nitocris* inaugurates this kind of hyper-spatial subjectivity with some force. Professor Marmion sits in his study with a mummy-case received that morning from Memphis pondering problems related to 'the forty-seventh proposition of Euclid'. The Professor's daughter, Nitocris, re-enters his study with a drink for him:

> Beside her stood another figure as familiar now to his eyes as her's [*sic*] was, dressed and tired and jewelled in a fashion equally familiar. Save for the difference in dress, Nitocris, the daughter of Rameses, was the exact counterpart in feature, stature, and colouring of Nitocris, the daughter of Professor Marmion. In her hands she carried a slender, long-necked jar of brilliantly enamelled earthenware and a golden flagon richly chased, and glittering with jewels, and these she put down on the table in exactly the same place as the other Nitocris had put her tray on, and as she did so he heard her voice again saying:
> 'Time was, is now, and ever shall be for those for whom Time has ceased to be— which is a riddle that Ma-Rimon may even now learn, since his soul has been purified and his spirit strengthened by earnest devotion through many lives to the search for True Knowledge.'[60]

Nitocris and her double speak simultaneously. The second chapter cuts across the first. It locates itself immediately in ancient Egypt, and begins in a spatial register: 'The City of a Hundred Kings, vast and sombre, stretched away into the dim, soft distance of the moonlit night to right and left and far behind him.' The Professor's consciousness is now in a different subjectivity: 'He was standing in the dark shadows of a higher pylon at one end of the broad white terrace of the Palace of Pepi in Memphis—he, Ma-Rimon, Priest of Amen-Ra and Initiate of the Higher Mysteries.'[61] Nitocris is 'standing beside him'. The trans-temporal multiplication of subjectivity is increased when Nitocris suggests that her dead lover resides also

[59] George du Maurier, *The Martian* (London and New York: Harper & Bros, 1897), p. 369.
[60] Griffith, *The Mummy*, p. 4.   [61] Griffith, *The Mummy*, p. 10.

in Ma-Rimon: 'Nefer is dead, yet is not Nefer re-incarnated in another form, another man of another build, but yet Nefer that was—and is beside me now?' The Professor was his daughter's lover in a previous life: 'The words of blasphemy came hot and fast between his kisses, and she heard them unresisting in his arms, giving him back kiss for kiss, and looking into his eyes under the dark lashes which half hid hers.'[62]

The incestuous union of the Professor and his daughter, barely buried beneath layers of reincarnated subjectivity, brings the truth of co-location's transgressive embodiment of spatial intimacy intensely to bear. A similar incestuous discomfort with higher spatial possession runs through *The Inheritors*. The narrator, Granger, meets, and finds himself irresistibly attracted towards, a Dimensionist woman who assumes the identity of his sister. It emerges that not only can she read minds— 'She seemed to divine my thoughts, to be aware of their very wording'—but that she can control them.[63] In the chapter entitled 'Murder by Suggestion', Phadrig, the oriental villain of *The Mummy and Miss Nitocris*, controls a Jewish gemstone dealer like 'a mechanical doll'.[64]

The permeability of the physical body, its susceptibility to higher spatial posses- sion, occurs as the spatial aspects of novelistic voice, or mood, tend towards a simi- lar form of co-location. This is perfectly exemplified in Henry James's use of the metaphor of possession: 'A beautiful infatuation this, always, I think, the intensity of the creative effort to get into the skin of the creature; the act of personal posses- sion of one being by another at its completest.'[65]

That James figures possession as the essence of his theory of the novel alerts us to his concern with spatiality and a responsiveness to the imagined spatiality of the period. To get into the skin of the character is to occupy the same space, primarily. Only then can the consciousness be shared: it cannot be shared from above, or below, or to the side. Thinking the same thoughts is part of the matter, but seeing with the same eyes, hearing with the same ears, sharing the sensible experience of the subjectivity is crucial: spatial identity and identification, the co-location made possible by higher-dimensional theories.

In the preface to *What Maisie Knew* James declared himself

> addicted to seeing 'through'—one thing through another, accordingly, and still other things through that—he takes, too greedily perhaps, on any errand, as many things as possible by the way.[66]

Here we have things in chains—things through things. James's 'through' identifies a variant meaning: not necessarily permeability but vicariance: seeing via. This is a position of mobility, as illustrated by the preceding sentence, describing 'his love,

---

    [62] Griffith, *The Mummy*, p. 13.
    [63] Joseph Conrad and Ford Madox Hueffer, *The Inheritors* (London: Dent, 1923), p. 11. All further references to this edition are given in the text using the abbreviation *I*.
    [64] Griffith, *The Mummy*, p. 141.
    [65] Henry James, *The Art of the Novel: Critical Prefaces*, ed. Richard P. Blackmur (New York and London: Charles Scribner's Sons, 1934), p. 37.
    [66] James, *The Art of the Novel*, pp. 153–4.

when it is a question of a picture, of anything that makes for proportion and perspective, that contributes to a view of all the dimensions'.

James's metaphor of 'the house of fiction' at the window of which stands the artist has become a critical commonplace: 'The spreading field, the human scene, is the "choice of subject"; the pierced aperture, either broad or balconied or slit-like and low-browed, is the "literary form".' That the form James favoured was broad and balconied was signalled in another window metaphor from another preface. Of Newman in *The American* he wrote:

> for the interest of everything is all that it is his vision, his conception, his interpretation: at the window of his wide, quite sufficiently wide consciousness we are seated, from that admirable position we 'assist'. He therefore supremely matters; all the rest matters only as he feels it, treats it, meets it.[67]

James's metaphorical windows are like the architectural things of higher spatial fictions: transparent dividers of space, permeable to vision and psychological ingress. Like doorways and mirrors they are portals, imaginative openings, structures that allow us from one space into another. These metaphorical structures indicate that James figures his imaginative practice, his theory of the novel, in spatial terms.

James's intensely spatialized thought about his practice contrasted with the superficial approaches of realism. In his 'A Plea for Romance', Frank Norris argued:

> Realism stultifies itself: It notes only the surface of things. For it, Beauty is not even skin deep, but only a geometrical plane, without dimensions and depth, a mere outside. Realism is very excellent as far as it goes, but it goes no further than the Realist himself can actually see.[68]

Norris's assessment is suspect: realists frequently went further than seeing. Armstrong and Scarry alert us to the information of the other senses as productive of space in George Eliot.

Higher spatial fictions intensify sense information until it is excessive or confused. Wheatcroft's blind fumblings in 'The Hall Bedroom' lead him into a succession of intense sensory experiences. In his second journal entry he reports that: 'One of my senses was saluted, nay, more than that, hailed, with imperiousness, and that was, strangely enough, my sense of smell, but in a hitherto unknown fashion. It seemed as if the odor reached my mentality first' (THB, 1287). A sequence of odours 'salutes' him in this way; roses are followed by lilacs by mignonette. 'I can not describe the experience, but it was a sheer delight, a rapture of sublimated sense. I groped further and further, and always into new waves of fragrance' (THB, 1288). He panics and uses a box of matches brought for the purpose to return him to his room.

The following night the experience is repeated but his sense of taste is stimulated. 'It was never cloying, though of such sharp sweetness that it fairly stung. It was the merging of a material sense into a spiritual one' (THB, 1289). An auditory

---

[67] James, *The Art of the Novel*, p. 37.
[68] Frank Norris, 'A Plea for Romantic Fiction', in *The Responsibilities of the Novelist and Other Literary Essays* (London: Grant Richards, 1903), p. 215.

experience follows the next night: 'The song had to do with me, but with me in unknown futures for which I had no images of comparison in the past; yet a sort of ecstasy as of a prophecy of bliss filled my whole consciousness' (THB, 1290). The following night he feels out wonderfully carved objects, a window through which a breeze blows and through this a tactile space evoking seance and sepulchre:

> Living beings, beings in the likeness of men and women, palpable creatures in palpable attire. I could feel the soft silken texture of their garments which swept around me, seeming to half infold me in clinging meshes like cobwebs.   (THB, 1296)

Finally he experiences excessive vision: 'I saw more than I can describe, more than is lawful to describe' (THB, 1297). Wheatcroft encounters first-hand the unrepresentability of higher space.

It is notable that Wheatcroft experiences the proximity senses first, before vision, the meta-sense, requiring distance. The experience of the sensations of odour, sound, taste, and touch directly in the mind—'the odor reached my mentality first'—compares to the readerly imagination. Blackwood's Victim is prompted in his higher spatial fugues by the sound of music, a particular sensory cue opening up the fourth dimension.

FORM: SPACE AND STRUCTURE

The things and bodies of higher spatial narratives reflect back to us a post-Kantian spatial imaginary. Might we also discern such evidence in their form? Attention to form reveals that what was reflected at the level of content in popular fiction nourished the writing of Modernist novelists at the level of technique.

Some distinctions will be useful. In his 1945 essay, Joseph Frank identified spatial form as the combination of effects that worked to make the reader 'apprehend [the] work spatially, in a moment of time, rather than as a sequence'. These included 'word-clusters' in poetry and the cultivation of 'the mythical imagination for which historical time does not exist—the imagination which sees the actions and events of a particular time merely as the bodying forth of eternal prototypes'.[69]

The editors of a collection of essays responding to Frank, wary, no doubt, of criticisms levelled at the theory, broadened this scope:

> 'Spatial form' in its simplest sense designates the techniques by which novelists subvert the chronological sequence inherent in narrative [...] Also, portions of a narrative may be connected without regard to chronology through such devices as image patterns, leitmotifs, analogy and contrast.[70]

Some respondents read under this definition aspects of fictional space we have already discussed—the spatial relationships of point of view or focalization, for example, are considered by Joseph Kestner as demonstrative of the category of

---

[69] Frank, 'Spatial Form in Modern Literature', p. 653.
[70] Jeffrey R. Smitten and Ann Daghistany, 'Editor's Preface', in *Spatial Form in Narrative* (Ithaca and London: Cornell University Press, 1981), pp. 13–15 (p. 13).

'sculptural secondary illusion': the attribution of sculptural volume to the novel that can also be achieved through comparisons of characters to sculpture or descriptions of sculptural works.[71] Both the sculptural secondary illusion and the pictorial secondary illusion, argues Kestner, serve to subvert the successive temporality of narrative; the latter through binding, delaying, and making static; the former through this filling out of volume.

This is perhaps the most useful insight provided by analyses of spatial form: the way in which they attune us to the complex interrelations of narrative time and narrative space. Frank wants us to read oscillation 'between these two poles', but we may benefit from making the oscillation smoother and the poles less extreme; transitions along a spectrum. Higher space certainly confuses the matter of where we locate time in relation to space, even before relativity.

Fiction's most famous Time Traveller, for example, was an expert in *n*-dimensional geometry. Wells's character argued that *'there is no difference between time and any of the three dimensions of space except that our consciousness moves along it'*.[72] He explained that there were attempts by mathematicians to describe geometry of four dimensions and noted, accurately, Simon Newcomb's involvement in such scholarly activity. The Time Traveller himself had 'been at work on this geometry of Four Dimensions for some time' and had produced some 'curious results'.[73] He described portraits of the same man at different ages as 'sections'. The story proper begins when the Time Traveller is asked to provide some 'experimental verification' to support his theoretical disquisition, and shows his audience a model time machine that vanishes upon operation. It ends with the narrator witnessing the Time Traveller himself disappear:

> I seemed to see a ghostly, indistinct figure sitting in a whirling mass of black and brass for a moment—a figure so transparent that the bench behind with its sheet of drawings was absolutely distinct; but this phantasm vanished as I rubbed my eyes.[74]

There are those who would read Wells's Time Traveller as a prophet of relativity but the broad church of speculative thought had long entertained the idea of considering time as a fourth dimension. Alfred Bork has drawn attention to the 'S' letter sent to the journal *Nature*, in which the correspondent discussed this very possibility. Bork acknowledges earlier speculations along similar lines in the work of D'Alembert.[75] We can't entirely trust Wells's own remembrances but perhaps we should credit him with some imaginative input into this idea. The lost essay 'The Universe Rigid', submitted to the *Fortnightly Review*, proposed 'a completely comprehended system of causation' which 'should admit of exact prophecy'. It was written in conjunction with

> a rather elaborate joke going on with Jennings and the others, about a certain 'Universal Diagram' I proposed to make, from which all phenomena would be derived

---

[71] Joseph Kestner, 'Secondary Illusion: The Novel and the Spatial Arts', in *Spatial Form in Narrative* (Ithaca and London: Cornell University Press, 1981), pp. 100–30 (p. 115).

[72] H.G. Wells, *The Time Machine* (London: Phoenix, 2004), p. 4. The italics are in the source.

[73] Wells, *The Time Machine*, p. 5.   [74] Wells, *The Time Machine*, p. 81.

[75] See Alfred M. Bork, 'The Fourth Dimension in Nineteenth-Century Physics', *Isis*, 55 (1964), 326–38; 'S', 'Four Dimensional Space', *Nature*, 31 (1885), 481.

by a process of deduction. (One began with a uniformly distributed ether in the infinite space of those days and then displaced a particle. If there was a Universe rigid, and hitherto uniform, the character of the consequent world would depend entirely, I argued along strictly materialist lines, upon the velocity of this initial displacement. The disturbance would spread outward with ever increasing complication.)[76]

These draft descriptions of a 'four dimensional space-time universe' resulting in 'a completely comprehended system of causation' show Wells's interest in mechanics and read like a Lucretian system of chance atomic flux.

The fifth 'time dimension' of the automatically written *I Awoke!* anticipated the narrative mechanics of *The Mummy and Miss Nitocris*, which gleefully exploited mobile temporality to pile reincarnated subjectivities into each other—'he might be both alive and dead at the same time—since Past, Present, and Future would be all one to him; the world without beginning or end'—to cut between ancient Egypt and present-day London and to repeat, with variations, previously narrated days: 'It was the 6th of June again. Once more Prince Zastrow rode with Ulik von Kessner and Alexis Vollmar [...] but now accompanied by two unseen Presences which belonged at once to their own world and also to another and wider one.'[77]

The narrative of *The Mummy* uses unannounced switches of voice and temporality. In his study the Professor recalls some lines from a poem: 'Was it hundreds of years ago, my love, / Was it thousands of miles away?' A new voice intrudes on the narrative without introduction:

'And why should it not be? Why should you, who were once Ma-Rimon, priest of Amen-Ra, in the City of Memphis—you who almost stood upon the threshold of the Inmost Sanctuary of Knowledge: you who, if your footsteps had not turned aside into the way of temptation and trodden the black path of Sin, might even now be dwelling on the Shores of Everlasting Peace in the Land of Amenti—dost thou dare to ask such a question?'[78]

The Professor is as jolted by this intrusion as is the reader: 'The sudden change of pronoun seemed to him to put the Clock of Time back indefinitely.' The pronoun may be what shocks the Professor but the sudden and rapid use of names of Egyptian origin, of mythical-sounding locations also disrupts the flow of the narrative, while the evocation of thresholds and footsteps makes space ambiguous. The Professor is returned to relational space just as suddenly: 'He was standing by his desk still facing the Mummy just as his daughter had left him after saying "goodnight."' The second interruption of a new voice explicitly disrupts the spatiality of the narrative.

The voice, strangely like his daughter's and his dead wife's also, appeared to come from nowhere and yet from everywhere, and it had a faint and far-away echo in it which

---

[76] H.G. Wells, *Experiment in Autobiography: Discoveries and Conclusions of a Very Ordinary Brain (Since 1866)*, 2 vols (London: Gollancz, 1934), I, pp. 214–15.

[77] Griffith, *The Mummy*, p. 161.   [78] Griffith, *The Mummy*, pp. 2–3.

harmonised most marvellously with other echoes which seemed to come up out of the depths of his own soul.[79]

The memories this voice provokes are described in terms that recall James's metaphor of the writer at the house of fiction: 'Why did they instantly draw before the windows of his soul a long panorama of vast cities, splendid palaces, sombre temples, and towering tombs.'

Other spatial effects may be performed upon time. It may be displaced or thickened, as well as disrupted. Dane, the troubled author who is the central character of Henry James's short story 'The Great Good Place', finds himself in a place he characterizes as 'a current so slow and so tepid that one floated practically without motion and without chill'.[80] Here, hours are described as 'melted'.[81] He is surprised to realize that

> whenever he chose to listen with a certain intentness he made out, as from a distance, the sound of slow, sweet bells. How could they be so far and yet so audible? How could they be so near and yet so faint? How, above all, could they, in such an arrest of life, be, to *time* things, so frequent?[82]

James's 'Great Good Place' is nowhere and everywhere—suggested, variously, as in Surrey, Hampshire, or near Bradford, or simply 'near town', it is described by Dane's companion as 'nearer everything—nearer every one'.[83] Dane has been substituted from his unbearably busy celebrity writer's life in London in which he was suffering from 'submersion by our eternal too much' by a young man who envies his life. 'It meant that he should live with my life, and think with my brain, and write with my hand, and speak with my voice.'[84] This is not a possession of spatial doubling but of displacement.

Spatially, the Great Good Place is like 'some great abode of an Order', a villa with a

> great cloister, inclosed externally on three sides and probably the largest, lightest, fairest effect, to this charmed sense, that human hands could ever have expressed in dimensions of length and breadth, opened to the south its splendid fourth quarter, turned to the great view an outer gallery that combined with the rest of the portico to form a high dry loggia.[85]

This description gestures towards dimensionality, and leaves the architectural structure open in one direction. Dane finds that he returns most frequently to the library. 'There were times when he looked up from his book to lose himself in the mere tone of the picture that never failed at any moment or at any angle.'[86] This is the pictorial secondary illusion, in Kestner's terms, the description of a scene as a picture: Kestner writes that 'the painterly quality of a free and static scene is apparent

---

[79] Griffith, *The Mummy*, p. 3.

[80] Henry James, 'The Great Good Place', in *The Complete Tales of Henry James*, ed. Leon Edel, 12 vols (London: Rupert Hart-Davis, 1962–4), XI, pp. 13–42 (p. 20). See also Elizabeth Lea Throesch, 'The *Scientific Romances* of Charles Howard Hinton: The Fourth Dimension as Hyperspace, Hyperrealism and Protomodernism', doctoral thesis, University of Leeds, 2007, pp. 207–17.

[81] James, 'The Great Good Place', p. 34. [82] James, 'The Great Good Place', p. 21.

[83] James, 'The Great Good Place', p. 22. [84] James, 'The Great Good Place', p. 28.

[85] James, 'The Great Good Place', pp. 20–1. [86] James, 'The Great Good Place', p. 34.

when the reader is encouraged to recognize the characters as existing beyond the canvas or to feel he has entered a picture'.[87] The Great Good Place is devoid of adult responsibility, and shares something with the more childlike aspects of Hinton's vision of higher spatial unity. Indeed, in a final conversation with a new companion the Place is referred to as a 'kindergarten'.

Kestner alerts us to a 'highly specialized form of this pictorial illusion' that seems to tend towards an opposite spatial effect: the framing of narratives within one another, a technique termed 'encadrement' by Victor Shklovsky.[88] Yet such frame narratives also cause a narrative 'delay', creating temporal space between sections of narrative. They also tend to emphasize space by making us aware of the materiality of the text within the text: the manuscript or journal that recounts the 'intra-diegetic' tale. 'The Hall Bedroom' is a most vividly spatially formed narrative. George Wheatcroft's story is told in a journal discovered in the hall bedroom; the frame narrative is the transcription of Elizabeth Jennings, the lessee of the guest-house in which the bedroom is located; lessee within lessee.

Inside the physical hall bedroom there is indeed a framed picture: an oil paint-ing hanging on this same wall, an old-fashioned and conventional landscape depicting lovers in a boat on a winding river. 'For some inexplicable reason the picture frets me,' writes Wheatcroft. 'I find myself gazing at it when I do not wish to do so. It seems to compel my attention like some intent face in the room' (THB, 1280).

As the story unfolds, so does the space of the room. Wheatcroft suspects that 'there is something very singular about this room' (THB, 1281–2). Attempting to reach the bottle of water he has put on a dresser in the middle of the night, he finds it beyond his reach: 'To my utter amazement, the steps which had hitherto sufficed to take me across my room did not suffice to do so' (THB, 1283). Despite a ninety-degree change of direction in the pitch dark, he again fails to reach the expected obstacle. 'I am telling the unvarnished truth when I say that I began to count my steps and carefully measure my paces after that, and I traversed a space clear of furniture at least twenty feet by thirty—a very large apartment' (THB, 1284). Stranger still, 'it was as if my feet pressed something as elastic as air or water, which was in this case unyielding to my weight. It gave me a curious sensation of buoyancy and stimulation.' On successive nights he ventures further and further into this space. In his final journal entry Wheatcroft records 'that doors and win-dows open into an out-of-doors to which the outdoors which we know is but a vestibule' (THB, 1297). This is the space depicted in the painting and here he meets the others who have disappeared from the room: 'It is true that the girl who disappeared from the hall bedroom was very beautiful.'

The extradiegetic narrative returns to describe the investigation into Wheatcroft's disappearance which reveals a hidden room behind a wall, 'a long narrow one, the length of the hall bedroom, but narrower, hardly more than a closet' (THB, 1298). Inside this doorless, windowless space a sheet of paper covered with figures is dis-covered and the investigators speculate a fifth-dimensional proof.

---

[87] Kestner, 'Secondary Illusion', p. 109.   [88] Kestner, 'Secondary Illusion', p. 110.

'The Hall Bedroom' is an insistently spatial piece, a story about a room containing a painting containing extrasensory spaces of succession. Armstrong alerts us to the containers of space—the walls, boundaries, and horizons that enable the production of three-dimensioned space. These containers are first established in order to be permeated in the production of higher space. Wheatcroft's personal 'limitations' are mirrored by the spatial limitations of his lodging quarters:

> Therefore behold me in my hall bedroom, settled at last into a groove of fate so deep that I have lost the sight of even my horizons. Just at present, as I write here, my horizon on the left, that is my physical horizon, is a wall covered with cheap paper. (THB, 1279)

Metaphorical and physical horizons are transcended by Wheatcroft's night-time journeys. Professor Marmion reports at the beginning of *The Mummy* a colleague's assertion that Pythagoras 'almost saw across the horizon of the world that we live in [...] almost reached the border which divides the world of three dimensions from the world of four'.[89] Marmion does cross that horizon.

The *encadrement* of a frame narrative gives a structure of constraint while the nesting of spaces within each other in the narrative telescopes inwards into 'an out-of-doors to which the outdoors which we know is but a vestibule'. Within the frame we experience *mise en abîme*. Wheatcroft's disappearance keeps the narrative mysteriously open, suspended in time. The painting on the wall is a spatial portal, like that in *Lilith*.

In William Hope Hodgson's *The House on the Borderland*, inverted Neverland is enclosed in framing narratives. Hodgson's spaces are the vast spaces of the physical cosmos, full of 'unnameables' and 'indescribables' described in long passages. In *The House on the Borderland* his narrative is enclosed within a prologue, an author's introduction, and a first chapter that recounts the discovery of an ancient manuscript. This manuscript tells of a man confronted with a descriptive rather than a formal *abîme*, a pit which functions as portal into the borderlands. The borderlands leak into architectural space:

> The light came from the end wall, and grew ever brighter until its intolerable glare caused my eyes acute pain, and involuntarily I closed them. It may have been a few seconds before I was able to open them. The first thing I noticed was that the light had decreased, greatly; so that it no longer tried my eyes. Then, as it grew still duller, I was aware, all at once, that, instead of looking at the redness, I was staring through it, and through the wall beyond.
>
> Gradually, as I became more accustomed to the idea, I realized that I was looking out on to a vast plain, lit with the same gloomy twilight that pervaded the room. The immensity of this plain scarcely can be conceived. In no part could I perceive its confines. It seemed to broaden and spread out, so that the eye failed to perceive any limitations.[90]

---

[89] Griffith, *The Mummy*, p. 2.

[90] William Hope Hodgson, *The House on the Borderland and Other Novels* (London: Gollancz, 2002), pp. 118–19.

Cosmological time is accelerated at a giddying rate:

> To see the sun rise and set, within a space of time to be measured by seconds; to watch
> (after a little) the moon leap—a pale, and ever growing orb—up into the night sky,
> and glide, with a strange swiftness, through the vast arc of blue; and, presently, to see
> the sun follow, springing out of the Eastern sky, as though in chase; and then again the
> night, with the swift and ghostly passing of starry constellations, was all too much to
> view believingly. Yet, so it was—the day slipping from dawn to dusk, and the night
> sliding swiftly into day, ever rapidly and more rapidly.[91]

Hodgson's work could not come before the *n*-dimensional turn and accordingly
'other' dimensional spaces recur in his work, signifying the cosmological immensity
of space and represented as a source of terror. In *The Night Land*, his flawed masterpiece depicting a place of the distant future, Hodgson dragged monstrous creatures
out of higher-dimensional space:

> And then, so it would seem, as that Eternal Night lengthened itself upon the world,
> the power of terror grew and strengthened. And fresh and greater monsters developed
> and bred out of all space and Outward Dimensions, attracted, even as it might be
> Infernal sharks, by that lonely and mighty hill of humanity, facing its end—so near to
> the Eternal, and yet so far deferred in the minds and to the senses of those humans.[92]

Hodgson's writing is characterized by his staggeringly detailed and lengthy descriptions of these alternate spaces. Genette observes that such detailed description also
serves as a retarding device: 'because it lingers over objects and beings considered
in their simultaneity and because it envisages the actions themselves as scenes,
[description] seems to suspend the flow of time and to contribute to spreading out
the narrative in space'.[93] Hodgson's spaces are narratively slow, spatially excessive,
and highly visual.

Wells's narrative structures frequently use a less formal version of *encadrement*, a
third-person relation of factual account. Plattner's first-person account as related to
the narrator is framed within a third-person description of events, what the narrator describes as 'the exterior version of the Plattner story—its exoteric aspect'. The
framing is rigid: immediately following the relation of the intrinsic version of
events the narrator returns to inform the reader that

> I have resisted, I believe successfully, the natural disposition of a writer of fiction to
> dress up incidents of this sort. I have told the thing as far as possible in the order in
> which Plattner told it to me. I have carefully avoided any attempt at style, effect, or
> construction.[94]

William J. Scheick has defined what he terms an 'aesthetic fourth dimension' in
H.G. Wells's later writing. He analyses Wells's fiction of the 1930s to identify what
he sees as a consistently deployed structural technique, the creation of

[91] Hodgson, *The House on the Borderland*, p. 165.
[92] Hodgson, *The House on the Borderland*, p. 329.
[93] Gérard Genette, 'Boundaries of Narrative', trans. Ann Levonas, *New Literary History*, 8 (1976),
1–13 (p. 7).
[94] Wells, 'The Plattner Story', p. 211.

an emergent interior structure which dialectically engages (in a mutually constitutive opposition) a more apparent exterior structure; in the process the exterior structure of his novels is expanded and is reformed/re-formed. The emergent inner structure is, for Wells, artistically equivalent to time, dream, or a dimensionality encompassing countless alternatives of human possibility. The emergence of the interior structure splinters the frame of the exterior structure in a late Wells novel even as (in Wells's view) the insight given by World War 1 fractured the extrinsic framework of social values in Western civilisation.[95]

When Scheick refers to 'fictional "space"' he refers not only to the space of structural organization, a space of relation within the text, but also to a space encompassing the reader, who is drawn into the text by the splintering of the narrative 'frame'. 'Ideally both reformed structure and modified reader sensibility become four dimensional or, in other words, open to ever-expanding indeterminate possibilities.'[96] This metaphorical usage of higher dimensionality resembles McGurl's.

Delayed decoding, Ian Watt's formulation for the aspect of impressionist literary technique that 'combines the forward temporal progression of the mind, as it receives messages from the outside world, with the much slower reflexive process of making out their meaning', also disrupts narrative succession.[97] Ford welcomed the designation of impressionism for his collaborations with Conrad: 'we saw that Life did not narrate, but made impressions on our brains'.[98] Granger's first impressions of higher space and early responses to the Dimensionist woman activate delayed decoding. The reader is plunged directly into Granger's conversation with this stranger and it is some pages before she tells her story. 'The Hall Bedroom', with its sequential unfolding of sensory experience and final reveal of a concealed room, is an extended exercise in delayed decoding; we might choose to read Fleda's ultimate realization of what the 'something' is that makes Mrs Gereth's things greater than the sum of their parts in the same way.

What is clear is that higher space nourished the formal, spatializing techniques read as definitive of canonical Modernist novels. Henry James engaged with the metaphysical version of the fourth dimension while Ford and Conrad were interested in the technical implications. For James, as for D.H. Lawrence later, there was optimism. For Ford and Conrad the implications were full of dread. Higher space is an engine behind spatial and temporal effects of both form and voice. Texts push and pull: we read effects such as the *mise en abîme*, but often bound within *encadrement*; we read the relocations of consciousness, temporal disturbances, doublings, and disappearances. We might perhaps be on safest ground observing that higher space produces an immense turbulence in the spatiality and temporality of fin de siècle fiction. This is also a period of enormous generic turbulence, driven by

[95] William J. Scheick, *The Splintering Frame: The Later Fiction of H.G. Wells* (Victoria, BC: University of Victoria English Literary Studies, 1984), pp. 24–5.

[96] Scheick, *The Splintering Frame*, p. 25.

[97] Ian Watt, *Conrad in the Nineteenth Century* (Berkeley and Los Angeles: University of California Press, 1979), p. 175.

[98] Ford Madox Ford, *Joseph Conrad: A Personal Remembrance* (London: Duckworth and Co., 1924), p. 182.

a myriad of factors: rapidly changing conditions of production, dizzying scientific discoveries, and massive geopolitical change. It was not only canonical Modernists who engaged with these ideas, however, but popular writers whose work was often as innovative, if less carefully worked through or theorized.

FEAR: DIANOIA, PARANOIA, AND AGORAPHOBIA

Higher space had flourished as a source of psychological threat in the years imme-diately after the turn of the century. Newspaper reports record that on the night of 21 April 1900 Sutherland Street Macklem, a 22-year-old undergraduate of Christ Church College, Oxford, committed suicide by jumping out of a window at his father's house in London. According to his father, Macklem had been 'overworking himself'. On the evening of the 20th, after going to the theatre, he had been 'unwell' and prescribed medicine by a doctor. To both his father and his brother, Macklem had raved about the fourth dimension. There was a connection between religion and science, but he could not discuss it, he said. A note discovered in his pocket following his fatal jump described the revelation of 'a fact': 'religion is con-cerned with the ultimate ideas of the brain and science is concerned with the body. We must study each equally.'[99]

The *Newcastle Weekly Courant* of 28 April, the same day in which it ran the report on the coroner's inquiry into Sutherland's suicide, ran the fourth instalment of a serialization of Griffith's novel *The Justice of Revenge*. Having suffered 'a terrible shock of such intensity that it would have driven most men insane', the novel's hero Stephen Lorne was experiencing traumatic after-effects:

> Sitting in the little room that was half study and half laboratory, the words and formu-
> lae on the sheet before him began to move about and form themselves into strange
> shapes and combinations, which gave him visions of amazing possibilities, and even
> glimpses into that fourth dimension of space which has been the dream of philo-
> sophers for ages.[100]

The excess of higher space began to threaten psychological damage in earnest: the attempt to realize the unrealizable, to think the impossible thought, could lead to madness, while madness could lead to visions of higher space. The utopian vision of Hinton, the optimistic millennialism of Stead, and the appropriation of the idea into the applied meditative techniques of C.W. Leadbeater's Theosophical work were inverted. The excess of higher spatial perception was presented as damaging, and its paradoxical pull psychologically irreconcilable. In the cases cited above we read the inability to reconcile the empirical and the ideal, the unhappy marriage of word and number, body and brain, laboratory and study.

---

[99] Anon., 'An Oxford Undergraduate's Suicide', *Jackson's Oxford Journal*, 7676 (1900), 8.
[100] Griffith, 'The Justice', 2.

This current in higher spatial thought was already present in *Flatland*. When A Square is first raised into Spaceland he finds the experience intensely disturbing:

> An unspeakable horror seized me. There was a darkness; then a dizzy, sickening sensation of sight that was not like seeing; I saw a Line that was no Line; Space that was not Space: I was myself, and not myself. When I could find voice, I shrieked aloud in agony, 'Either this is madness or it is Hell.' 'It is neither,' calmly replied the voice of the Sphere, 'it is Knowledge; it is Three Dimensions: open your eye once again and try to look steadily.' (*F*, 64)

A Square feels the horror of unrepresentability. He has neither the words nor the images with which to describe his radically new experience. His habitual mode of seeing is also insufficient and his existential position ambivalent: 'I was myself and not myself.'

In *The Inheritors*, Granger experiences something similar to A Square's horror:

> I felt a kind of unholy emotion [...] What had happened? I don't know. It all looked contemptible. One seemed to see something beyond, something vaster—vaster than cathedrals, vaster than the conception of the gods to whom cathedrals were raised. The tower reeled out of the perpendicular. One saw beyond it, not roofs, or smoke, or hills, but an unrealized, an unrealizable infinity of space. (*I*, 8–9)

Like A Square, Granger finds language insufficient to account for his experience: the space is unrealized and unrealizable. His feeling of contempt is a presentiment of how the Dimensionists regard the inhabitants of lower space. As his walking partner baldly explains to him that she comes from the fourth dimension, he senses the risks to sanity posed by this concept but finds himself seduced. He remarks that 'there was a touch of the incongruous, of the mad, that appealed to me' (*I*, 9); 'there was a newness, a strangeness about her; sometimes she struck me as mad, sometimes as frightfully sane' (*I*, 11).

As *The Inheritors* develops, paranoia of penetrability becomes a fear of a very real threat. Granger has a history of neurasthenic illness: 'I had been ill; trouble of the nerves, brooding, the monotony of life in the shadow of unsuccess' (*I*, 11). Given his condition, he is particularly unfortunate to fall under the sway of the manipulative Dimensionist who uses her telepathic gifts to exert psychological control. As she pursues her ends her puppets are left psychologically weakened:

> 'That De Mersch was crumbling up,' she suddenly completed my unfinished sentence; 'oh, that was only a grumble, premonitory. But it won't take long now. I have been putting on the screw. Halderschrodt will...I suppose he will commit suicide, in a day or two. And then the—the fun will begin.' (*I*, 140)

Halderschrodt does commit suicide; Granger's aunt 'break[s] up'; and when the Dimensionist's machinations are revealed, Granger himself feels 'a sudden mental falling away' (*I*, 150).

Blackwood's 'Victim' makes explicit the mechanisms behind this paranoid fear of observation and control enabled by access to higher space. Dr Silence

> realised—and this was most unusual—that this individual whom he desired to watch knew that he was being watched. And, further, that the stranger himself was also

watching! In fact, that it was he, the doctor, who was being observed—and by an observer as keen and trained as himself.[101]

It is the eye of the Victim that enables Silence to first recognize his corporeality. We are reminded of the evil eye of mythology, a panoptical malevolence and recurrent symbol of Egyptian curse narratives of the Edwardian period. The mechanism of paranoia is further extended when Mudge warns Silence that 'anything you think of vividly will reach my mind'. In *The Mummy and Miss Nitocris* this paranoid fear is extended to outright mind control, of the kind suggested in *The Inheritors*, as Phadrig disposes of several spies by means of hypnotic murder.

Wells's Invisible Man, Griffin, is considered a madman for his violent and socially aberrant behaviour, while Davidson's condition is described as a 'transitory mental aberration'.[102] In *Stella* a committee of the Society for Psychical Research investigating Beechwood House address Hugh Churton:

> They talked to me with charming frankness, assuring me that if my insanity was caused by a spectre it was a very interesting case; but if, on the other hand, the spectre originated in my deranged mental condition, it was of no moment to the Society. (*S*, 44)

The earlier history of higher spatial thought suggests one facet to this particular concretion. The inflection of psychological disorder had been externally glossed onto fourth-dimensional thinking by the Seybert Commission's post-mortem investigations into Zöllner's trials with Slade and its diagnosis of insanity. Had this become internalized in higher space by the fin de siècle? That the accounts of Frances Sedlak's use of Hinton's cubes oscillate between admonition and celebration suggests the persistent influence of the Zöllner event.

Yet the period in which higher space is developed in genre fictions is concurrent with the 'belle époque' of phobias, in David Trotter's phrase.[103] The standard account of spatial phobias, indeed the account developed by the psychologists who first observed them—Westphal, Legrand, and Ball—was that these pathologies were by-products of urbanization. As Vidler writes: 'The resonance of a sickness associated with closed or open spaces, of symptoms that whatever their cause seemed to be triggered by the new configurations of urban space introduced by modernization, was irresistible to critics and sociologists alike.'[104] Ford Madox Ford had not yet been diagnosed with agoraphobia by the time *The Inheritors* was published but as his biographer Max Saunders points out, 'Granger is [...] Ford's first fictional agoraphobic.'[105]

---

[101] Blackwood, 'A Victim of Higher Space', p. 233.

[102] Wells, 'The Plattner Story', p. 174.

[103] David Trotter, 'The Invention of Agoraphobia', *Victorian Literature and Culture*, 32 (2004), 463–74 (p. 474).

[104] Anthony Vidler, *Warped Space: Art, Architecture, and Anxiety in Modern Culture* (Cambridge, MA and London: MIT Press, 2000), p. 31.

[105] Max Saunders, *Ford Madox Ford: A Dual Life*, 2 vols (Oxford: Oxford University Press, 1996), I, p. 132.

The fundamental openness of higher space, its absolute interpenetration, is Janus-faced. The same polarization as observed in the altruism/egoism opposition recurs: one man's utopian dissolution of ego is another's terrifying loss of identity; one man's glory in absolute truthfulness, another's paranoid fear of observation. Dianoia flips easily into paranoia. Bruce Clarke's observation that Hinton's manipulation of cubes is a means of control against the incipient paranoia of observation may have some truth to it, but the phobics of higher spatial fictions find little respite through their manipulation of things. Blackwood's Victim slips away; Granger falls off.

The source for which Frank reached in an attempt to explain why a shift towards spatial form might have come about proves informative. Wilhelm Worringer's *Abstraction and Empathy*, published in German in 1908 and a source for T.E. Hulme, among others, opposed the urge to abstraction to the urge to empathy:

> Whereas the precondition for the urge to empathy is a happy pantheistic relationship of confidence between man and the phenomena of the external world, the urge to abstraction is the outcome of a great inner unrest inspired in man by the phenomena of the outside world [...] We might describe this state as an immense spiritual dread of space.[106]

Worringer reads abstraction as 'the suppression of the representation of space'. For Frank, the same urge was behind the suppression of time aspects in the modern novel. The idea of 'an immense spiritual dread of space' experienced by 'the man disquieted by the obscurity and entanglement of phenomena' acquires increasing interest as higher-dimensional fictions themselves become more concerned with dread and terror. The search for the source of such fear can begin with the entanglement of phenomena in the things of higher spatial fiction and there are many places in which to locate them.

This representation of the psychological threat of higher space became a tool for the provocation of fear in the body of fiction that most systematically worked with the unrepresentability of higher space. In his canon-making essay 'Supernatural Horror in Literature', H.P. Lovecraft praised the work of many of the authors discussed here and indicated the central role of aberrant spatiality in the creation of his 'cosmic horror':

> A certain atmosphere of breathless and unexplainable dread of outer, unknown forces must be present; and there must be a hint, expressed with a seriousness and portentousness becoming its subject, of that most terrible conception of the human brain—a malign and particular suspension or defeat of those fixed laws of Nature which are our only safeguard against the assaults of chaos and the daemons of unplumbed space.[107]

The 'unplumbed space' that produced daemons might be immeasurable in scale or might confound the notions of up and down founded on standard space. Both

---

[106] Wilhelm Worringer, *Abstraction and Empathy: A Contribution to the Psychology of Style*, trans. Michael Bullock (Chicago: Ivan R. Dee, 1997), p. 15.

[107] H.P. Lovecraft, 'Supernatural Horror in Literature', in *At the Mountains of Madness: The Definitive Edition* (New York: The Modern Library, 2005), pp. 103–73 (p. 107).

n-dimensional and non-Euclidean spaces were favoured Lovecraft tropes. In his classic short story 'The Call of Cthulhu', the account of second-mate Johansen of R'lyeh indicates how Lovecraft deployed the language of the new geometries:

> He dwells only on broad impressions of vast angles and stone surfaces—surfaces too great to belong to anything right or proper for this earth [...] I mention this talk about angles because it suggests something Wilcox had told me of his awful dreams. He said that the geometry of the dream place he saw was abnormal, non-Euclidean, and loathsomely redolent of spheres and dimensions apart from ours.[108]

The story 'The Dreams in the Witch House' continues the theme of 'The Hall Bedroom'. Walter Gilman, who is studying 'non-Euclidean calculus' at Miskatonic university, rents a room of 'queerly irregular shape' that was previously inhabited by the witch Keziah Mason. During her trial Mason 'had told Judge Hathorne of lines and curves that could be made to point out directions leading through the walls of space to other spaces beyond'.[109]

Lovecraft works directly with the crisis of representation set in chain by non-Euclidean and *n*-dimensional geometries. We might even read his style as spatially responsive. For Roger Luckhurst, Lovecraft's style is characterized by the use of catachresis:

> Adjectives move in packs, flanked by italics and exclamation marks that tell rather than show. His horror is always premised on a contradiction: the indescribable is always exhaustively described [...] The power of the weird crawls out of these sentences because of the awkward style. These repetitions build an incantatory rhythm, tying baroque literary form to philosophical content. Conceptually, breaking open the world requires the breaking open of language and the conventions of realism [...] This rhetorical device is known as catachresis, the deliberate abuse of language, such as mixed metaphors [...] Lovecraft's horror fictions employ a language that continually stumbles against the trauma of the unrepresentable Thing, the shards of the sublime falling back into the debris of his busted sentences.[110]

The philosopher Graham Harman also lights upon this aspect of Lovecraft's style, which he reads as a form of linguistic 'cubism'. For Harman, Lovecraft's work describes a Husserlian rift between objects and their qualities. Lovecraft's things are not Henry James's things, they are not possessed with a spirit and do not become vessels for consciousness, but are fundamentally repellent. He writes: 'Language is overloaded by a gluttonous excess of surfaces and aspects of the Thing.'[111] His favoured example of this is a passage from *The Mountains of Madness* describing the architecture of an ancient city in the heart of Antarctica:

[108] H.P. Lovecraft, 'The Call of Cthulhu', in *The Call of Cthulhu and Other Weird Tales* (London: Vintage, 2001), pp. 61–98 (p. 93).

[109] H.P. Lovecraft, 'The Dreams in the Witch House', in *The Call of Cthulhu and Other Weird Tales* (London: Vintage, 2001), pp. 235–80 (pp. 236–7).

[110] Roger Luckhurst, 'Introduction', in H.P. Lovecraft, *The Classic Horror Stories* (Oxford: Oxford University Press, 2013), pp. vii–xxviii (pp. xix–xx).

[111] Graham Harman, *Weird Realism: Lovecraft and Philosophy* (Winchester and Washington: Zero Books, 2012), p. 25.

There were truncated cones, sometimes terraced or fluted, surmounted by tall cylindrical shafts here and there bulbously enlarged and often capped with tiers of thinnish scalloped discs; and strange, beetling, table-like constructions suggesting piles of multitudinous rectangular slabs or circular plates or five-pointed stars.[112]

Harman writes of this passage that 'the near-incoherence of such descriptions undercuts any attempt to render them in visual form. The very point of the descriptions is that they fail, hinting only obliquely at some unspeakable substratum of reality.'[113]

This description of an excess of surface, as Harman's use of the term 'cubist' suggests, is enabled by a spatial imagination in which one might observe from higher dimensions: Leadbeater seeing all sides of a cube at once on the astral plane; Picasso's reading in the work of Jouffret.[114] Lovecraft's letters make clear why his space was no a priori condition for human thought:

Now all my tales are based on the fundamental premise that common human laws and interests and emotions have no validity or significance in the vast cosmos-at-large. To me there is nothing but puerility in a tale in which the human form—and the local human passions and conditions and standards—are depicted as native to other worlds or other universes. To achieve the essence of real externality, whether of time or space or dimension, one must forget that such things as organic life, good and evil, love and hate, and all such local attributes of a negligible and temporary race called mankind, have any existence at all.[115]

The *n*-dimensional and non-Euclidean spaces informing the physics of the twentieth century, 'the utterly unplumbed gulfs still farther out—the nameless vortices of never-dreamed-of strangeness, where form and symmetry, light and heat, even matter and energy themselves, may be unthinkably metamorphosed or totally wanting', were emblematic of this real externality.[116]

It would be wrong to describe this Worringerian spiritual dread of space as a recent tendency in higher spatial thought: its seed was there from the beginning. A decade before even A Square's fictional experience C.M. Ingleby wrote to his friend C.J. Monro: 'Your last has made me giddy and perplexed, and brought back once more to me the horrors of my childhood when I endeavoured to trace out immensity. I dare say my space, in those dreadful days of dawning intuition, was of four dimensions, and cursed.'[117]

---

[112] H.P. Lovecraft, 'At the Mountains of Madness', in *At the Mountains of Madness: The Definitive Edition* (New York: The Modern Library, 2005), pp. 1–102 (p. 29).

[113] Graham Harman, 'On the Horror of Phenomenology: Lovecraft and Husserl', *Collapse*, 4 (2009), 333–64 (p. 339).

[114] See Linda Dalrymple Henderson, *The Fourth Dimension and Non-Euclidean Geometry in Modern Art* (Princeton: Princeton University Press, 1983), p. 138.

[115] H.P. Lovecraft, *Selected Letters 1923–1929*, ed. August Derleth and Donald Wandrei (Wisconsin: Arkham House, 1968), vol. 2, p. 150.

[116] Lovecraft, *Selected Letters*, pp. 150–1.

[117] MS Monro correspondence, Acc 1063/2457.

THE GLOBE: HIGHER SPACE AND EMPIRE

The fear of the immensity of space mapped onto the immense space sensed and sometimes experienced in modern life. In an influential discussion of 'Imperialism and Modernism', Fredric Jameson describes the incommensurability of the experience of 'a global space that like the fourth dimension somehow constitutively escapes you':

> Colonialism means that significant structural segment[s] of the economic system as a whole [are] now located elsewhere, beyond the metropolis, outside of the daily and existential experience of the home country [...] unknown and unimaginable for the subjects of the imperial power.[118]

In Jameson's reading it is the reality of physical, global space that becomes 'unthinkable' and 'unrepresentable'. Anxieties over colonial space are repeatedly evident in higher spatial fictions.

With his vision warped by a freakish fourth-dimensional accident, H.G. Wells's Davidson witnesses the exercises of the British navy at a hemisphere removed. Davidson rambles about a beach and ship while bumping into various pieces of laboratory equipment as if blinded: ' "We seem to have a sort of invisible bodies," said he. "By Jove! there's a boat coming round the headland." '[119] His colleagues take him to their Dean's office where they discuss the nature of his hallucinations, of a beach, ships, and a penguin. The narrator continues: 'For three weeks Davidson remained in this singular state, seeing what at the time we imagined was an altogether phantasmal world.'[120]

Some years later a friend who is a lieutenant in the navy shows Davidson a photograph of the *Fulmar*, a ship that has been in the South Seas for the past six years. Davidson recognizes it as the ship from his hallucination and is able to describe details that could not have been known to anyone who had not seen the ship during its mooring at Antipodes Island. The narrator concludes:

> It's perhaps the best authenticated case in existence of real vision at a distance. Explanation there is none forthcoming, except what Professor Wade has thrown out. But his explanation invokes the Fourth Dimension, and a dissertation on theoretical kinds of space. To talk of there being 'a kink in space' seems mere nonsense to me; it may be because I am no mathematician. When I said that nothing would alter the fact that the place is eight thousand miles away, he answered that two points might be a yard away on a sheet of paper, and yet be brought together by bending the paper round [...] He thinks, as a consequence of this, that it may be possible to live visually in one part of the world, while one lives bodily in another.[121]

This is the space of the colonial experience at its greatest, trans-hemispheric extent. Britain, and the reach of its military, did extend around the globe, as dizzying as

---

[118] Fredric Jameson, 'Imperialism and Modernism', in *Nationalism, Colonialism and Literature* (Minneapolis: University of Minnesota Press, 1990), pp. 43–68 (pp. 50–1).
[119] Wells, 'The Plattner Story', p. 176.   [120] Wells, 'The Plattner Story', p. 180.
[121] Wells, 'The Plattner Story', pp. 182–3.

that was for the reader in his Hampstead bedsit, or indeed the scientist in his laboratory. The making proximal of distant and exotic locations was a trick Wells also used in the short story 'The Door in the Wall', in which a door in Kensington opens into an Eden-like, oriental garden. We recall that before such global, spatial short-circuits were featured in Wells's fiction they were mystically enacted for Theosophists like Massey and Leadbeater.

Critics have been drawn to figuring Wells's Time Traveller as a colonial explorer and so, in his way, is Algernon Blackwood's Racine Mudge who is 'so confused in geography as to find myself one moment at the North Pole, and the next at Clapham Junction—or possibly at both places simultaneously'.[122] Mudge explores the extended geographical space of the polar expeditionary pushing back the edges of the mapped world, an adventurer in the psychic domain, questing for the north pole of the occult. His higher space is similarly mapped onto geographic space.

Seemingly the higher spatial text least amicable to such a reading, MacDonald's *Lilith*, has been read in these terms by Kelly Shearsmith. Intriguingly, Shearsmith makes Vane's journey read like a quasi-objectival counterpart to that of Marlow in *Heart of Darkness*:

> Although Vane goes native, MacDonald is able to reinscribe the late-Victorian colonial narrative by directing his journey within rather than without, taking him into the nature of things instead of into the things of nature and the differences in men those may produce.[123]

The idea that Vane might be a colonial explorer of the interior is particularly allusive. The colonial extensions of higher space explicit in other texts are here internalized. We might choose to read Vane's allegorical journey as a distorted-mirror version of the more explicitly geopolitical moves in the higher spatial fictional canon.

The Inheritors makes these moves explicit. Conrad and Ford's novel carries with it echoes of both *Heart of Darkness*—Granger notes towards the end of the novel 'the real horrors of the Systeme Groënlandais—flogged, butchered, miserable natives, the famines, the vices, diseases, and the crimes' (*I*, 183)—and Conrad's own biography; his affair, in Nice, with a Spanish Legitimist echoed in Granger's infatuation with a Dimensionist.

The relationship between international space and higher space in *The Inheritors* is clear: the Dimensionists are expansionists. Those previously 'crowded out' of global space are returning to recolonize it. Evolutionary divergence has taken place long before the start of the narrative and the more highly evolved Dimensionists are now returning from the transitional space a changed species:

> Your ancestors were mine, but long ago you were crowded out of the Dimension as we are to-day, you overran the earth as we shall do to-morrow. But you contracted diseases as we shall contract them,—beliefs, traditions; fears; ideas of pity...of love.   (*I*, 10)

---

[122] Blackwood, 'A Victim of Higher Space', p. 241.
[123] Kelly Shearsmith, 'Chiasmatic Christianity: Lilith's Sense of an Ending', in *Lilith in a New Light: Essays on the George MacDonald Fantasy Novel*, ed. Lucas H. Harman (Jefferson and London: McFarland, 2008), pp. 143–60 (p. 158).

The very modern experience of threatening news from the inconceivable imperial margins is made present; for the colonies, read the fourth dimension. The 'anxiety of reverse colonisation' described by Stephen Arata in the context of Stoker's *Dracula* is a useful concept here. Arata writes that a narrative informed by this anxiety 'expresses both fear and guilt. The fear is that what has been represented as a "civilised" world is on the point of being colonized by "primitive" forces.'[124] The Dimensionists of *The Inheritors* are no longer primitive, but they once were; they are mankind's evolutionary cousins, themselves 'crowded out' by three-dimensional man. Arata continues: 'They are also responses to cultural guilt. In the marauding, invasive Other, British culture sees its own imperial practices mirrored back in monstrous forms.'[125] In *The Inheritors* higher space is made the source of this mirroring.

The tangle of specific period concerns in *The Mummy and Miss Nitocris* complicates this picture while emphasizing the fictional connection between international geographical space and the fourth dimension. Contemporary occult accents (consistent with Theosophical thought) are recurrent in the text: Phadrig is an 'Adept'; the Marmions are adherents of 'the doctrine' of reincarnation and believe ghosts to be 'astral' bodies. Most striking is the mummy-case in the Professor's study, through which the narrative is instigated.

The second wave of Egyptophilia, dating from the 1881 discovery of a cache of royal mummies concealed by a priesthood nervous of grave-robbers, was still going strong at the time of publication, buoyed by the popularity of exhibits of antiquities at the British Museum and stories and rumours concerning the occult resonances of such objects. The Russian Revolution of 1905 was dominating European news, and strikes and protests were spreading through the Baltic nations. In Latvia and Estonia workers were involved in armed conflict with Baltic German aristocracy. *The Mummy* works all these current political concerns through its plot. Its specific interest in Egyptology, and the priesthood of Amen-Ra, was yet more focused in current events.

The mummy-case of a priestess of Amen-Ra was held at the British Museum and was by 1907 already at the centre of rumours of occult happenings. In 1904 the then editor of the *Daily Express* and friend of Arthur Conan Doyle, Bertram Fletcher Robinson, had been researching the rumours of a curse surrounding the case and the family of Thomas Douglas Murray, who had brought it back from Egypt in the mid-1860s.[126] Fletcher Robinson died suddenly in 1907 and an article about 'The Mysterious Mummy' appeared in *Pearson's Weekly*, Griffith's erstwhile publisher, in 1909. This article noted that Madame Blavatsky had encountered the mummy in the possession of a previous owner and had asked for it to be removed from the room 'declaring it to be a thing of utmost danger'. W.T. Stead and Murray's brother, meanwhile, were later reported to have been performing psychic

[124] Stephen D. Arata, 'The Occidental Tourist: *Dracula* and the Anxiety of Reverse Colonization', *Victorian Studies*, 33 (1990), 621–45 (p. 623).

[125] Arata, 'The Occidental Tourist', 623.

[126] See Roger Luckhurst, 'The Mummy's Curse: A Study in Rumour', *Critical Quarterly*, 52 (2010), 6–22.

experiments with the mummy in 1906 and the palmist Cheiro wrote about it in his undated book *True Ghost Stories* around the same time.

Griffith's story is not a curse tale but it does contain elements consistent with curse rumours and narratives: his interest in the mummy emerges from a nexus of occult rumour with a notable centre in the London periodical press in which he worked. Why the combination of four-dimensional and mummy narratives already sketched out in separate stories? *The Mummy* suggests some theorization in terms of shared liminality: mummies are undead creatures, sexually ambiguous, as emphasized by the passage in the text in which Oskar Oskarovitch is revenged by being united with his true bride, the corpse of the mummy, laid on his bridal bed. The text refers throughout to the crossing of the threshold between three- and four-dimensional existence—one chapter even titled 'Beyond the Threshold'.

The most compelling connection, though, is colonial. As Roger Luckhurst remarks about mummy stories, 'the remarkable overdetermination of factors in Egypt tends to be downplayed, reduced to sketchy background, just when Egyptian specificity needs to be emphasised'.[127] The factors Luckhurst urges us to 'bring back into play' include the British occupation of Cairo in September 1882 following the rebellion in June 1882 after ten years of worsening debt in Egypt. This marked a turning point of British imperial policy and the beginning of a religious war fought at the southern border of Egypt with the Sudanese which witnessed the botched evacuation of Gordon in 1884 and Kitchener's massacre of rebel forces at Omdurman in 1898. Luckhurst argues that 'wherever there is imperial occupation, there is a reserve of supernaturalism, an occult supplement to allegedly enlightened rule that becomes one of the popular currencies for acknowledging and perhaps even beginning to negotiate the consequences of colonial violence and guilt'.[128]

Intriguingly, *The Mummy and Miss Nitocris* responds to imperial occupation not only with an exploration of supernaturalism but also with a jingoistic restatement of the superiority of British foreign policy. The doublings of reincarnation give historical and ethical authority and justification to the Marmions: the Queen and the priest of Amen-Ra, 'initiate of the higher mysteries', are reincarnated as British, while the murderous warrior Menkau-Ra and his priest are reincarnated as a Russian and an Egyptian. The Professor uses his powers responsibly while the foreigners would use theirs for aggression: the dastardly plotter Oskar Oskarovitch cannot understand why Professor Marmion, who has solved the mystery of powered flight, is determined to keep it to himself. *The Mummy* does not so much represent the anxiety of reverse colonization as respond to it with a counterblast. It takes hold of the supernatural supplement and attempts whiggishly to reclaim imperial superiority. In so doing it works against the current of higher spatial fictions.

This reading certainly appeals in terms of its spatial congruence, the way abstract space becomes a metaphor for lived space. It can give us insight into the generic turbulence we are in the midst of when we read canonical Modernists writing SF.

---

[127] Luckhurst, 'The Mummy's Curse', p. 16.
[128] Luckhurst, 'The Mummy's Curse', p. 17.

Christina Britzolakis reads the 'generic instability' of *The Inheritors* as a response to 'a historically specific metropolitan experience of cognitive dissonance in the face of violently reconfigured relations among urban, national, and global space'.[129] In this reading, fictional genre responds to actual spatial experience.

We must be wary, though, of reducing the picture to a single reading because it is neat. Extended space was not always a source of dread and neither was Worringer's 'entanglement of phenomena'. Euphoric communion with objects both human and natural was an essential aspect of the utopian higher spatial visions of Hinton, Stead, and Leadbeater, themselves owing much to the immersion in nature described in James Hinton or the thought of Goethe and the German Romantics. Describing Fechner's psychophysical project, Jonathan Crary describes well how entanglement could be a

> delirious merging of the interiority of a perceiver into a single charged and unified field, every part of it vibrating with the same forces of repulsion and attraction, an infinite nature [...] where life and death are simply different states of a primal energy.[130]

There is something of this delirium in Henry James's carefree and otherworldly vision of the fourth dimension and the intense joy he took in his 'possession' of characters. There is at the same time something more measured about James's theorization of his practice and this carefully detailed appreciation of connectivity in space might be figured through a productive distance. Anthony Vidler contrasts Worringer's *Raumscheu* with Alby Warburg's *Denkraum*, a space of reason providing distance from the natural, phenomenal world:

> In this ascription, space was beneficent, and the more the better. Indeed, the 'progress' that Warburg measured seemed to increase in direct proportion to the amount of mental and physical space that might be conquered by society in order to create a sufficient barrier between nature and civilisation.[131]

In a similar spirit, we might set against Hodgson's terrifying abysses the panoramic views of Wells's utopian fictions such as that described from a tower in *In the Days of the Comet*: 'There for one clear moment I saw it; its galleries and open spaces, its trees of golden fruit and crystal waters, its music and rejoicing, love and beauty without ceasing flowing through its varied and intricate streets.'[132]

Distance within an extended space could be seen to allow the subject a privileged position from which to observe, a removal from entanglement. The utopian wing of higher spatial fiction could also offer an alternative to the spatial theory at the heart of Worringer's thesis. Worringer argued that:

[129] Christina Britzolakis, 'Pathologies of the Imperial Metropolis: Impressionism as Traumatic Afterimage in Conrad and Ford', *Journal of Modern Literature*, 29 (2005), 1–20 (p. 2).

[130] Jonathan Crary, *Techniques of the Observer: On Vision and Modernity in the Nineteenth Century* (Cambridge, MA: MIT Press, 1992), p. 147.

[131] Vidler, *Warped Space*, p. 45.

[132] H.G. Wells, *In the Days of the Comet* (London: Macmillan, 1912), p. 305.

The artist was forced to approximate the representation to a plane because three-dimensionality, more than anything else, contradicted the apprehension of the object as a closed material individuality, since perception of three-dimensionality calls for a succession of perceptual elements that have to be combined; in this succession of elements the individuality of the object melts away [...] Suppression of representation of space was dictated by the urge to abstraction through the mere fact that it is precisely space which links things to one another, which imparts to them their relativity in the world-picture and because space is the one thing it is impossible to individualise.[133]

Hinton's *Stella* contains a passage, quoted from the journals of Michael Graham, of a strikingly different view from a higher spatial theorist:

'Compare the work of the sculptor and that of the painter. The sculptor makes an object resembling much more that which he represents than does a painter.

'But the painter has a greater scope. In the thinness of his medium lies his greater power; in the remoteness of his representation of the thing represented he gains his power of wide pourtrayal [*sic*]. And Thought is an art, a representation of the world in mind-stuff, in ideas of space, motion, matter. To say that the world is made of space, motion, matter, is as absurd as to say that it is made of paint.

'The more that can be represented in any medium the less that medium resembles the things represented, and thought which pictures all reality uses a medium which is most unreal. Its power lies in the nothingness of its means.'   (*S*, 14–15)

In this transcendent view, higher spatial thought gave the widest 'pourtrayal' because the most remote; and that 'pourtrayal' pictured all reality in the immaterial.

---

[133] Worringer, *Abstraction and Empathy*, pp. 986–7.

Conclusion

It might be said that higher space becomes what it does through productive mistranslation. In the history of higher space we are always dealing with languages: algebra, the language of mathematics; geometry, the language of space; rhetorics, the language of philosophical explanation; metaphor, the language of the imagination.

The story begins with an algebraic speculation but this is soon translated into the language of projective geometry; in order to explain the concept as demonstrated in projective geometry, it is then translated into the rhetorical construction of an analogy; from both analogy and projective geometry, physical theses are speculated, concealing the linguistic construction of rhetoric and the descriptive function of geometry; these physical theses are then hypostatized, assumed concrete rather than abstract; both the hypostatized versions of these theses and their speculative forebears become once again linguistic constructions, as allegories or metaphors in fiction.

What happens most consistently is that the translations are accomplished without acknowledgement that they are translations: the different languages, each subject to its own grammar, each incapable of mimesis but concealing its failings in different ways, is assumed to be directly translatable into the next. This process is itself generative, catalysing new forms, new languages, and new ideas and inserting these into new contexts. Zöllner's misreading of analogy was crucial to the cultural development of the idea of higher space. His experiments produced a potent symbol for how this space was thought: the knot, binding together disparate discourses, binding together world and idea, suggesting ways that we too can approach higher space through its loops and inversions.

Flatland exploited the imaginative mobility of higher space in accordance with this very model. Its intra- and meta-textuality unknotted and retied the complexities of the dimensional analogy, and, crucially, personified the process of thinking it. *Flatland* presented this thought in imaginative representations that proliferated through literary continuations. Charles Howard Hinton's mediation of oscillating higher space through basic material objects provided detail and practice, grounding his visionary promises in tangible things. His hybridization of ethical philosophy and speculative mathematical and physical ideas made clear the philosophical stake of the subject in the higher spatial game, the co-constitutive relations of subject and object after *n*-dimensions. Hinton's work was, also, crucially, utopian, casting the no-place of the fourth dimension as a quintessentially good place of progressive communion.

The utopianism and ambivalence of Hinton's work was continued in occult engagements with higher space that hypostatized once again the abstractions of the fourth dimension coincident with supportive scientific advance: the granular material permeability of Theosophy was seemingly suggested by scientific discoveries concerning the nature of matter, and radiation in particular. Higher space truly became an analogue for global space under the aegis of the Theosophical Society, an organization whose network expanded within a subsection of global colonial space.

The counter-intuitive tendency of the Theosophical imagination towards rendering such dissipating ideas in planar, hierarchical form tilts the same notion towards a distinction between perception and conception. As one Theosophical commentator put it:

> The sensory surfaces of the body, and hence, our sensations, are two-dimensional, our perceptions of objects are three-dimensional while our conceptions are four-dimensional.[1]

This distinction, with its own appropriately prepositional bias, certainly seems to grasp a core element of what is at stake, and the continuity between conception and the productive imagination makes clear again the loops between thought and matter occasioned by higher spatial thought. Looking for 'a counterpart of this [the fourth dimension] in common life', the mathematician William Spottiswoode lit upon a literary example: the representation of past, present, and future in one 'common focus'. He continued:

> Or once more, when space already filled with material substances is mentally peopled with immaterial beings, may not the imagination be regarded as having added a new element to the capacity of space, a fourth dimension of which there is no evidence in experimental fact?[2]

Spottiswoode's remark demonstrates the extent to which the fourth dimension was figured as the creative imagination.

In an essay of 1990 Gillian Beer asked whether scientific ideas were translated or transformed in their passage into literary culture, arguing that neither term adequately described the complex shifts of register that take place:

> It turns out not always to be a simple matter to re-distil ideas absorbed into other formations. The implications of scientific ideas may manifest themselves in narrative organizations. They may be borne in the fleeting reference more often than in the expository statement, condensed as metaphors or skeined out as story, alive as joke in the discordances between diverse discursive registers. Lightness and suspicion may tell more than scrutiny and exposition.
>
> Scientific material does not have clear boundaries once it has entered literature. Once scientific arguments and ideas are read outside the genre of the scientific paper and the institution of the scientific journal, change has already begun.[3]

This account hopes to add mistranslation and hybridization to the array of possibilities for what can happen to scientific ideas in literature. It argues that not just

[1] Charles Johnston, 'Psychism and the Fourth Dimension', *The Theosophist*, 9 (1888), 423.
[2] William Spottiswoode, 'Presidential Address', *Report of the Forty-Eighth Meeting of the BAAS Held at Dublin in August 1878* (London: John Murray, 1879) p. 23.
[3] Gillian Beer, 'Translation or Transformation? The Relations of Literature and Science', *Notes and Records of the Royal Society of London*, 44 (1990), 81–99 (p. 90).

language, metaphor, and narrative form might respond in the encounter, but so too might technique. The idea of higher space emerged with permeability as an essential feature; this permeability was enacted at the metaphorical level, where it lent itself to combination.

In the fin de siècle, topology was theorized concurrently with topophobia. That the theorization and description of spatial fears was simultaneous with the development of the new geometries and their spatial correlates is not coincidence, but it is important to maintain the flux and ambiguity of this moment and complicate the critical commonplaces of Modernism. Rapid changes in the spatial imaginary of the fin de siècle informed cultural production. The position of the *n*-dimensional turn in the context of the field of spatial developments should be amplified. Higher space provided resources that radically altered imaginative possibility and out-stripped the physical resources of technological and communicative advance. Higher space made possible spatial union, instant communication, corporeal inversion, co-location, bi-location, possession. Reflection of these possibilities occurred at the levels of both form and content and the results were ambiguous. To quote Beer once again: 'the transformed materials of scientific writing become involved in social and artistic questioning. That questioning is enacted sometimes at the level of semantics, sometimes of form or of broken story. Transformations and imbalances reveal as much as congruities.'[4]

Looking forward to Modernism, the extent to which visual artists worked with *n*-dimensional ideas has been thoroughly described. Modernist literary engagements with higher space have been identified in disparate works—Bruce Clarke has written on D.H. Lawrence, Linda Henderson on Gertrude Stein—but a broader, unifying account would be fruitful. Research that considers how the altered spatial imaginary described here informed also Dora Marsden's writing on space in *The Egoist*, W.B. Yeats's Theosophically inspired poems, Pound's 'vortex', and Mary Butts's 'vision', for example, and contrasting this with the way the same spatial thought was dealt with in the work of non-canonical 'pulp' Modernists, briefly approached here, would both build on extant work on Modernism and open up fresh channels. From such foundations we might be able to work towards an account of the spatial imaginary of the twentieth century, and how we arrive at responses to the contemporary higher spaces of string theory, via SF hyperdrives and New Age alternate universes. The development of topology and morphology, and negotiations of Relativity Theory, will provide useful levers for prising open the convoluted abstract spaces of the twentieth century. With regard to the contem-porary recurrence of higher space in the context of string theories, the most useful set of questions might revolve around where we might locate the intense, visionary, and generative cultural work that such spaces spawned in the fin de siècle. It seems likely that we are all now higher spatial subjects.

---

[4] Beer, 'Translation or Transformation', p. 97.

Bibliography

PRIMARY SOURCES

A Collection of Prospectuses of the Educational Exhibition of 1854, 3 vols (London: Royal Society of Arts, 1854)

A Square, 'The Metaphysics of Flatland', *Athenaeum*, 2980 (1884), 733

Abbott, Edwin A., *A Shakespearean Grammar* (London and New York: Macmillan and Company, 1870)

Abbott, Edwin A., *Hints on Home Teaching* (London: Seeley, Jackson, and Halliday, 1883)

Abbott, Edwin A., *Flatland: A Romance of Many Dimensions*, 2nd edn (London: Seeley and Co., 1884; repr. New York: Dover Publications, 1992)

Abbott, Edwin A., *The Kernel and the Husk: Letters on Spiritual Christianity* (London: Macmillan, 1886)

Abbott, Edwin A., *The Annotated Flatland: A Romance of Many Dimensions*, ed. Ian Stewart (Reading, MA and Oxford: Perseus, 2002)

Abbott, Edwin A., *Flatland: An Edition with Notes and Commentary*, ed. William F. Lindgren and Thomas F. Banchoff (Cambridge: Cambridge University Press; Washington, DC: Mathematical Association of America, 2010)

Abbott, Edwin A., and J.R. Seeley, *English Lessons for English People* (London: Seeley, Jackson and Halliday, 1871)

Aksakow, Aleksander, 'A New Manifestation with Dr. Slade at Leipzig University', *Spiritualist*, 12 (1878), 78

Anon., 'The Political Euclid', *Punch*, 1 (1841), 149

Anon., 'Most Natural Selection', *Punch, or the London Charivari*, 1 April 1871, 127

Anon., 'Mr. Darwin on the Descent of Man', *The Times*, 8 April 1871, 5

Anon., 'A New Philosophy', *City of London School Magazine*, 1 (1877), 277–81

Anon. [probably William H. Harrison], 'On Space of Four Dimensions', *Daily Telegraph*, 2 April 1878, 2

Anon., 'Proceedings of Societies', *Chemical News and Journal of Industrial Science*, 37 (1878), 271–2

Anon., 'Modern Spiritualism', *Spectator*, 2739 (1880), 1661–2

Anon., *Astronomical Society Monthly Notices*, February 1883, 185

Anon., 'Flatland', *Athenaeum*, 2977 (15 November 1884), 622

Anon., 'Flatland', *Literary World*, 15 (1884), 389–90; reproduced on Flatweb [http://library.brown.edu/cds/flatweb/1884litworld.html] [17 January 2013] (para. 1)

Anon., 'Flatland: A Romance of Many Dimensions', *Spectator*, 2944 (1884), 1583–4

Anon., 'Literary Gossip', *Athenaeum*, 2978 (1884), 660

Anon., 'Flatland', *Literary World*, 16 (1885), 93

Anon., 'Flatland', *New York Times*, 23 February 1885, 3

Anon., 'Humor and Satire', *Literary News*, 6 (1885), 85

Anon., 'Scientific Romances. No. II', *Mind*, 10 (1885), 613

Anon., 'Societies and Academies', *Nature*, 31 (1885), 328–32

Anon., 'Civilization: Its Cause and Cure and Other Essays', *Lucifer*, 6 (1890), 159–60

Anon., 'What is the Nature of Intuition, and Which are the Best Means of Developing this Faculty?', *Vahan*, 2 (1892), 5

Anon., *I Awoke! Conditions of Life on the Other Side Communicated by Automatic Writing* (London: Simpkin and Marshall, 1893)

Anon., 'An Oxford Undergraduate's Suicide', *Jackson's Oxford Journal*, 7676 (1900), 8

Anon., 'Summer Meeting of the London Branch: Flatland', *Mathematical Gazette*, 7 (1914), 228–31

Aristotle, 'Physics', in *The Complete Works of Aristotle: The Revised Oxford Translation*, ed. Jonathan Barnes, 2 vols (Princeton: Princeton University Press, 1984), I, pp. 315–446

Aristotle, 'Poetics', in *The Complete Works of Aristotle: The Revised Oxford Translation*, ed. Jonathan Barnes, 2 vols (Princeton: Princeton University Press, 1984), II, pp. 2316–40

Aristotle, 'Rhetoric', in *The Complete Works of Aristotle: The Revised Oxford Translation*, ed. Jonathan Barnes, 2 vols (Princeton: Princeton University Press, 1984), II, pp. 2152–269

Barrett, William, 'Invisible Beings', *Nonconformist and Independent*, 4 (1881), 16–17

Barrett, William, *On the Threshold of the Unseen* (London: Kegan Paul, 1918)

Bartlett, Rev. J.B., 'A Glimpse of the "Fourth Dimension"', *Boy's Own Paper*, 12 (1890), 462

Besant, Annie, and C.W. Leadbeater, *Thought Forms* (London and New York: The Theosophical Publishing House, 1905)

Bierce, Ambrose, *The Collected Works*, 12 vols (New York: The Neale Publishing Company, 1909–12)

Blackwood, Algernon, 'A Victim of Higher Space', in *The Complete John Silence Stories*, ed. S.T. Joshi (New York: Dover Publications, 1998), pp. 230–46

Blavatsky, H.P., *The Secret Doctrine*, 2 vols (London: The Theosophical Publishing Company, 1888)

Blavatsky, H.P., *Isis Unveiled*, 2 vols, 2nd edn (Point Loma: Theosophical Publishing Co., 1910)

Busch, Wilhelm, *Edward's Dream*, ed. and trans. Paul Carus (Chicago: Open Court Publishing Company, 1909)

Canning, G., John Hookham Frere, and G. Ellis, 'The Loves of the Triangles', in *Poetry of The Anti-Jacobin*, ed. Charles Edmonds, 3rd edn (London: Sampson Low, Marston, Searle & Rivington, 1890), pp. 151–64

Carpenter, Edward, 'Underneath and After All', *Lucifer*, 6 (1890), 248

Carpenter, Edward, *From Adams Peak to Elephanta* (London: Swan Sonnenschein, 1892)

Carpenter, Edward, *My Days and Dreams* (London: G. Allen & Unwin, 1916)

Carroll, Lewis, *Through the Looking Glass and What Alice Found There* (London: Academy Editions, 1977)

Cayley, Arthur, 'Chapters in the Analytical Geometry of (n) Dimensions', *Cambridge Mathematical Journal*, 4 (1843), 119–27

Cayley, Arthur, 'Sur quelques théorêmes de la géometrie de position', *Crelle's Journal*, 31 (1846), 213–27

Cayley, Arthur, 'A Memoir on Abstract Geometry', *Philosophical Transactions of the Royal Society of London*, 160 (1870), 51

Child, Lydia Maria, *Philothea: A Romance*, 2nd edn (Boston: Otis, Broaders and Company, 1839)

Clifford, W.K., 'On the Space-Theory of Matter', *Transactions of the Cambridge Philosophical Society*, 2 (1876), 157–8 (repr. in *Mathematical Papers*, ed. Robert Tucker (London: Macmillan and Co., 1882), pp. 21–2)

Clifford, W.K., 'The Philosophy of the Pure Sciences III: The Postulates of the Science of Space', in *Lectures and Essays*, ed. Leslie Stephen and Sir Frederick Pollock, 2 vols (London: Macmillan and Co., 1879), I, pp. 295–323

Clifford, W.K., 'Problems and Solutions from The Educational Times', in *Mathematical Papers*, ed. Robert Tucker (London: Macmillan and Co., 1882), pp. 565–627

Clifford, W.K., *The Common Sense of the Exact Sciences* (London: Kegan, Paul, Trench & Co., 1886 [1885])

Collins, Frank S., 'The Fourth Dimension, a Paper read by Frank S. Collins, Part 1', *The Path*, 4 (1889), 17–19

Conan Doyle, Arthur, *The History of Spiritualism*, 2 vols (London: Cassell and Company, 1926)

Conrad, Joseph, and Ford Madox Hueffer, *The Inheritors* (London: Dent, 1923)

Coryn, Herbert, 'The Fourth Dimension', *Lucifer*, 12 (1893), 326–32

Crosland, Newton, C.C.M., and G.W., MD, 'The Fourth Dimension', *Light*, 1 (1881), 31

Crosland, Newton and E.T.B., 'The Fourth Dimension', *Light*, 1 (1881), 23

Darwin, Charles, *The Descent of Man, and Selection in Relation to Sex* (London: John Murray, 1871)

Darwin, Charles, *The Origin of Species by Means of Natural Selection, or the Preservation of Favoured Races in the Struggle for Life*, 6th edn (London: John Murray, 1872)

Darwin, Charles, *The Formation of Vegetable Mould, Through the Action of Worms* (London: John Murray, 1881)

De Cyon, E., 'The Anti-Vivisectionist Agitation', *Contemporary Review*, 43 (1883), 498–510

Dick, Frederick J., 'The Meaning of Separated Life—A Mathematical Story of 2, 3 & 4 Dimensions', *Lucifer*, 6 (1890), 243–5

Diogenes Laertius, 'Xenophanes', in *The Lives of Eminent Philosophers*, trans. R.D. Hicks, 2 vols (London: William Heinemann, 1950), II, pp. 425–9

Dodgson, Charles Lutwidge, *The Dynamics of a Parti-cle* (Oxford: James Parker and Co., 1874)

Du Maurier, George, *The Martian* (London and New York: Harper & Bros, 1897)

Eddington, Arthur, *Space, Time and Gravitation* (Cambridge: Cambridge University Press, 1920)

Edge, H.T., 'Popular Misconceptions about the Fourth Dimension', *The Path*, 4 (1889), 252

Ellis, H. Havelock, 'Hinton's Later Thought', *Mind*, 9 (1884), 384–405

Engels, Friedrich, 'Natural Science and the Spirit World', Marxists.org, http://www.marxists.org/archive/marx/works/1883/don/ch10.htm [accessed 9 February 2012] (repr.of *Illustrierter Neue Welt-Kalender für das Jahr 1898*)

Euclid, *Euclid's Elements of Geometry: The First Six Books and the Portions of the Eleventh and Twelfth Books Read at Cambridge*, ed. Robert Potts (London and New York: Longmans, Green and Co., 1895)

Euclid, *The Elements*, trans. Sir Thomas Heath, 3 vols (New York: Dover Publications, 1956)

Fechner, Gustav Theodor, 'The Comparative Anatomy of Angels', trans. Hildegard Corbet and Marilyn E. Marshall, *Journal of the History of the Behavioral Sciences*, 5 (1869), 135–51

Fechner, Gustav Theodor, 'Zöllner's Mediumistic Experiments: Extracts from the Diary of Gustav Theodor Fechner, Late Professor in Vienna, Died November 19th, 1887', *Light*, 8 (1888), 256–7

Ford, Ford Madox, *Joseph Conrad: A Personal Remembrance* (London: Duckworth and Co., 1924)

Freeman, Mary E. Wilkins, 'The Hall Bedroom', in *Short Story Classics (American)*, ed. William Patten (New York: P.F. Collier and Son, 1905), IV, pp. 1275–99

Fullerton, George S., 'On Space of Four Dimensions', *Journal of Speculative Philosophy*, 18 (1884), 113–21

Fullerton, George S., 'A Letter from Professor Fullerton to Mr. C.C. Massey', *Light*, 7 (1887), 451

Fullerton, George S., and others, *Preliminary Report of the Commission Appointed by the University of Pennsylvania to Investigate Modern Spiritualism in Accordance with the Request of the Late Henry Seybert* (Philadelphia: J.B. Lippincott Co., 1887)

Funk, Isaac K., *The Widow's Mite and Other Psychic Phenomena* (New York: Funk and Wagnall's Company, 1904)

Gordon, E.A. Hamilton, 'The Fourth Dimension', *Science Schools Journal*, 5 (1887), 145–51

Griffith, George, 'The Justice of Revenge', *Newcastle Weekly Courant*, 28 April 1900, 2

Griffith, George, *The Mummy and Miss Nitocris* (Milton Keynes: Tutis Digital Publishing, 2008; repr. of London: T. Werner Laurie, 1906)

Grove, William Robert, *Address to the British Association for the Advancement of Science* (London: Longmans, Green and Co., 1867)

Halsted, George Bruce, 'Bibliography of Hyper-Space and Non-Euclidean Geometry', *American Journal of Mathematics*, 1:3 (1878), 261–76

Hartmann, Franz, *The Talking Image of Urur* (New York: John W. Lovell and Company, 1890)

Helmholtz, Hermann von, 'Integrals of the Hydrodynamical Equations, which Express Vortex-motion', trans. P.G. Tait, *Philosophical Magazine*, 33 supplement (1867), 485–511

Helmholtz, Hermann von, 'The Axioms of Geometry', *Academy*, 1 (1870), 128–31

Helmholtz, Hermann von, 'The Axioms of Geometry', *Academy*, 3 (1872), 52–3

Helmholtz, Hermann von, 'On the Use and Abuse of the Deductive Method in Physical Science', *Nature*, 11 (1874), 149–51

Helmholtz, Hermann von, 'The Origin and Meaning of the Axioms of Geometry', *Mind*, 1 (1876), 301–21

Heun, Karl, 'Science Note-Book', *Nature*, 31 (1884), 51–2

Hinton, C.H., 'What is the Fourth Dimension?', *The University Magazine*, 96 (1880), 15–34

Hinton, C.H., 'What is the Fourth Dimension?', *Cheltenham Ladies College Magazine*, 8 (1883), 31–52

Hinton, C.H., 'Fourth Dimension', in *Hazell's Annual Cyclopaedia, 1886* (London: Hazell, Watson and Viney, 1886), pp. 183–5

Hinton, C.H., 'A Picture of Our Universe', in *Scientific Romances*, 2 vols (London: Swan Sonnenschein, 1886), I, pp. 161–204

Hinton, C.H., 'The Persian King', in *Scientific Romances*, 2 vols (London: Swan Sonnenschein, 1886), I, pp. 33–128

Hinton, C.H., 'What is the Fourth Dimension?', in *Scientific Romances*, 2 vols (London: Swan Sonnenschein, 1886), I, pp. 3–32

Hinton, C.H., *A New Era of Thought* (London: Swan Sonnenschein, 1888)

Hinton, C.H., *Stella and an Unfinished Communication: Studies in the Unseen* (London: Swan Sonnenschein, 1895)

Hinton, C.H., 'Many Dimensions', in *Scientific Romances: Second Series* (London: Swan Sonnenschein, 1896), pp. 28–44

Hinton, C.H., 'On the Education of the Imagination', in *Scientific Romances: Second Series* (London: Swan Sonnenschein, 1896), pp. 3–22

Hinton, C.H., *An Episode of Flatland: or How a Plane Folk Discovered the Third Dimension* (London: Swan Sonnenschein, 1907)

Hinton, James, *Man and his Dwelling Place* (London: John W. Parker and Son, 1859)

Hinton, James, *Life in Nature* (London: Smith, Elder and Co., 1862)

Hinton, James, *The Mystery of Pain* (London: Smith, Elder and Co., 1866)

Hodgson, Shadworth, 'Introduction', in *Chapters on the Art of Thinking*, ed. C.H. Hinton (London: C. Kegan Paul and Co., 1879)

Hodgson, William Hope, *The House on the Borderland and Other Novels* (London: Gollancz, 2002)

Hopkins, Ellice, *Life and Letters of James Hinton* (London: C. Kegan Paul, 1878)

Huxley, T.H., 'The Scientific Aspects of Positivism', *Fortnightly Review* (1869), 653–70

Ingleby, Dr. C.M., 'Transcendent Space', *Nature*, 1 (1870), 289

Jacob, Alexander, *Henry More's Manual of Metaphysics: A Translation of the Enchiridium Metaphysicum (1679) with an Introduction and Notes*, 2 vols (Zurich: Georg Olms Verlag Hildesheim, 1995)

James, Henry, *The Art of the Novel: Critical Prefaces*, ed. Richard P. Blackmur (New York and London: Charles Scribner's Sons, 1934)

James, Henry, 'The Great Good Place', in *The Complete Tales of Henry James*, ed. Leon Edel, 12 vols (London: Rupert Hart-Davis, 1962–4), XI, pp. 13–42

James, Henry, *The Spoils of Poynton* (Oxford: Oxford University Press, 1982)

James, Henry, 'The Art of Fiction', in *The Art of Criticism: Henry James on the Theory and Practice of Fiction*, ed. William Veeder and Susan M. Griffin (Chicago and London: University of Chicago Press, 1986), pp. 165–83; repr. of *Longman's Magazine*, 4 (1884), 502–21.

Jevons, William Stanley, 'Helmholtz on the Axioms of Geometry', *Nature*, 4 (1871), 482

Jinarajadasa, C., *The Golden Book of the Theosophical Society: A Brief History of the Society's Growth from 1875–1925* (Adyar, Madras: Theosophical Publishing House, 1925)

Jinarajadasa, C., 'Introduction', in C.W. Leadbeater, *The Astral Plane* (Adyar: Theosophical Publishing House, 1970), pp. vi–xx

Johnston, Charles, 'Psychism and the Fourth Dimension', *The Theosophist*, 9 (1888), 423

Jung, C.G., *The Zofingia Lectures*, trans. Jan van Huerck (London: Routledge and Kegan Paul, 1983)

Kandinsky, Wassily, *On the Spiritual in Art*, trans. Hilla Rebay (New York: The Solomon R. Guggenheim Foundation, 1946)

Kant, Immanuel, 'Thoughts on the True Estimation of Living Forces (Selected Passages)', in *Kant's Inaugural Dissertation and Early Writings on Space*, ed. and trans. John Handyside (Westport, CT: Hyperion Press, 1979), pp. 3–18

Kant, Immanuel, 'Concerning the Ultimate Ground of the Differentiation of the Directions in Space', in *Cambridge Edition of the Works of Immanuel Kant: Theoretical Philosophy 1755–1770*, ed. and trans. David Walford in collaboration with Ralf Meerbote (Cambridge: Cambridge University Press, 1992), pp. 361–72

Kant, Immanuel, 'On the Form and Principles of the Sensible and the Intelligible World [Inaugural Dissertation]', in *Cambridge Edition of the Works of Immanuel Kant: Theoretical Philosophy 1755–1770*, ed. and trans. David Walford in collaboration with Ralf Meerbote (Cambridge: Cambridge University Press, 1992), pp. 373–416

Kant, Immanuel, *Critique of Pure Reason*, ed. and trans. Paul Guyer and Allen W. Wood (Cambridge: Cambridge University Press, 1998)

Kant, Immanuel, 'Prolegomena to Any Future Metaphysics', in *Cambridge Edition of the Works of Immanuel Kant: Theoretical Philosophy after 1781*, ed. Henry Allison and Peter Heath, trans. Gary Hatfield, Michael Friedman, Henry Allison, and Peter Heath (Cambridge: Cambridge University Press, 2002), pp. 29–170

Kendall, May, *Dreams to Sell* (London: Longmans, Green and Co., 1887)

Kepler, Johannes, *The Harmony of the World*, trans. E.J. Aiton, A.M. Duncan, and J.V. Field (Philadelphia: The American Philosophical Society, 1997)

Kepler, Johannes, *Mysterium Cosmographicum*, trans. A.M. Duncan (Norwalk, CT: Abaris Books, 1999 [1981])

Klein, Felix, 'Bemerkungen über den Zusammenhang der Flächen', *Mathematische Annalen*, 9 (1876), 476–82

Klein, Felix, *Development of Mathematics in the 19th Century*, trans. M. Ackerman (Brookline: Math Sci Press, 1979)

Land, J.P.N., 'Kant's Space and Modern Mathematics', *Mind*, 2 (1877), 38–46

Leadbeater, C.W., *The Astral Plane: Its Inhabitants and Phenomena* (London: The Theosophical Society Publishing Company, 1895)

Leadbeater, C.W., *The Devachanic Plane* (London: The Theosophical Publishing Society, 1896)

Leadbeater, C.W. [C.W.L.], 'Does a Highly Developed Ego, That of a Master, for Instance, Put on the Limitations of the Physical Brain When it Descends to Work on the Physical Plane?', *Vahan*, 8 (1898), 7

Leadbeater, C.W., *Clairvoyance* (London: The Theosophical Publishing Society, 1899)

Leadbeater, C.W., *The Fourth Dimension (A lecture given by C.W. Leadbetter [sic] before the Amsterdam Lodge T.S. in April, 1900. Stenographic notes in Dutch by J.J. Hallo, Jr.; translated into English by Mrs. Marie Knothe)* (San Francisco: Mercury Publishing Office, 1900)

Leadbeater, C.W., *The Other Side of Death* (London: Theosophical Publishing Society, 1903)

Leadbeater, C.W., *How Theosophy Came to Me* (Adyar: Theosophical Publishing House, 1930)

Lewes, G.H., 'Kant's View of Space', *Nature*, 1 (1870), 289

Lovecraft, H.P., *Selected Letters 1923–1929*, ed. August Derleth and Donald Wandrei (Wisconsin: Arkham House, 1968), vol. 2

Lovecraft, H.P., 'The Call of Cthulhu', in *The Call of Cthulhu and Other Weird Tales* (London: Vintage, 2001), pp. 61–98

Lovecraft, H.P., 'The Dreams in the Witch House', in *The Call of Cthulhu and Other Weird Tales* (London: Vintage, 2001), pp. 235–80

Lovecraft, H.P., 'At the Mountains of Madness', in *At the Mountains of Madness: The Definitive Edition* (New York: The Modern Library, 2005), pp. 1–102

Lovecraft, H.P., 'Supernatural Horror in Literature', in *At the Mountains of Madness: The Definitive Edition* (New York: The Modern Library, 2005), pp. 103–73

MacDonald, George, *Lilith* (London: Chatto & Windus, 1895; repr. Grand Rapids, MI: Wm. B. Eerdmans Publishing Company, 2000)

MacDonald, Greville, 'Introduction', in *Lilith* (London: George Allen and Unwin, 1924), pp. ix–xx

Manning, Henry Parker, *Geometry of Four Dimensions* (New York: Dover Publications, 1956; repr. of Macmillan, 1914)

Marenholtz-Bülow, Baroness, *Child and Child-Nature: Contributions to the Understanding of Fröbel's Educational Theories*, trans. Alice M. Christie (London: Swan Sonnenschein, 1879)

Massey, C.C., 'Letter', *Bombay Gazette*, 13 March 1879, 3

Massey, C.C., 'The Fourth Dimension', *Light*, 1 (1881), 15

Massey, C.C., 'M.E. de Cyon and the Late Professor Zöllner', *Light*, 3 (1883), 188

Massey, C.C., 'Zöllner: An Open Letter to Professor George S. Fullerton', [supplement to] *Light*, 7 (1887), 375–84

Massey, C.C., 'Translator's Preface', in Carl du Prel, *The Philosophy of Mysticism*, trans. C.C. Massey, 2 vols (London: George Redway, 1889), I, pp. ix–xxii

Möbius, August, 'On Higher Space', in *Sourcebook in Mathematics*, ed. D.E. Smith, trans. Henry P. Manning (New York: McGraw Hill Book Company, 1929), pp. 525–6 (first published in *Der barycentrische Calcul* (Leipzig: [no publisher] 1827))

Monck, W.H. Stanley, 'Kant's View of Space', *Nature*, 1 (1870), 386

More, Henry, *The Immortality of the Soul*, ed. Alexander Jacob (Dordrecht and Lancaster: Nijhoff, 1987)

Newcomb, Simon, 'Note on a Class of Transformations which Surfaces May Undergo in Space of More Than Three Dimensions', *American Journal of Mathematics*, 1 (1878), 1–4

Nichols, T.L., MD, 'Remarkable Physical Manifestations', *Spiritualist*, 12 (1878), 174–5

Olcott, Henry Steel, *Old Diary Leaves*, 6 vols (New York and London: G. Puttnam and Sons, 1895)

Oliver, Rev. George, *The Revelations of A Square* (London: Richard Spencer, 1855)

Oliver, Rev. George, *The Pythagorean Triangle, or the Science of Numbers* (London: John Hogg and Co., 1875)

Payne, Joseph, *Fröbel and the Kindergarten System of Elementary Education* (London: Henry S. King and Co., 1874)

Pearson, Karl, *The Grammar of Science*, 2nd edn (London: Adam and Charles Black, 1900)

Plato, *The Dialogues of Plato*, ed. and trans. Benjamin Jowett, 5 vols (Oxford: The Clarendon Press, 1871)

Plato, 'Timaeus', in *Timaeus and Critias*, trans. Desmond Lee (Harmondsworth: Penguin, 1987), pp. 27–124

'Puzzled', 'Questions and Answers', *Light*, 1 (1881), 7

Raine, Kathleen, and George Mills Harper (eds), *Thomas Taylor the Platonist: Selected Writings* (New York: Bollingen Foundation, 1969)

Randolph, B., *Seership! The Magnetic Mirror* (Toledo, OH: K.C. Randolph, 1884)

Reid, Thomas, *An Inquiry into the Human Mind on the Principles of Common Sense*, ed. Derek R. Brookes (Edinburgh: Edinburgh University Press, 1997)

Reimers, C., 'Vindication of Dr Slade', *Medium and Daybreak*, 9 (1878), 232

Riemann, Bernhard, 'On the Hypotheses which Lie at the Bases of Geometry', trans. W.K. Clifford, *Nature*, 8 (1873), 14–17

Robert-Houdin, Jean Eugene, *The Secrets of Stage Conjuring*, ed. and trans. Professor Hoffmann [pseud. Angelo John Lewis] (London: G. Routledge & Sons, 1881)

Rodwell, G.F., 'On Space of Four Dimensions', *Nature,* 8 (1873), 8–9

'S', 'Four Dimensional Space', *Nature,* 31 (1885), 481

Sambourne, Linley, 'Man is but a Worm', in *Punch's Almanack for 1882* (London: Punch Office, 1882)

Schofield, A.T., *Another World* (London: Swan Sonnenschein, 1888)

Schubert, Hermann, 'The Fourth Dimension: Mathematical and Spiritualistic', *Monist*, 3 (1893), 435–49

'Scientific and Literary Societies', *Leeds Mercury*, 1 April 1899, 9

Shaw, Nellie, *A Czech Philosopher on the Cotswolds; Being an Account of the Life and Work of Francis Sedlak* (London: C.W. Daniel Co., 1940)

Shelley, Percy, *Prometheus Unbound*, ed. Lawrence John Zillman (Seattle: University of Washington Press, 1959)

Sidgwick, Mrs. Henry, 'Results of a Personal Investigation into the Physical Phenomena of Spiritualism with Some Critical Remarks on the Evidence for the Genuineness of Such Phenomena', *Proceedings of the Society for Psychical Research*, 4 (1887), 45–74

Simmons, J., 'Slade in Europe', *Spiritualist*, 12 (1878), 186

Sinnett, A.P., *The Occult World*, 3rd edn (London: Trübner and Co., 1883)

Sinnett, A.P., *The Early Days of Theosophy in Europe* (London: Theosophical Publishing House, 1922)

Skilin, C.W., 'Zöllner', *Medium and Daybreak* (1882), 803–4

Sommerville, Duncan M.Y., *Bibliography of Non-Euclidean Geometry* (London: Harrison & Sons, 1911)

Spottiswoode, William, 'Presidential Address', *Report of the Forty-Eighth Meeting of the BAAS Held at Dublin in August 1878* (London: John Murray, 1879)

Stead, W.T., 'On the Eve of the Fourth Dimension, Mathematical and Spiritualistic', *Review of Reviews*, 7 (1893), 542

Stead, W.T., 'Some Books of the Month', *Review of Reviews*, 7 (1893), 325

Stead, W.T., 'Throughght: Or, On the Eve of the Fourth Dimension: A Record of Experiments in Telepathic Automatic Handwriting', *Review of Reviews*, 7 (1893), 426–32

Steiner, Rudolf, *The Fourth Dimension* (Great Barrington, MA: Anthroposophic Press, 2001)

Stevenson, R.L., 'A Gossip on Romance', *Longman's Magazine*, 1 (1882), 69–79

Stevenson, R.L., *The Letters of Robert Louis Stevenson*, ed. Sidney Colvin, 3 vols, 10th edn (London: Methuen and Co., 1911)

Stringham, W.I., 'Regular Figures in n-Dimensional Space', *American Journal of Mathematics*, 3 (1880), 1–14

Sylvester, J.J., 'A Plea for the Mathematician', *Nature*, 1 (1869), 237–9, 261–3 (abridged with supplemental notes); repr. in full in Sylvester's *The Laws of Verse* (London: Longmans, Green, and Co., 1870), pp. 109–12

Tait, P.G., *Lectures on Some Recent Advances in Physical Science* (London: Macmillan, 1876)

Tait, P.G., 'On Knots', *Transactions of The Royal Society of Edinburgh*, 28 (1877), 145–90

Tait, P.G., 'Zöllner's Scientific Papers', *Nature*, 17 (1878), 420–2

Tait, P.G., 'Johann Benedict Listing', *Nature*, 27 (1883), 316–17

Tait, P.G., and Balfour Stewart, *The Unseen Universe or, Physical Speculations on a Future State*, 4th edn (London: Macmillan and Co., 1876)

Thomson, Sir William, 'On Vortex Atoms', in *Mathematical and Physical Papers* (Cambridge: Cambridge University Press, 2011), pp. 1–12 (repr. of 'On Vortex Atoms', *Proceedings of the Royal Society of Edinburgh*, 6 (1867), 94–105)

Tucker, R., 'Flatland: A Romance of Many Dimensions', *Nature* (1884), 76

Tyndall, John, *Fragments of Science: A Series of Detached Essays, Addresses and Reviews*, 6th edn, 2 vols (London: Longmans, Green and Co., 1879)

Uspensky, P.D., *Tertium Organum*, trans. Nicholas Bessaraboff and Claude Bragdon (London: Kegan Paul & Co., 1923)

Wadsworth, Edward, 'Inner Necessity', *Blast*, 1 (1914), 119–25

Waltershausen, Wolfgang Sartorius von, *Gauss: A Memorial*, trans. Helen Worthington Gauss (Colorado Springs, CO: self-published, 1966)

Wells, H.G., *In the Days of the Comet* (London: Macmillan, 1912)

Wells, H.G., *Experiment in Autobiography: Discoveries and Conclusions of a Very Ordinary Brain (Since 1866)*, 2 vols (London: Gollancz, 1934)

Wells, H.G., 'The Plattner Story', in *Selected Short Stories* (London: Penguin, 1968), pp. 193–211

Wells, H.G., 'The Remarkable Tale of Davidson's Eyes', in *Selected Short Stories* (London: Penguin, 1968)

Wells, H.G., *The Invisible Man: A Grotesque Romance* (London: J.M. Dent, 1995)

Wells, H.G., *The Time Machine* (London: Phoenix, 2004)

Wilde, Oscar, 'The Canterville Ghost', in *The Canterville Ghost, The Happy Prince and Other Stories* (London: Penguin, 2010), p. 197 (repr. from *Court and Society Review*, 23 February 1887)

Willink, Arthur, *The World of the Unseen: An Essay on the Relation of Higher Space to Things Eternal* (London: Macmillan, 1893)

Worringer, Wilhelm, *Abstraction and Empathy: A Contribution to the Psychology of Style*, trans. Michael Bullock (Chicago: Ivan R. Dee, 1997)

Wundt, Wilhelm, 'Spiritualism as a Scientific Question: An Open Letter to Prof. Hermann Ulrici of Halle', *Popular Science Monthly*, 15 (1879), 578–93 (repr. and trans. of *Der Spiritismus: Eine Sogenannte Wissenschaftliche Frage* (Leipzig: Engelmann, 1879))

Zimmermann, Robert, *Henry More und die vierte Dimension des Raumes* (Vienna: Carl Gerold's Sohn, 1881)

Zöllner, Johann Carl Friedrich, *Über die Natur der Cometen: Beiträge zur Geschichte und Theorie der Erkenntnis* (Leipzig: Engelmann, 1872)

Zöllner, Johann Carl Friedrich, 'On Space of Four Dimensions', *Quarterly Journal of Science*, 8 (1878), 227–37

Zöllner, Johann Carl Friedrich, *Transcendental Physics*, trans. Charles Carleton Massey, 4th edn (Boston: Banner of Light, 1901)

SECONDARY SOURCES

Arata, Stephen D., 'The Occidental Tourist: *Dracula* and the Anxiety of Reverse Colonization', *Victorian Studies*, 33 (1990), 621–45

Armstrong, Isobel, 'Spaces of the Nineteenth-Century Novel', in *The Cambridge History of Victorian Literature*, ed. Kate Flint (Cambridge: Cambridge University Press, 2012), pp. 575–97

Bakhtin, Mikhail, *Rabelais and his World*, trans. Helene Iswolsky (Bloomington: Indiana University Press, 1984)

Beer, Gillian, *Darwin's Plots: Evolutionary Narrative in Darwin, George Eliot and Nineteenth-Century Fiction* (London: Routledge & Kegan Paul, 1983)

Beer, Gillian, 'Translation or Transformation? The Relations of Literature and Science', *Notes and Records of the Royal Society of London*, 44 (1990), 81–99

Beer, Gillian, *Open Fields: Science in Cultural Encounter* (Oxford: Clarendon Press, 1996)

Bell, Ian F.A., *Critic as Scientist: The Modernist Poetics of Ezra Pound* (London: Methuen, 1981)

Bergonzi, Bernard, *The Early H.G. Wells* (Manchester: Manchester University Press, 1961)

Bork, Alfred M., 'The Fourth Dimension in Nineteenth-Century Physics', *Isis*, 55 (1964), 326–38

Bortoft, Henry, *Goethe's Scientific Consciousness* (London: Institute for Cultural Research, 1986)

Brake, Laurel, and Marysa Demoor (eds), *Dictionary of Nineteenth-Century Journalism* (Ghent: Academia Press; London: British Library, 2009)

Bringmann, Wolfgang G., Norma J. Bringmann, and Norma L. Medway, 'Fechner and Psychical Research', in *G.T. Fechner and Psychology*, ed. Josef Brožek and Horst Gundlach (Passau: Passavia, 1988), pp. 243–56

Britzolakis, Christina, 'Pathologies of the Imperial Metropolis: Impressionism as Traumatic Afterimage in Conrad and Ford', *Journal of Modern Literature*, 29 (2005), 1–20

Brooker, Joseph, 'Satire Bust: The Wagers of Money', *Law and Literature*, 17 (2005), 321–44

Brown, Alan Willard, *The Metaphysical Society: Victorian Minds in Crisis, 1869–1880* (New York: Columbia University Press, 1947)

Brown, Bill, 'Thing Theory', *Critical Inquiry*, 28 (2001), 1–22

Brown, Daniel, *The Poetry of Victorian Scientists: Style, Science and Nonsense* (Cambridge: Cambridge University Press, 2013)

Buchwald, Jed Z., 'Electrodynamics in Context: Object States, Laboratory Practice, and Anti-Romanticism', in *Hermann von Helmholtz and the Foundations of Nineteenth Century Science*, ed. David Cahan (Berkeley, Los Angeles, and London: University of California Press, 1993)

Butor, Michel, 'The Space of the Novel' and 'The Book as Object', in *Inventory*, ed. Richard Howard (London: Jonathan Cape, 1961), pp. 31–8, 39–56

Cahan, David, 'Anti-Helmholtz, Anti-Zöllner, Anti-Duhring: The Freedom of Science in Germany during the 1870s', in *Universalgenie Helmholtz: Rückblick nach 100 Jahren*, ed. Lorenz Kruger (Berlin: Akademie Verlag, 1994), pp. 330–44

Cajori, Florian, 'Origins of Fourth Dimension Concepts', *The American Mathematical Monthly*, 33 (1926), 397–406

Cassirer, Ernst, *The Platonic Renaissance in England*, trans. James P. Pettegrove (Edinburgh: Nelson, 1953)

Clarke, Bruce, 'A Scientific Romance: Thermodynamics and the Fourth Dimension in Charles Howard Hinton's "The Persian King"', Weber Studies, 14 (1997), http://www.altx.com/ebr/w%28ebr%29/essays/clarke.html [accessed 24 February 2010] (para. 1 of 28)

Clarke, Bruce, *Energy Forms: Allegory and Science in the Era of Classical Thermodynamics* (Ann Arbor: University of Michigan Press, 2001)

Cohn, Dorrit, *Transparent Minds: Narrative Modes for Presenting Consciousness in Fiction* (Princeton: Princeton University Press, 1978)

Colman, Andrew M., *A Dictionary of Psychology* (Oxford: Oxford University Press, 2008)

Connor, Steven, 'Making an Issue of Cultural Phenomenology', *Critical Quarterly*, 42 (2000), 2–6.

Connor, Steven, 'Afterword', in *The Victorian Supernatural*, ed. Nicola Bown, Carolyn Burdett, and Pamela Thurschwell (Cambridge: Cambridge University Press, 2004), pp. 258–77

Connor, Steven, *The Matter of Air: Science and Art of the Ethereal* (London: Reaktion Books, 2010)

Connor, Steven, 'Thinking Things', *Textual Practice*, 24 (2010), 1–20

Connor, Steven, 'Pregnable of Eye: X-Rays, Vision and Magic', http://www.stevenconnor.com/xray/ [accessed 22 October 2012]

Connor, Steven, 'Thinking Things', http://www.stevenconnor.com/thinkingthings [accessed 22 October 2012]

Coxeter, H.S.M., 'Alicia Boole Stott', in *Women of Mathematics: A Bibliographic Sourcebook*, ed. Louise S. Grinstein and Paul J. Campbell (Westport, CT and London: Greenwood Press, 1987), pp. 220–4

Crary, Jonathan, *Techniques of the Observer: On Vision and Modernity in the Nineteenth Century* (Cambridge, MA: MIT Press, 1992)

Cuddon, J.A., *A Dictionary of Literary Terms* (Harmondsworth: Penguin, 1982)

Daly, Nicholas, *Modernism, Romance and the Fin de Siècle: Popular Fiction and British Culture, 1880–1914* (Cambridge: Cambridge University Press, 1999)

Dixon, Joy, *Divine Feminine: Theosophy and Feminism in England* (Baltimore: Johns Hopkins University Press, 2001)

Dixon, Thomas, *The Invention of Altruism: Making Moral Meanings in Victorian Britain* (Oxford: Oxford University Press, 2008)

Dodds, E.R., 'The Astral Body in Neoplatonism', in Proclus, *The Elements of Theology*, ed. and trans. E.R. Dodds (Oxford: Clarendon Press, 1933)

Draznin, Yaffa Claire, *My Other Self: The Letters of Olive Schreiner and Havelock Ellis, 1884–1920* (New York: Peter Lang, 1992)

Faivre, Antoine, *Theosophy, Imagination, Tradition: Studies in Western Esotericism* (New York: State University of New York Press, 2000)

Fischer, Gerd, *Mathematical Models* (Braunschweig, Wiesbaden: Friedr. Vieweg & Sohn, 1986)

Frank, Joseph, 'Spatial Form in Modern Literature: An Essay in Three Parts', *Sewanee Review*, 53 (1945), 221–40, 433–56, 643–53

Frank, Joseph, 'Spatial Form: Thirty Years After', in *Spatial Form in Narrative*, ed. Jeffrey R. Smitten and Ann Daghistany (Ithaca and London: Cornell University Press, 1981), pp. 202–43

Friedländer, Paul, *Plato: The Dialogues, Second and Third Periods*, trans. Hans Meyerhoff (London: Routledge and Kegan Paul, 1969)

Funkenstein, Amos, *Theology and the Scientific Imagination from the Middle Ages to the Seventeenth Century* (Princeton: Princeton University Press, 1989)

Gardner, Martin, *Mathematical Carnival* (Washington, DC: The Mathematical Association of America, 1989 [1965])

Geduld, Harry M., *The Definitive Time Machine* (Indianapolis: Indiana University Press, 1987)

Genette, Gérard, 'Complexe de Narcisse', in *Figures I* (Paris: Editions de Seuil, 1966), pp. 21–8

Genette, Gérard, 'Boundaries of Narrative', trans. Ann Levonas, *New Literary History*, 8 (1976), 1–13

Genette, Gérard, *Narrative Discourse: An Essay in Method*, trans. Jane E. Lewin (Ithaca, NY: Cornell University Press, 1983)

Gentner, Dedre, 'Analogy in Scientific Discovery: The Case of Johannes Kepler', in *Model-based Reasoning: Science, Technology, Values*, ed. Lorenzo Magnani and Nancy J. Nersessian (New York: Kluwer Academic/Plenum, 2002), pp. 21–40

Gersh, Stephen, 'The Medieval Legacy from Ancient Platonism', in *The Platonic Tradition in the Middle Ages*, ed. Stephen Gersh and Maarten J.F.M. Hoenen (Berlin and New York: Walter de Gruyter, 2002), pp. 3–30

Gibbons, Tom H., 'Cubism and "The Fourth Dimension", in the Context of the Late Nineteenth-Century and Early Twentieth-Century Revival of Occult Idealism', *Journal of the Warburg and Courtauld Institutes*, 44 (1981), 130–47

Gilbert, Elliott L., ' "Upward, not Northward": *Flatland* and the Quest for the New', *English Literature in Transition*, 34 (1991), 391–404

Grant, Edward, *Much Ado About Nothing: Theories of Space and Vacuum from the Middle Ages to the Scientific Revolution* (Cambridge: Cambridge University Press, 1981)

Gray, Jeremy, *Plato's Ghost: The Modernist Transformation of Mathematics* (Princeton: Princeton University Press, 2008)

Hall, G.S., *The Founders of Modern Psychology* (New York and London: D. Appleton and Company, 1924)

Harman, Graham, 'On the Horror of Phenomenology: Lovecraft and Husserl', *Collapse*, 4 (2009), 333–64

Harman, Graham, *Weird Realism: Lovecraft and Philosophy* (Winchester and Washington: Zero Books, 2012)

Hatfield, Gary C., *The Natural and the Normative* (Cambridge, MA and London: MIT Press, 1990)

Henderson, Andrea, 'Math for Math's Sake: Non-Euclidean Geometry, Aestheticism, and "Flatland" ', *PMLA*, 124 (2009), 455–71

Henderson, Linda Dalrymple, 'Italian Futurism and "The Fourth Dimension"', *Art Journal*, 41 (1981), 317

Henderson, Linda Dalrymple, *The Fourth Dimension and Non-Euclidean Geometry in Modern Art* (Princeton: Princeton University Press, 1983)

Henderson, Linda Dalrymple, 'Mysticism as the "Tie That Binds": The Case of Edward Carpenter and Modernism', *Art Journal*, 46 (1987), 29–37

Henderson, Linda Dalrymple, 'X Rays and the Quest for Invisible Reality in the Art of Kupka, Duchamp, and the Cubists', *Art Journal*, 47 (1988), 323–40

Henderson, Linda Dalrymple, 'Re-Introduction', in *The Fourth Dimension and Non-Euclidean Geometry in Modern Art* (Cambridge, MA and London: MIT Press, 2013), pp. 1–96

Herbert, Christopher, *Victorian Relativity: Radical Thought and Scientific Discovery* (Chicago: University of Chicago Press, 2001)

Hermann, D.B., 'Zöllner Studies at Archenhold Observatory 1974–1994', in *Karl Friedrich Zöllner and the Historical Dimension of Astronomical Photometry*, ed. Christiaan Sterken and Klaus Staubermann (Brussels: VUB University Press, 2000), pp. 151–9

Hessenbruch, Arne, 'Science as Public Sphere: X-Rays Between Spiritualism and Physics', in *Wissenschaft und Öffentlichkeit in Berlin, 1870–1930*, ed. Constantin Goschler (Stuttgart: Franz Steiner, 2000), pp. 89–126

Hoffman, Banesh, 'Introduction', in *Flatland* (New York: Dover Publications, 1952), pp. iii–iv

Innes, Shelley, 'Mary Boole and Curve Stitching: A Look into Heaven', *Endeavour*, 28 (2004), 36–8

Jameson, Fredric, 'Imperialism and Modernism', in *Nationalism, Colonialism and Literature* (Minneapolis: University of Minnesota Press, 1990), pp. 43–68

Jann, Rosemary, 'Abbott's *Flatland*: Scientific Imagination and "Natural Christianity"', *Victorian Studies*, 28 (1985), 473–90

Jenkins, Alice, 'Spatial Imagery in Nineteenth Century Representations of Science: Faraday and Tyndall', in *Making Space for Science: Territorial Themes in the Shaping of Knowledge*, ed. Crosbie Smith and Jon Agar (New York: Palgrave Macmillan, 1998), pp. 181–91

Jenkins, Alice, *Space and the 'March of Mind'* (Oxford: Oxford University Press, 2007)

Johnson, K. Paul, *The Masters Revealed: Madame Blavatsky and the Myth of the Great White Lodge* (Albany: State University of New York Press, 1994)

Jones, Harry C., 'Jacobus Henricus Van't Hoff', *Proceedings of the American Philosophical Society*, 50 (1911), iii–xii

Kaplan, Walter J., and Alison Chaiken, 'Flatland Fans', *Science News*, 126 (1984), 355

Kestner, Joseph, 'Secondary Illusion: The Novel and the Spatial Arts', in *Spatial Form in Narrative* (Ithaca and London: Cornell University Press, 1981), pp. 100–30

Koven, Seth, *Slumming: Sexual and Social Politics in Victorian London* (Princeton: Princeton University Press, 2004)

Koyré, Alexander, *From the Closed World to the Infinite Universe* (Baltimore: Johns Hopkins Press, 1979)

Kragh, Helge, 'Zöllner's Universe', *Physics in Perspective*, 14 (2012), 392–420

Latour, Bruno, *We Have Never Been Modern*, trans. Catherine Porter (Cambridge, MA: Harvard University Press, 1993)

Latour, Bruno, *Pandora's Hope* (Cambridge, MA: Harvard University Press, 1999)

Latour, Bruno, *Reassembling the Social: An Introduction to Actor Network Theory* (Oxford: Oxford University Press, 2008)

Latour, Bruno, *On the Modern Cult of the Factish Gods* (Durham, NC: Duke University Press, 2010)

Luckhurst, Roger, *The Invention of Telepathy* (Oxford: Oxford University Press, 2002)

Luckhurst, Roger, 'The Mummy's Curse: A Study in Rumour', *Critical Quarterly*, 52 (2010), 6–22

Luckhurst, Roger, 'Introduction', in H.P. Lovecraft, *The Classic Horror Stories* (Oxford: Oxford University Press, 2013), pp. vii–xxviii

McGurl, Mark, 'Social Geometries: Taking Place in Henry James', *Representations*, 68 (1999), 59–83

Materer, Timothy, *Modernist Alchemy: Poetry and the Occult* (Ithaca, NY: Cornell University Press, 1995)

Miller, J. Hillis, *Ariadne's Thread: Story Lines* (New Haven: Yale University Press, 1992)

Mortimer, Joanne Stafford, 'Annie Besant and India 1913–1917', *Journal of Contemporary History*, 18 (1983), 61–78

Mumby, F.A., and Frances H.S. Stallybrass, *From Swan Sonnenschein to George Allen and Unwin Ltd.* (London: George Allen and Unwin, 1955)

Murray, Paul, *A Fantastic Journey: The Life and Literature of Lafcadio Hearn* (London: Routledge, 1993)

Mussell, James, *Science, Time and Space in the Periodical Press: Movable Types* (Aldershot and Burlington, VT: Ashgate, 2007)

Norris, Frank, 'A Plea for Romantic Fiction', in *The Responsibilities of the Novelist and Other Literary Essays* (London: Grant Richards, 1903)

Owens, Alex, *The Place of Enchantment: British Occultism and the Culture of the Modern* (Chicago: University of Chicago Press, 2004)

Parshall, Karen Hunger, *James Joseph Sylvester: Jewish Mathematician in a Victorian World* (Baltimore: Johns Hopkins University Press, 2006)

Plotnitsky, Arkady, *The Knowable and the Unknowable: Modern Science, Nonclassical Thought, and the 'Two Cultures'* (Michigan: University of Michigan Press, 2002)

Poortman, J.J., *Vehicles of Consciousness*, 4 vols (Utrecht: The Theosophical Society in the Netherlands, 1978)

Poovey, Mary, *Making a Social Body: British Cultural Formation, 1830–1864* (Chicago: University of Chicago Press, 1995)

Poovey, Mary, *A History of the Modern Fact: Problems of Knowledge in the Sciences of Wealth and Society* (Chicago: University of Chicago Press, 1998)

Porter, Theodore M., *Karl Pearson: The Scientific Life in a Statistical Age* (Princeton: Princeton University Press, 2004)

Prothero, Stephen, *The White Buddhist: The Asian Odyssey of Henry Steel Olcott* (Bloomington and Indianapolis: Indiana University Press, 1996)

Richards, Joan L., *Mathematical Visions: The Pursuit of Geometry in Victorian England* (Boston: Academic Press, 1988)

Rocke, Alan J., *Image and Reality* (Chicago: University of Chicago Press, 2010)

Rotman, Brian, *Signifying Nothing: The Semiotics of Zero* (Basingstoke: The Macmillan Press, 1987)

Rowbotham, Sheila, *Edward Carpenter: A Life of Liberty and Love* (London: Verso, 2009)

Rucker, Rudolf v. B., 'Introduction', in *Speculations on the Fourth Dimension: Selected Writings of Charles H. Hinton* (New York: Dover Publications, 1980), pp. v–xix

Saunders, Max, *Ford Madox Ford: A Dual Life*, 2 vols (Oxford: Oxford University Press, 1996)

Scheick, William J., *The Splintering Frame: The Later Fiction of H.G. Wells* (Victoria, BC: University of Victoria English Literary Studies, 1984)

Serres, Michel, *The Parasite*, trans. Lawrence R. Schehr (Baltimore and London: Johns Hopkins University Press, 1982)

Serres, Michel, 'Gnomon: The Beginnings of Geometry in Greece', in *A History of Scientific Thought*, ed. Michel Serres (Oxford: Blackwell, 1995), pp. 73–123

Shearsmith, Kelly, 'Chiasmatic Christianity: Lilith's Sense of an Ending', in *Lilith in a New Light: Essays on the George MacDonald Fantasy Novel*, ed. Lucas H. Harman (Jefferson; London: McFarland, 2008), pp. 143–60

Silver, Daniel S., 'The Last Poem of James Clerk Maxwell', *Notices of the AMS*, 55 (2008), 1266–70

Smith, Jonathan, *Fact and Feeling: Baconian Science and the Nineteenth-century Literary Imagination* (Madison: University of Wisconsin Press, 1994)

Smith, Jonathan, Lawrence I. Berkove, and Gerald A. Baker, 'A Grammar of Dissent: *Flatland*, Newman, and the Theology of Probability', *Victorian Studies*, 39 (1996), 129–50

Smitten, Jeffrey R., 'Introduction: Spatial Form and Narrative Theory', in *Spatial Form in Narrative*, ed. Jeffrey R. Smitten and Ann Daghistany (Ithaca and London: Cornell University Press, 1981), pp. 15–36

Smitten, Jeffrey R., and Ann Daghistany, 'Editor's Preface', in *Spatial Form in Narrative* (Ithaca and London: Cornell University Press, 1981), pp. 13–15

Staubermann, Klaus B., 'Tying the Knot: Skill, Judgement and Authority in the 1870s Leipzig Spiritistic Experiments', *British Journal for the History of Science*, 34 (2001), 67–79

Staubermann, Klaus B., *Astronomers at Work: A Study of the Replicability of 19th Century Astronomical Practice* (Frankfurt am Main: Verlag Harri Deutsch, 2007)

Stewart, Ian, *The Annotated Flatland* (New York: Perseus Publishing, 2002)

Stromberg, Wayne H., 'Helmholtz and Zöllner: Nineteenth-century Empiricism, Spiritism, and the Theory of Space Perception', *Journal of the History of the Behavioral Sciences*, 25 (2006), 371–83

Surette, Leon, *The Birth of Modernism: Ezra Pound, T.S. Eliot, W.B. Yeats, and the Occult* (Montreal/Kingston: McGill-Queen's University Press, 1994)

Suvin, Darko, 'Victorian Science Fiction, 1871–85: The Rise of the Alternative History Sub-Genre', *Science Fiction Studies*, 10 (1983), 148–69

Suvin, Darko, *Victorian Science Fiction in the UK: The Discourses of Knowledge and Power* (Boston: G.K. Hall & Co., 1983)

Taliaferro, R. Catesby, 'Foreword', in *The Timaeus and the Critias, or Atlanticus: The Thomas Taylor Translation* (New York: Pantheon Books, 1952), pp. 9–36

Throesch, Elizabeth Lea, 'The *Scientific Romances* of Charles Howard Hinton: The Fourth Dimension as Hyperspace, Hyperrealism and Protomodernism', doctoral thesis, University of Leeds, 2007

Treitel, Corinna, *A Science for the Soul: Occultism and the Genesis of the German Modern* (Baltimore: Johns Hopkins University Press, 2004)

Trotter, David, 'The Invention of Agoraphobia', *Victorian Literature and Culture*, 32 (2004), 463–74

Valente, K.G., 'Transgression and Transcendence: *Flatland* as a Response to "A New Philosophy"', *Nineteenth Century Contexts*, 26 (2004), 61–77

Valente, K.G., '"Who Will Explain the Explanation?": The Ambivalent Reception of Higher Dimensional Space in the British Spiritualist Press, 1875–1900', *Victorian Periodicals Review*, 41 (2008), 124–49

Vasco, Gerhard M., *Diderot and Goethe: A Study in Science and Humanism* (Geneva: Librairie Slatkine, 1978) (trans. of Goethe, *Werke* (Weimar: Hrsg. Im Auftrage der Grossherzogin Sophie von Sachsen, 1877–1919), sec. 4, XXXIV, pp. 136–7)

Veeder, William, and Susan M. Griffin, 'Commentary on The Art of Fiction', in *The Art of Criticism: Henry James on the Theory and Practice of Fiction*, ed. William Veeder and Susan M. Griffin (Chicago and London: University of Chicago Press, 1986), pp. 184–8

Vidler, Anthony, *Warped Space: Art, Architecture, and Anxiety in Modern Culture* (Cambridge, MA and London: MIT Press, 2000)

Viswanathan, Gauri, 'The Ordinary Business of Occultism', *Critical Inquiry*, 27 (2000), 1–20

Walkowitz, Judith R., *City of Dreadful Delight: Narratives of Sexual Danger in Late-Victorian London* (London: Virago, 1992)

Washburne, Carleton W., and Sidney P. Marland, *Winnetka: The History and Significance of an Educational Experiment* (Englewood Cliffs, NJ: Prentice-Hall, 1963)

Watt, Ian, *Conrad in the Nineteenth Century* (Berkeley and Los Angeles: University of California Press, 1979)

Weedon, Alexis, *Victorian Publishing: Book Publishing for the Mass Market 1836–1916* (Aldershot: Ashgate, 2003)

White, Roger M., *Talking about God: The Concept of Analogy and the Problem of Religious Language* (Farnham: Ashgate, 2010)

Yates, Frances A., *The Art of Memory* (London: Peregrine Books, 1969)

UNPUBLISHED MANUSCRIPTS

London, London Metropolitan Archives, MS Monro correspondence, ACC/1063/2109a

London, London Metropolitan Archives, MS Monro correspondence, Acc 1063/2457

London, Royal Institution, MS John Tyndall, RI MS JT/1/H/48

Reading, Reading University Library, MS Swan Sonnenschein and Co., 3282

Reading, Reading University Library, MS Swan Sonnenschein and Co., 4058, Charles Howard Hinton to William Swan Sonnenschein, 22 February 1887

FILMS

Flatland: The Film, dir. Ladd Ehlinger Jr (F.X. Vitolo, 2007)

Flatland: The Movie, dir. Dano Johnson and Jeffrey Travis (Flat World Productions, 2007)

Name Index

Abbott, Edwin 44 n. 11, 133
  English Lessons for English People 91
  Flatland 6, 10, 72–102, 111–12, 132, 179,
    195, 206
  Hints on Home Teaching 98
  Kernel and the Husk, The 91, 101, 111
  Shakespearean Grammar, A 96
Aksakov, Aleksander 53–4
Ames, Julia 154
Anaximenes of Miletus 22
Arata, Stephen D. 202
Aristotle 23, 71, 90, 92, 137
  Physics 21–2
  Poetics 33–4
  Rhetoric 34
Armstrong, Isobel 8, 167–9, 185, 191
A Square *see* Abbott, Edwin

Baeyer, Adolf 70
Bakhtin, Mikhail 85
Banchoff, Thomas 75, 79, 84–5, 95
Baraduc, Hippolyte 161
Barrett, William 103
Bartlett, Rev. J.B. 11, 146
Beer, Gillian 5, 13, 207, 208
  Darwin's Plots 12, 32–3, 39
Bergonzi, Bernard 171
Berkeley, Bishop George 51, 61
Besant, Annie 143–4, 148, 152, 160,
  163, 165
  Thought Forms 160–2
Besant, Walter 99
Bierce, Ambrose 182
  'Mysterious Disappearance' 181
  'Science to the Front' 172, 181
Blackwood, Algernon
  'A Victim of Higher Space' 126–7, 172, 177,
    182, 186, 195–6, 197, 201
Blavatsky, Helena Petrova 138–40, 142, 143,
  145, 149, 152, 163, 202
  Isis Unveiled 135, 140, 158
  Key to Theosophy 158
  Secret Doctrine, The 134, 140, 147, 149,
    152, 157
Boehme, Jakob 176
Bork, Alfred M. 187
Bortoft, Henry 130
Britten, Emma Hardinge 65
Britzolakis, Christina 204
Brooker, Joseph 86
Brown, Alan Willard 109
Brown, Bill 178

Brown, Daniel 9
Bruno, Giordano
  De umbris idearum 123
Buchwald, Jed Z. 51
Burgess, Gelett 140
  'The Ghost Extinguisher' 172
Busch, Wilhelm
  Eduards Traum (*Edward's Dream*) 179
Butor, Michael 166–7, 169, 173, 174
Butts, Mary 208

Cajori, Florian 21
Candler, Howard 112
Canning, G.
  'Loves of the Triangles, The' 80–1
Carpenter, Edward 151, 152, 164
  From Adams Peak to Elephanta 150
  Civilisation: Its Cause and Cure 149
  My Days and Dreams 149
  'Underneath and After All' 149
Carroll, Lewis 78, 79, 81
  *Through the Looking Glass and What Alice
    Found There* 174
Cassirer, Ernst 25
Cayley, Arthur 15, 28, 88, 117
  'Chapters in the Analytic Geometry of (n)
    Dimensions' 2, 26
  'A Memoir on Abstract Geometry' 27
  'On Some Theorems of Geometry of
    Position' 26–7
Child, Lydia M.
  Philothea: A Grecian Romance 157–8
Clarke, Bruce 10, 40, 113, 118, 132, 157,
  197, 208
Clausius, Rudolf 49
Clifford, William Kingdon 14
  Common Sense of the Exact Sciences 36
  'On the Space-Theory of Matter' 28–9
  Postulates of the Science of Space, The 29
Cohn, Dorrit 170
Collins, Frank S. 145
Connor, Steven 13, 44, 48, 160–1
  'Afterword' in *The Victorian Supernatural* 7,
    11, 101
  'Thinking Things' 3–4, 165
Conrad, Joseph 193
  Inheritors, The 172, 184, 195, 201
Coryn, Herbert 157
Crary, Jonathan 204
Crookes, William 41, 50, 54, 56, 110
Crosland, Newton 61
Cyon, Emil de 61–2, 64 n. 77

Daly, Nicholas 98, 99
Darwin, Charles 39
  Descent of Man, The 37–8
  Origin of Species, The 38
Darwin, Erasmus
  'Loves of the Plants, The' 80
Davenport Brothers 66–7
Descartes, René 1, 9, 21, 23, 24, 25
Dick, Frederick J. 146
Dixon, Joy
  Divine Feminine 141, 147–8
Dixon, Thomas 107–8
Dodds, E. R. 157
Dodgson, Charles Lutwidge
  The Dynamics of a Part-icle 82
Du Maurier, George
  Martian, The 183
Du Prel, Carl
  Philosophy of Mysticism, The 151
Dühring, Eugene 51, 69

Eddington, Arthur
  Space, Time and Gravitation 36–7
Edge, Henry T. 145
Ellis, Henry Havelock 107, 108
Engels, Friedrich 43
  'Natural Science in the Spirit World' 43
Euclid 25, 102, 183
  Elements, The 33, 87
  satire of 81

Falk, Herman John 119, 121, 122, 124, 140
Fechner, Gustav Theodor 52, 63, 65–6, 151,
  204
  'The Comparative Anatomy of Angels'
  (Dr Mises) 79–80
Felt, George Henry 135, 136, 147
Fludd, Robert 123
Ford, Ford Madox *see* Hueffer, Ford Madox
Frank, Joseph
  'Spatial Form in Modern Literature' 170–1,
  186
Freeman, Mary Wilkins
  'Hall Bedroom, The' 172, 177, 181, 182,
  185–6, 190–1, 193, 198
Friedländer, Paul 136
Fröbel, Friedrich 121–2
Fullerton, George 65, 67 n. 88
  Preliminary Report 62–4
Funkenstein, Amos 23

Galton, Frances 86
Gardner, Martin
  Mathematical Carnival 127–8
Garnett, William 78, 84 n. 34
Gauss, Carl Friedrich 2, 6, 14, 30, 34, 41,
  51, 126
Genette, Gérard 169–70, 171, 174–5, 192
Gentner, Derdre 70
Gibbons, Tom H. 42

Gibson, William
  Neuromancer 6
Gilbert, Elliot L. 98
Goethe, Johann Wolfgang von 120, 129–30,
  136, 204
Gordon, E.A. Hamilton 133
Graham, Michael 116, 176, 182, 205
Grant, Edward 22
Gray, Jeremy 7
Griffith, George
  Angel of the Revolution, The 172
  Justice of Revenge, The 177, 179, 194
  Mummy and Miss Nitocris, The 172, 174,
  183–4, 188–9, 191, 196, 202–3
Grove, William Robert 45, 46, 107

Hall, G.S.
  Founders of Modern Psychology, The 69
Halsted, George Bruce 11
  'Bibliography of Hyper-Space and Non-
  Euclidean Geometry' 26
Harman, Graham 198–9
Harrison, Andrea 96
Harrison, William 42, 58
Hartmann, E. von
  Spiritismus, Der 64
Hartmann, Franz
  Talking Image of Urur, The 175
Hatfield, Gary 30, 32
Hellenbach, L. B. von 59, 64
Helmholtz, Hermann von 21, 37, 42, 44,
  49–51, 55, 69, 79
  'Axioms of Geometry, The' 29–31
  'Origin and Meaning, The' 31–2
Henderson, Andrea 75
Henderson, Linda Dalrymple 10, 12, 150,
  161, 208
  *Fourth Dimension and Non-Euclidean
  Geometry in Modern Art, The* 6, 105,
  125, 140
Herbert, Christopher 5
Hering, Ewald 30, 32
Hertz, Heinrich 70
Hessenbruch, Arne 58
Hinton, Charles Howard 10–11, 74, 79, 80,
  97, 100, 102, 106, 112, 133–4, 150–1,
  155, 171, 190, 194, 197, 206
  'Casting out the Self' 119
  Episode of Flatland, An 117–18
  influence on Theosophy 139–41, 162–5
  'Many Dimensions' 116
  New Era of Thought, A 105, 109, 114–15,
  118–19, 121, 124, 130, 145, 162
  'On the Co-ordination of Space' 109–10
  On the Education and Imagination 119–21,
  129
  Persian King, The 113
  'Picture of Our Universe, A' 113–14, 180
  'Plane World, A' 111, 113, 179
  Science Notebook 110

Scientific Romances 73, 103, 111, 113, 119, 152
Stella and an Unfinished Communication 116–17, 171, 176–7, 182, 196, 205
  use of cubes 122–8, 130, 132
  'What is the Fourth Dimension?' 103–4, 110, 123–4
Hinton, James 11, 108–9, 115, 204
  Life in Nature 106
  Man and his Dwelling Place 106
  Mystery of Pain, The 107
Hodgson, Shadworth 106, 109
Hodgson, William Hope 172, 204
  House on the Borderland, The 191–2
  Night Land, The 192
Hoff, Jacobus Henricus Van't
  Chimie dans l'espace, La 53
Hoffmann, Banesh 77
Hoffmann, Oscar von 52
Home, Daniel 67
Hueffer, Ford Madox 193, 196
  Inheritors, The 172, 184, 195, 201
Hume, A.O. 143
Hutton, R.H. 99

I Awoke! Conditions of Life on the Other Side Communicated by Automatic Writing (Anon.) 155–6, 188
Ingleby, C.M. 15, 199

James, Henry 99, 184–5, 193, 198, 204
  American, The 185
  'Great Good Place, A' 172, 189
  narrative perspective 169–70
  Spoils of Poynton, The 172, 178
Jameson, Fredric 200
Jann, Rosemary 75, 78, 86, 91, 96, 98
Jarry, Alfred 79
Jenkins, Alice 2, 25, 141
Jevons, William Stanley 31
Jinarajadasa, Curuppumullage 147 n. 39, 159
Johnson, Paul K. 142–3
Johnson, Samuel 86
Jowett, Benjamin 136–7
Judge, William Q. 145
Jung, Carl Gustav 43

Kandinsky, Wassily
  'Concerning the Spiritual in Modern Art' 164
  On the Spiritual in Art 43
Kant, Immanuel 8, 11, 14–16, 21, 26, 30, 32, 33, 41, 55, 117, 167–8, 182
  'Concerning the Ultimate Ground of Differentiation of the Directions in Space' 18
  Critique of Pure Reason 14, 17, 19–20
  'Thoughts on the True Estimation of Living Forces' 1, 17–18

Kekulé, August 70
Kendall, May
  'A Pure Hypothesis' 133
Kepler, Johannes 11
  Cosmographicum Mysterium 119–20
Kern, Stephen
  Culture of Time and Space, The 12
Kestner, Joseph 186–7, 189–90
Kingsland, W. 146
Klein, Felix 27–8, 42, 45, 47, 52, 53, 54, 69
Koven, Seth 108

Lamarck, Jean-Baptiste 38
Land, J.P.N. 32, 79
Lang, Andrew 98, 99
Latour, Bruno 66, 68, 109, 172–3, 177
  On the Modern Cult of the Factish Gods 3
  Pandora's Hope 5
  We Have Never Been Modern 3–4
Leadbeater, Charles W. 140–1, 142, 144, 147, 165, 194, 199, 204
  Astral Plane, The 158–9
  Clairvoyance 162–3
  How Theosophy Came to Me 158–9
  Other Side of Death, The 163–4
  Thought Forms 160–2
Leibniz, Gottfried Wilhelm 1, 18, 157
Lewes, G.H. 16, 21
  'Kant's View of Space' 15
Lewis, Wyndham 164
Lindgren, William F. 79, 84–5, 95
Listing, Johann Benedict 44, 68
Lovecraft, H.P. 172
  'Call of Cthulhu, The' 198
  'Dreams in the Witch House, The' 198
  'Supernatural Horror in Literature' 197
  'At the Mountains of Madness' 198–9
Luckhurst, Roger 67 n. 88, 198, 203
  Invention of Telepathy, The 7, 9, 138, 152–4
Ludwig, Carl 52

MacDonald, George
  Lilith 171–4, 176, 201
MacDonald, Greville 176
McGurl, Mark 166, 193
Manning, Henry Parker 43
Marsden, Dora 208
Maskelyne, John Nevil 67
Massey, Charles Carleton 42, 58, 60–2, 64–5, 67 n. 88, 68, 103, 138–9, 151
Maxwell, James Clerk 9, 37, 40, 44, 60, 70, 84 n. 34
  Paradoxical Ode, A 46–8
Miller, J. Hillis 94, 95
Mivart, Sir George 38
Möbius, Augustus 6, 34, 52, 53, 80, 93, 94, 146, 169
  'On Higher Space' 1–2, 26, 33
Monck, W.H. Stanley 15

Monro, Cecil James 37, 60, 199
More, Henry 21, 23
  Immortality of the Soul, The 24
Mumby, F.A. 110
Murray, Thomas Douglas 202
Myers, Frederick 150, 154
  'Subliminal Consciousness' 153

Nettleship family 108
Newcomb, Simon 28, 56, 125, 187
Norris, Frank
  'Plea for Romantic Fiction, A' 185

Oliver, Rev. George
  Pythagorean Triangle, The 135–6, 137
  Revelations of A Square, The 82
Owens, Alex 135

Payne, Joseph
  *Fröbel and the Kindergarten System of
    Elementary Education* 121
Pearson, Karl
  Grammar of Science, The 36, 101
Picasso, Pablo 6, 125, 199
Plato 22, 102, 158
  Meno 93, 131
  Timaeus 11, 79, 113, 120, 136–7
Plotnitsky, Arkady 7
Poovey, Mary 2–3, 4–5
Pound, Ezra 164, 208
Proclus 137
  Elements of Theology, The 157
Proctor, R. A. 95, 111, 133

Randolph, Paschal Beverly 175
Reichenbach, Baron von 52
Reimers, C.
  'Vindication of Dr Slade' 56
Richards, Joan L. 2, 25–6, 30
Ricoeur, Paul 39
Riemann, Bernhard 2, 14, 28, 29, 30, 31, 32,
    41, 58
Robert-Houdin, Jean Eugene 67
Robinson, Fletcher 202
Rocke, Alan J. 70
Rodwell, G.F.
  'On Space of Four Dimensions' 35
Rotman, Brian 7–8
Rowbotham, Sheila 149
Rucker, Rudolf v. B. 10

Sarasvati, Swami Dayananda 143
Saunders, Max 196
Scheick, William J. 192–3
Schläfli, Ludwig 125
Schlegel, Viktor 125
Schofield, A.T.
  Another World 100–1
Schopenhauer, Arthur
  Metaphysics 50

Schreiner, Olive 101 n. 72, 108
Schubert, Hermann 156
Sedlak, Francis 127–8, 129, 196
Seeley, J.R. 72, 91, 97
Serres, Michel 4, 130
  'Gnomon: The Beginnings of Geometry in
    Greece' 131–2
Seybert, Henry 62
Shakespeare, William 102
  Hamlet 72
  Tempest, The 72
Shaw, Nellie
  Czech Philosopher on the Cotswolds, A 128–9
Shearsmith, Kelly 201
Shelley, Percy
  Prometheus Unbound 46–8
Shklovsky, Victor 190
Sidgwick, Eleanor 68
Sinnett, A.P. 138, 141, 148, 158
Slade, Henry 41, 42, 43, 51–60, 62, 63, 64,
    66–8, 196
Smith, Jonathan 75, 77, 78, 91
Sommerville, Duncan 42
Sonnenschein, William Swan 73, 74, 100, 104,
    110–11, 113, 115, 117, 121–2, 140
Spottiswoode, William 56–7, 207
Stanton, Edward 155
  Dreams of the Dead 153
Staubermann, Klaus B. 50
Stead, William T. 11, 152–3, 155–6, 164, 194,
    202, 204
  Letters from Julia 154
  Real Ghost Stories 153–4
Steiner, Rudolf 21, 24, 105
Stevenson, Robert Louis
  'Gossip on Romance, A' 98
Stewart, Balfour 46, 83–4
Stewart, Iain 75
Stott, Alicia Boole 124–5
Stringham, W.I. 125
Suvin, Darko 78, 83, 85, 99
Sylvester, James Joseph 6, 8, 26, 28, 31, 35, 36
  'Plea for the Mathematician, A' 14–16, 34

Tait, Peter Guthrie 40, 44–5, 48, 49, 50, 54,
    69, 83
  'Johann Benedict Listing' 68
  *Lectures on Some Recent Advances in Physical
    Science* 35
  Unseen Universe, The 46, 83–4
Taliaferro, R. Catesby 137
Taylor, Thomas 137
Thiersch, Geheimrath 52
Thomson, William 37, 40, 46, 49, 50
  'On Vortex Atoms' 44–5
Throesch, Elizabeth 10, 11, 105, 172 n. 23
Trotter, David 196
Tupper, J.L. 31
Tyndall, John 48, 49–50
  Fragments of Science 46

Ulrici, Hermann 57

Valente, K.G. 66, 75, 82
Vidler, Anthony 196, 204
Viswanathan, Gauri 148, 156

Wadsworth, Edward 164
Warburg, Alby 204
Watt, Ian 193
Weber, Wilhelm Eduard 49, 51, 52, 54, 64
Wedmore, Basil 125
Weedon, Alexis 110–11
Wells, H.G. 132–3, 171, 187–8, 193
  'Chronic Argonauts, The' 133
  'Door in the Wall, The 201
  In the Days of the Comet 204
  Invisible Man: A Grotesque Romance, The 177
  'Plattner Story, The' 180, 182, 192,
    196, 200
  'Remarkable Case of Davidson's Eyes,
    The' 176
  Time Machine, The 187
Wichelhaus, Hermann 70

Wilde, Oscar
  'Canterville Ghost, The' 134
Willink, Arthur
  World of the Unseen, The 153, 155
Wolff, Christian 17
Worringer, Wilhelm
  Abstraction and Empathy 197, 204–5
Wundt, Wilhelm 52, 57–8, 63, 65

Yates, Frances
  Art of Memory, The 122–3
Yeats, W.B. 164, 208

Zimmermann, Robert 21
Zöllner, Johann Carl Friedrich 9, 40, 43, 45,
    48, 57–8, 67–8, 69, 71, 103, 105, 130,
    134, 145, 161, 167, 196, 206
  criticism of 61–6
  mediumistic experiments 50–4, 56
  'On Space of Four Dimensions' 41
  Transcendental Physics 42, 50, 55, 58–60,
    67 n. 88
  Über die Natur der Cometen 49–50

Subject Index

Academy 29, 31, 79, 97
Actor Network Theory (ANT) 3–4
altruism, and spatial voiding 107, 115, 116, 182, 197
American Journal of Mathematics 28
 analogy 9–10, 32–3
 and the ambiguity of knowledge 70–1
 and biblical geometry 100–1
 and comparative anatomy 79–80
 continuity and vortex atoms 46
 and Darwinian thought 37–9
 flatfishes and worms 36–7
 in *Flatland* 75–6, 90–4, 112
 higher and colonial/global space 132, 165
 imagination and thought 129–30, 146
 and n-dimensions 14, 16, 33–5
 and proportional metaphors 33–4
 as a reification allegory 40
 and satire 83
 spiritualism and space 24
 Theosophical thoughts on 146
anatomy, comparative 79–80
Anti-Jacobin, The 80–1
astral space/plane 141, 157–64
Astronomical Society Monthly Notices 61
Athenaeum, The 76, 79, 94, 95, 97

Banner of Light 53
Blast 164
body
 and biblical geometry 101
 and co-location 184
 and dimensional analogy 35, 38, 166
 and incongruent counterparts 168–9, 180
 permeability 178–82
 and three-dimensionality 18
Bombay Gazette 138–9
Boy's Own Paper 146
British Association for the Advancement of Science (BAAS) 45

Cambridge Mathematical Journal 26
'casting out the self' 106, 119, 129, 152, 164
Chemical News 110
City of London School Magazine 75, 79, 82
co-location 24, 140, 168
 and possession 183–5
consciousness
 astral/planar vision 157–64
 and co-location 183 4
 cosmic, and Theosophy 141, 145, 149, 150–1
 and dimensional analogy 34, 59, 104, 146

 in narrative perspectives 170
 and psycho-physical sensibility 151
Contemporary Review 29, 61
continuity/discontinuity
 of existence, philosophy 114–15
 and knottedness 46–8
 satire on 83–4
 of spatial thought 89
Court and Society Review 134
Crelle's Journal 26
cubes
 and astral vision 162–3
 and dread of space 197
 and knowledge production 131
 pedagogical use of 112, 118–19, 120–2
 as quasi-objects 11, 130
 ratio and harmony of 119–20
 spatial philosophy on 110
 see also tesseracts

Daily Express 202
Daily Telegraph 42, 54
Devachanic Plane 160

Educational Times 28
ether 104
 and the astral plane 159–64
 knitted vortex 44, 47, 48
 and thought materialization 80
 and unity of thought 114–15

fear 194
 control of, and abstraction 197
 cosmic horror 197–8
 paranoia of penetrability 195–6
 phobias 196
 of unrepresentability 195
fiction 10, 12
 and body permeability 178–82
 books and readers as spatial objects 166
 fin de siècle, high-dimensional spaces in 171–2
 global space and colonialism 200–5
 and Kantian spatiality 167–8
 portals and spatial fluidity 172–6
 possession, and co-location 183–6
 sense embodiment in 169, 185–6
 and spatial fears 194–9
 and spatial form 186–93
 spatiality of narrative perspective 170–1
Flatland (Abbott) 72–5
 analogical thinking in 75–6, 90–4

classical references 84–5
gendered congruence in 179
Hintonian practices in 112
and the horror of unrepresentability 195
imagination in 97–100
influence on Christian theology 100–1
line measurement in 94
and multidimensionality 166
precursors 79–84
reviews of 76–9
satire in 76, 79, 85–8
sensual hierarchy in 88–9
spatial fear in 195
A Square as a character 95–6
stage adaptation 101–2
form
and dimensional analogy 37–8
and imagination 186–7
Kantian spatiality 18
and pictorial illusion 189–90
spatial limitations 191
and temporality 170–1, 187–9, 192–3
see also cubes; knots
Fortnightly Review 187

geometry
biblical 100–1
geometrical satire 75, 80–2, 179
impact of Modernism on 6–7
and memory 123
and pedagogy 112
projective 27–8
on symmetry of knowledge 4–5
Pythagorean 135–7
thoughts on three-dimensionality 1–2
see also *n*-dimension; non-Euclidean geometry
gnomons 131–2
Greece, ancient, and Victorian England
satire on 84–5, 86, 92–3

illusion 59, 91
pictorial 189–90
and spatial form 186–7
and thought, Theosophical 146
imagination 1, 69, 207
and cognition 78, 96–7
and continuity 48
and education 116, 119–20
and spatial form 17, 20, 24, 57, 186–7
and thought 129–30
Innatism
criticism 30–2
in *Flatland* 93, 95–6

Kantian space 8
criticism of 14–15, 30
and geometry 19–20
and incongruent counterparts 167–8

as a priori intuition 20–1
and three-dimensionality 17–18
kinaesthetics 121
knots 9
and analogies 69–71
as closed space curves 44–5
mediumistic experiments on 53–6, 66–8
and the Principle of Continuity 44–8
and spiritualism 41–4
and transcendental physics 59
and vortex atoms 44–6

Leeds Mercury 125
Light 60, 64, 65
lines 1, 27
and analogical thinking 90, 93
infinity of 84
measurability 94
permeability 44, 80
satire of Euclid 81
Literary News 77
Literary World, The 76, 77, 79, 87
Longman's Magazine 98, 99
Lucifer 126, 144, 146, 149, 157, 160

matter 28–9, 207
and continuity 45–6
permeability of 44, 59, 114, 157, 160–1, 175
and relationality 18
Theosophy on 138–9, 157
and thought 106, 115, 130
Medium and Daybreak 56, 61
memory and space 122–3
Mind 29, 79, 109, 113
mirrors and spatial fluidity 172–6
Modernism 3, 12
and Imperialism 200
influence on geometry 6–7
and narrative spatial form 171, 186–93, 208
occult imagery in 164–5
and x-rays 161
Monist, The 156

naturalism 45, 57, 109, 141
in Darwin 37–9
in Zöllner 49–50
Nature 6, 14, 35, 49, 112, 187
n-dimension 2, 6, 10, 12, 13
and analogy 14, 16, 33, 37, 90
in *Flatland* 90
in Kant 20
and knots 56
in Lovecraft 198–9
and Modernism 7, 208
and perception 30–2
and sense embodiment 169
thought development 25–8, 83, 167–8
in Wells's Time Traveller 187

Neoplatonism 16, 21, 85, 120
    and Hintonian cubes 122
    and planes 157–8
    and Theosophy 135–7
New York Times 77
Newcastle Weekly Courant 194
Nicol prisms 53
Nonconformist and Independent, The 103
non-Euclidean geometry 2, 14, 16, 20, 181
    criticism of 51, 69, 110
    development 5–7, 19–20, 25–6, 29–31, 82
    in *Flatland* 77, 88
    in Lovecraft 198–9
    and spiritualism 42

Occult Review 126
occultism 6–7, 9, 11–12, 25
    and global space 202–3
    hierarchical relations in 148
    Hinton's influence on 105, 126, 152, 164
    and mirrors 175–6
    possession, and co-location 183–6
    and space permeability 154–6, 160–5
    see also supernatural
orientability(non-), in *Flatland* 85, 87, 89, 92, 97

Pall Mall Gazette 64, 99, 152, 153
Path, The 145
Pearson's Weekly 202
permeability 133
    astral planes 158–64
    of bodies 178–85
    and portals 172–8
    Theosophy on 139–40, 152, 157
    and thought as a medium 154–6, 165
Philosophical Magazine, The 26
philosophy
    continuity of life 114–15
    eternal return 116
    mental 106, 109
    moral 107
    and physical models 113
Pioneer, The 138
planes
    astral 157–63
    in geometry 1, 15, 18, 26–7, 168
    and Neoplatonism 158
    and spirituality 142, 147
    see also *Flatland*
polytopes 124–5
portals 172–8, 185
projective geometry 7, 27–9, 40, 206, 208
    knots 45
Psychische Studien 53–4
Punch 38, 81, 133
pyramids 130–1
Pythagoreanism 23
    in *The Mummy* 191
    and Theosophy 135–6

Quarterly Journal of Science 41, 54, 58
quasi-objects 4, 5
    cubes as 11, 130
    knots as 68

refraction, and incorporeality 116, 182
relativity 5, 61
    in *Flatland* 87, 166
    Kant on 18
    of perception 161
Review of Reviews 152, 153, 154, 156

satire
    and analogical thinking 83
    about continuity 83–4
    in *Flatland* 76, 79, 85–8, 92–3
    geometrical 75, 80–2, 179
    on naturalism 38
science fiction 77–8
Science Schools Journal 133
Scientific American 42, 127
semiotics of mathematics 8
sense perception 31, 55
    embodiment of, in fiction 169, 185–6
    external 19
    hierarchy, in *Flatland* 88–9
    Theosophy on 139
Sewanee Review 170
Seybert Commission Report, The 62–6,
        196
Spectator, The 60, 77, 99
Sphinx 65
spiritualism 9, 11, 21
    and co-location 24, 140, 168, 183–5
    and ghost stories 134
    interest in, and *n*-dimensions 41–2
    interpenetration with space 24–5
    and knottedness 53–6
    mediumistic experiments 51–3, 60, 66–8
    mystical revival 135
    political implications 43
    scientific investigation of 62–6
    see also occultism; supernatural;
        Theosophy
Spiritualist, The 53, 56, 147
spissitude 21–5
    Henry More on 21–5
    theology on 23–4
subjectivity
    corporeal 178–82
    and dimensional analogy 34
    inter- 4
    negation of 106
    and possession 183–4
supernatural 7, 71
    and cosmic horror 197
    and Imperialism 203
    and mental illusions 146
    possession and co-location 183–4

tesseracts 124–5, 127
  and astral visions 162–3
  visualization of 127–9
  and vortex atoms 69
theology
  and continuity 46
  Flatland's influence on 100–1
  on omnipresence 23–5
Theosophist, The 144–5
Theosophy, Theosophical Society 207
  hierarchy in 140–1, 147–8, 152
  and Hintonian thought 145
  and 'illusion' 146
  and Pythagorean geometry 135–7
  social network 142–4
  on space and matter 138–9
  and space permeability 139–40,
    152, 157
  and tesseract visualizations 126–9
  and unity of consciousness 149–52
Times, The 38
thought-forms 161–2, 165

time, temporality
  delayed decoding 193
  dimensional progression of 156
  and spatial form 187–9, 187–90, 192
  suppression of 171, 197
topology *see* projective geometry

University Magazine 103

Vahan, The 152, 162
visualization
  in *Flatland* 89
  of mind processes 127–9, 130
  of perception, criticism 30–2
void, theories on 22
see also altruism
vorticism 164
  vortex atoms, and knots 44–6

windows 173, 174, 185, 186, 190

x-rays 160–1